Natural
Food
Colorants

ift Basic Symposium Series

Edited by
INSTITUTE OF FOOD TECHNOLOGISTS
221 N. LaSalle St.
Chicago, Illinois

Natural Food Colorants

Science and Technology

edited by

Gabriel J. Lauro

California State Polytechnic University, Pomona
Pomona, California

F. Jack Francis

University of Massachusetts
Amherst, Massachusetts

CRC Press
Taylor & Francis Group
Boca Raton London New York

CRC Press is an imprint of the
Taylor & Francis Group, an **informa** business

CRC Press
Taylor & Francis Group
6000 Broken Sound Parkway NW, Suite 300
Boca Raton, FL 33487-2742

First issued in paperback 2019

© 2000 by Taylor & Francis Group, LLC
CRC Press is an imprint of Taylor & Francis Group, an Informa business

No claim to original U.S. Government works

ISBN-13: 978-0-8247-0421-6 (hbk)
ISBN-13: 978-0-367-39823-1 (pbk)

Visit the Taylor & Francis Web site at
http://www.taylorandfrancis.com

and the CRC Press Web site at
http://www.crcpress.com

Preface

Responding to its membership, the Institute of Food Technologists' Continuing Education Committee sponsored a Basic Symposium on Natural Colorants on July 23 and 24, 1999. Together, the editors, as co-chairs, selected and coordinated the speakers, conducted the 2-day symposium, and assembled the manuscripts for this book. It was a labor of love.

Seventeen internationally renowned scientists and managers came together in Chicago to review and update scientific information relative to natural colorants. The topic was not only appropriate but timely, reflecting the increased consumer demand for "all-natural" products, which encourages the use of natural colorants. It was the aim of the symposium to improve, correct, and update the current knowledge base by covering all aspects of natural colorants from their chemistry, preparation, formulation, application, and measurement to safety, regulatory, and health considerations.

The chromophores reviewed were selected on the basis of commercial usage, availability, and potential future usage. Omitted were pigments that are rare, have limited distribution, or have a low probability of obtaining regulatory acceptance in both the United States and the European Union.

We would like to thank all of the speakers for their participation, and especially for preparation and submission of a chapter for this book, thus making available a permanent record of this historic event for those who could not attend and for the generation of scientists to follow. Unfortunately, two speakers, John Hallagan and Joseph F. Borzelleca, because of the press of current assignments, could not complete their chapters in time to meet the publication deadline. Abstracts of their talks are presented below.

Regulations in the United States (John Hallagan, IACM, Washington, DC): In the United States, color additives are regulated under the authority of the 1960 Color Additives Amendments to the Federal Food, Drug, and Cosmetic Act, which provides for the establishment of the pre-market approval program currently administered by the Food & Drug Administration. FDA regulates color additives as members of one of two classes of materials, "certified" color additives or color additives "exempt from certification." These two classes generally correspond to "synthetic" (certified) and "natural" (exempt) color additives. Both classes of color additives are subject to the same safety requirements. Different regulations govern the labeling of foods containing the two classes of color additives, with more specific disclosure required for certified color additives.

Determination of Safety (Joseph F. Borzelleca, Virginia Commonwealth University, Richmond, VA): The safety evaluation (SE) of food ingredients is an essential component of premarketing research and development and usually follows the demonstration of functionality. The extent of SE testing is a function of the chemical and/or physical nature of the ingredient and the extent of exposure (and appropriate regulations). Potentially sensitive segments of the consuming population are identified, and appropriate evaluations are conducted. SE testing in humans is highly desirable and should be conducted as soon as the animal data support it. For new macroingredients, postmarketing surveillance is recommended. The need for extensive animal testing may be modified by data documenting human exposure. A long history of safe use may support the safety of an ingredient but historical data alone are not an adequate basis for documenting safety. The extent of nonhuman data required to establish safe limits of human exposure will be determined by the quality and quantity of the available and properly documented human exposure data. For natural (noncertifiable) colors, the extent of animal testing should be less than that required for synthetic (certifiable) colors. A summary of the available SE data in animals and in humans for selected natural colors is presented and evaluated.

This book should extend the horizons of those using natural colorants for years to come.

Gabriel L. Lauro, Ph.D.
F. Jack Francis, Ph.D.

Contents

Contributors

Ron Buescher   Department of Food Science, University of Arkansas, Fayetteville, Arkansas

I. L. Goldman   Department of Horticulture, University of Wisconsin–Madison, Madison, Wisconsin

James A. Guzinski   KALSEC, Inc., Kalamazoo, Michigan

Bruce S. Henry   Phytone, Ltd., Burton-on-Trent, Staffordshire, England

George A. F. Hendry   Department of Biological Sciences, University of Dundee, Dundee, Scotland

Jennifer D. Houghton   Ard Sgoil Port Righ Biology, Portree, Isle of Skye, Scotland

William Kamuf   D.D. Williamson & Co., Inc., Louisville, Kentucky

Yoshiaki Kato   San-Ei Gen F.F.I., Inc., Osaka, Japan

Luis L. Levy   INEXA, Industria Extractora C.A., Quito, Ecuador

Carol L. Locey   Color Products Department, KALSEC, Inc., Kalamazoo, Michigan

Kevin Loughrey   Minolta Corporation, Ramsey, New Jersey

G. Mazza   Pacific Agri-Food Research Centre, Agriculture and Agri-Food Canada, Summerland, British Columbia, Canada

Richard E. Mudgett*   University of Massachusetts, Amherst, Massachusetts (Retired)

Alexander R. Nixon   D.D. Williamson & Co., Inc., Louisville, Kentucky

Minhthy L. Nguyen   Department of Food Science and Technology, The Ohio State University, Columbus, Ohio

Owen Parker   D.D. Williamson & Co., Inc., Louisville, Kentucky

Diana M. Rivadeneira   Research and Development, INEXA, Industria Extractora C.A., Quito, Ecuador

José Schul   Chemistry Institute of Caen, Caen, France

Steven J. Schwartz   Department of Food Science and Technology, The Ohio State University, Columbus, Ohio

Joachim H. von Elbe   Department of Food Science, University of Wisconsin–Madison, Madison, Wisconsin

Ronald E. Wrolstad   Department of Food Science and Technology, Oregon State University, Corvallis, Oregon

Luoqing Yang   Wild Flavors, Inc., Erlanger, Kentucky

---

* *Current affiliation:* Consultant, Technical Liaisons, Amherst, Massachusetts

Natural
Food
Colorants

1

Carmine

José Schul

Chemistry Institute of Caen
Caen, France

> Anyone wishing to gain insight into aspects of the commercial
> exploitation of products derived from cochineal for food and
> beverage uses, should be prepared to enter a field embracing
> not only the physical, chemical and biological sciences, but also
> a range of other disciplines including archaeology, economics,
> geography, history, linguistics and even visual arts. No short pa-
> per could hope to do justice to this intricate and fascinating sub-
> ject (Lloyd, 1979).

Cochineal is an extremely important product for the Peruvian econ-
omy, not just in the commercial value of the material itself but also for the
jobs that it creates. As a consequence, the Peruvian Society of Natural Colors has been conducting extensive research on improving all facets of the
cochineal-producing industry.

EARLY USE OF NATURAL COLORANTS

Colors have played a very important role in the history of humankind.
Think about the surprising colors of the paintings found in the caverns of
Altamira (Spain), Lascaux (France), and elsewhere around the world.

For colorants, early humans turned to nature. In the wrappings of a mummy dated around 3200 B.C. three colorants were found: two ochres and safflower yellow. A cloth from Theba, 3000 B.C. was found to be dyed with indigo (Schweppe, 1992).

The use of certain colors became symbolic due to the rarity and expense of their natural sources. In 1464, Pope Paul II introduced the red robes for Cardinals as a symbol of distinction. In oriental kingdoms only the rulers wore purple. Julius Caesar, after rising to full power, reserved for himself the right to wear purple robes. His highest officials were allowed to wear purple stripes. This tradition can still be seen today in the form of purple stripes on the trousers of some generals (Schweppe, 1992).

If you visit the Art Museum of Lima you may see woolen cloth from the Paracas Culture (1100 B.C.). These fabrics were colored with cochineal extract and look as if they could have been dyed and woven yesterday. Cochineal extract is one of the few organic colorants that resists degradation with time.

GENERAL INFORMATION

Cochineal red is extracted from the wingless female scale insect variously called cochineal, *Dactylopius coccus*, coccus Cacti, or *Dactilopius coccus* L. Costa. It is a parasitic insect whose host plants are cacti, principally the Opuntia and Nopalea. The dried insect bodies contain up to 24% carminic acid, a red hydroxyanthraquinone dye. The high concentration of carminic acid in the insect is a defense against predators that do not like the smell or taste.

Cochineal is not the only insect to contain colorants. Others include kermes (*Kermes vermilio*), Polish cochineal (*Porphirophora polonica*), and lac (*Kerria lacca*). Until the discovery of cochineal in the New World, Kermes, which is mentioned several times in the Old and New Testaments, was the most important of all red dyes. Kermes, however, contains much less colorant and gives a red shade that is less pure than cochineal.

HISTORICAL EVENTS

The story begins in 1512 with the Spanish conquest of Mexico. The Spaniards found the Aztecs using the extract of an insect called "Nochetzli" as a paint and as a dye. The Spaniards recognized the similarity to kermes and immediately recognized its commercial value.

Export to the Spanish court began in 1523, and cochineal became a very "hot" item. After New World gold and silver, it was the most valuable commodity.

Because the microscope had not yet been invented, Europeans believed that the dried insects were red berries. The Spaniards were not eager to correct the misunderstanding, and cochineal's insect origins remained a trade secret. Eventually, in 1694 Nicolaas Hartsocker, a lens, microscope, and telescope manufacturer living near Paris, demonstrated that cochineal was actually the dried body of a female insect filled with red eggs (Schweppe, 1992).

In 1785, Thiery de Menonville took cochineal from Oaxaca to Port au Prince, Haiti. He wrote a book about his adventures and a treatise on the cultivation of Nopalea and the coccus Cacti. Menonville tried to convince the French colonial authorities to introduce, particularly to Santo Domingo, the cultivation of the plant and the insect. Unsuccessful, he died in 1790.

In 1835, cochineal was introduced to the Canary Islands, where it is still cultivated. Soon European markets contained cochineal from a variety of sources: Honduras, Vera-Cruz, the Canary Islands, Java, and perhaps others. Many of the original types have since disappeared.

Up until the second half of the nineteenth century, the brilliant red of wool or cotton dyed with cochineal was impossible to obtain with any other dye. The development of synthetic dyes in 1856 resulted in a decline in the use of cochineal. Synthetics proved to be much less expensive, more consistently available, and offered a wide range of bright, attractive shades.

The 1970s and 1980s were difficult times for U.S. manufacturers and users of synthetic food colors. Food color regulations mandated that colorants could not be permitted if they "induced cancer in man or animal when fed at any level" (Food, Drug and Cosmetic Act, 1960). FDA interpreted this as a requirement that colors be tested to prove zero risk. Soon the colors were being used as a proving ground for new toxicology methods.

In 1976 FD&C Red #2 was "delisted" by FDA "out of concern for the public concern." The safety of Red #40 was successfully defended by industry but not before alarming the marketplace. When questions began to arise concerning the safety of FD&C Red #3, the industry was left with few options.

If FD&C Red #3 were banned, the only remaining synthetic red would be FD&C Red #40, and it was often not an acceptable replacement. The two colors were significantly different both in shade and in chemical properties.

The marketplace was changing too. U.S. food manufacturers were looking at export markets, the majority of which did not permit FD&C Red #40 at that time. There was a need for a universally permitted red colorant that was stable and attractive.

In anticipation of a potential FDA ban on Red #3, there was a resurgence of interest in cochineal-based colorants. Traditionally used as a cosmetic color in the United States, cochineal-based colorants could provide bright, attractive red shades similar to Red #3. The color was extremely stable to heat and light. Cochineal-based colorants were sensitive to low pH, but so was Red #3. Finally, almost every country in the world permitted it for food use.

In 1990 when FDA did eventually "delist" FD&C Red #3 (and the provisionally approved uses of the dye) [21 CFR §81.10(u)], the industry had already elected cochineal-based colorants as some of the best replacements.

But the food industry, accustomed to synthetic colors, asked for quality controls to which the carmine manufacturers were not accustomed. The manufacturing methods were old fashioned, the official control methods were out of date, and the published literature nonexistent. It was a challenge to adopt the requirements and to develop new colorants based on cochineal.

But the industry *did* react, and today many cochineal-based colorants are available for food use. These colors are no longer seen as simple replacements for synthetic dyes, but rather are selected for their own unique properties and shades.

One problem continued to limit the use of cochineal-based colorants in the United States—the lack of kosher certification. The author's company, Warner-Jenkinson, has recently been able to obtain Kosher certification through several Rabbinical authorities. Opinions vary on this issue, and a specific rabbinical service may or may not accept these certifications.

U.S. REGULATORY STATUS

Cochineal extract and carmine are regulated by FDA in 21CFR 73.100. This section in the Code of Federal Regulations provides definitions, chemical specifications, and allowable applications.

LIFE CYCLE AND PROPAGATION

Cochineal is an insect and a parasite that lives anchored on cactus leaves. The male has two small wings and the female none. The female attaches herself to the green part of the cactus leaf with her mouth and feeds on the nutritious juice. For protection from the weather, the female covers her body with a white waxy substance.

The propagation method is to put the female insects, which are ready to lay their eggs, in small, open-meshed fabric bags. The bags are fixed onto the cactus leaves using the plant's thorns. The infestation epoch in the *sierra* (the mountains) is during the months of April to October and in the spring on the *costa* (the coast).

Climatic factors greatly influence the growth of cochineal. Rains especially affect the young cochineal, washing them off the host plant after which they rapidly die. The best temperature is between 16 and 24°C with a relative humidity of 85%.

With a development cycle of approximately 120 days, two or three crops can be collected each year. The male life span is only 65 days; it dies shortly after breeding.

The collection of cochineal is done by hand. Only those insects that have completed their biological cycle are collected. The cochineal is brushed from the leaves, air-dried in the sun, and cleaned by screening.

Today, cochineal are produced mainly in Peru, Chile, and the Canary Islands, with some production in Bolivia. Cochineal production in Peru during 1996 was 645,000 kg, in 1997 607,000 kg, and in 1998 699,000 kg. Some of this was used domestically for the production of carminic colorants, but the majority was exported to Europe and the United States (Committee of Natural Colorants, 1999).

COCHINEAL EXTRACT

Cochineal extracted with water gives a solution of carminic acid, proteins, and minor quantities of several salts. This liquid is referred to as *cochineal extract*.

Cochineal extract is an acid-sensitive, magenta red, water-soluble color. It becomes orange at pH 4 and is very stable. Compared to FD&C Red #40, its heat and light stability would be rated as excellent. Applications for this colorant are typical for a water-soluble dye, but there are also some unique and traditional uses.

Fruit preparations often use this color. Although the shade is orange in acidic fruit, when the colored fruit is incorporated into a compound food, the shade changes to a bright red at higher pH.

At a beverage pH 2.5–3.5, cochineal extract produces a stable orange shade. New age beverages, in particular, utilize cochineal extract.

Finally, surimi, or mock crab legs, have traditionally used cochineal extract often in conjunction with other natural colors. The colorant was selected both for its shade and for its ability to migrate the appropriate distance through the meat, creating a realistic appearance.

CARMINIC ACID

Carminic acid is the root of all carmine colors. It is the active coloring agent in all cochineal-based colorants. It is one of the most light and heat stable

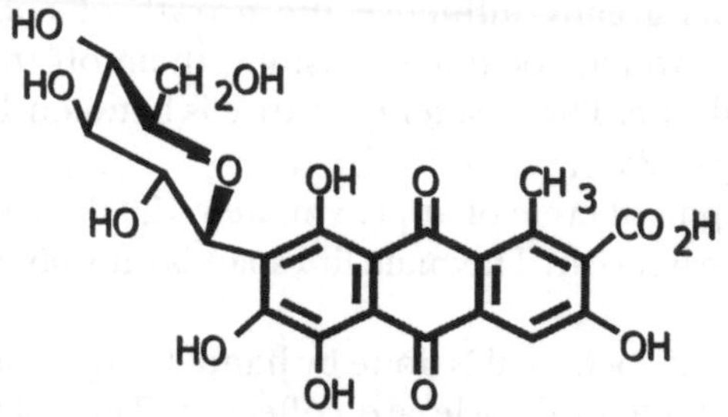

FIG. 1.1   The structure of carminic acid.

of all the natural colorants and is more stable than many synthetic food colors. At 135°C carminic acid becomes brownish, and it decomposes at 205°C.

Its definitive structural formula was determined by Overeem and van der Kerk in 1964 and confirmed in 1965 by Bhatia and Venkatamaran. It is a hydroxyanthraquinone linked to a glycosyl group (Fig. 1.1). It is interesting to note that kermesic acid, extracted from kermes, has the same chemical formula but lacks the glycosyl group.

Commercial carminic acid usually contains 50–95% carminic acid, the balance consisting of 3–5% water, 2–3% proteins, maltodextrin, and some salts.

The analytical methods used to calculate carminic acid content are the same ones used to analyze carminic acid in carmine pigments, soluble carmine, carmine pigment solutions, etc. These FCC II and FCC III methods are definitively inaccurate (Food Chemicals Codex, 1981).

If Merck Carminic Acid 97–98% is analyzed using FCC Method II, a result of 106% is obtained. FCC Method III gives a result of 89.5%. The best way to determine the carminic acid content in a commercial product may be to analyze the protein, ash, and moisture contents and then calculate the balance as carminic acid.

ACID-PROOF CARMINIC ACID

There has been much research and development on new forms of cochineal-based colorants. One of the biggest challenges has been to develop a bright red carmine product for beverages, whose usual pH is acidic. On September 15, 1992, U.S. patent 5,147,673 was awarded to the author for the product called *Carmacid R*, a red colorant based on carminic acid. It is substantially stable against color change when exposed to acidic media.

CARMINE

Water-soluble carminic acid has the ability to complex with a variety of metals to form an insoluble pigment called carmine. This pigment has various hues depending on the metal used. With mercury or tin, a scarlet shade is obtained, a crimson shade is provided by zinc or aluminum, a gray or purplish color with iron, dull violet with barium, green with uranium, and purple with chromium. The textile industry uses several of these metals as mordants to obtain different shades.

For food use, the substrates are aluminum and calcium. By careful process control, the shade of carmine can be varied from a bright magenta red to a purple shade.

Carmine is insoluble in alcohol and practically insoluble in water. It is extremely heat and light stable.

In the world market, carmine typically contains 27–65% carminic acid. In the United States, the specifications require a 50% minimum carminic acid content [21 CFR §73.100 (b)(2)].

Carmine can be used in a variety of food products—essentially anywhere that insoluble pigments are appropriate, including in dry or oil-based products where there is insufficient moisture for water-soluble dyes, or products where migration of a soluble dye would be a problem.

In principal, the preparation of a carmine pigment is extremely easy. Adding aluminum and calcium ions, in almost any proportion, to cochineal extract at a wide range of pH and temperature will result in a precipitate that *analytically* is carmine pigment. The secret is in making a quality carmine.

In the carmine business there are no "standard colors." There have been many carmine suppliers over the years, each one having its own unique shade. A user will standardize using that shade and insist on obtaining that exact hue and tinting strength. If a supplier wants to stay in the market, it must manufacture all the hues and all the tinting strengths made by all of the carmine manufacturers in the world.

Natural carmine is probably the only colorant in the world that, when prepared with the same ingredients, is expected to be redder, yellower, bluer, duller, have various tinting strengths, and still safeguard its 50% carminic acid content.

In practice this is almost possible by very careful adjustment of the cochineal extraction and laking process. In fact, carmine pigment manufacturing requires constant controls, step by step, in order to maintain consistency.

Two important features of carmine pigments (as for all pigments) are the hue and the tinting strength. *Hue* is the shade that the human eye perceives, for instance, red, green, yellow, or blue. *Tinting strength* is the relative quantity of colorant needed in a product to achieve a specific intensity.

The measurement of these attributes is usually done in a diluted form. In the case of carmine, the pigment is typically dispersed in talc at 1%. The L, a, b values are then determined using any good colorimeter. The human eye is still the most sensitive colorimeter, but it is impossible to quantify the results.

In the L, a, b system, L denotes lightness/darkness, a red/green, and b blue/yellow. In theory, one unit of difference in any dimension is enough to be perceptible to human color vision.

The following L, a, b measurements show the enormous differences between several carmines, each one having its own commercial value. All the samples were prepared at 1% in talc.

	L	a	b
Carmine 2XY (50%)	72.38	24.70	0.062
Carmine FGM (50%)	78.00	14.31	−0.490
Carmine PG (50%)	70.69	13.78	−3.810

Another set of data was generated from a single type of carmine manufactured with different carminic acid contents. It is obvious that the L, a, b values are not linear with the carminic acid concentration.

	L	a	b
Carmine 2XY (50%)	72.38	24.70	0.062
Carmine 2XY (35%)	74.90	21.92	−0.277
Carmine 2XY (30%)	74.44	18.58	−0.359
Carmine 2XY (25%)	77.91	17.74	−0.415

For the end user, the value of a carmine pigment is related to its tinting strength, i.e., how much pigment is needed in a food or cosmetic in order to obtain the desired shade.

For the carmine manufacturer, the costs are mainly related to the carminic acid content.

To further prove that the value of a carmine pigment should be based on its tinting strength and not on its carminic acid content, additional test methods were obtained from the French Centre Technique de la Salaison, de la Charcuterie et des Conserves de Viande, Maison Alfort, France. This is the main state-owned French institute that controls quality in the meat industry. Carmine is used extensively to color sausages in France.

The coloring strength of several carmines was tested using their methods. These involve dispersing the color in a variety of meat bases: surimi, sausage, salami, and liver. The end results show that the coloring strength of carmine does not necessarily relate to its carminic acid content.

	Usage level	L	a	b
E5 (40% c.a. content)	0.5 g/kg	36.4	42.1	4.5
F8 (40% c.a. content)	0.3 g/kg	40.5	44.5	3.3
B2 (40% c.a. content)	0.8 g/kg	40.5	44.5	3.3
F6 (30% c.a. content)	0.3 g/kg	40.3	43.0	3.4

Comparing the samples B2 and F8 in salami paste, it takes 2.66 times as much of B2 as F8 to achieve the same shade. It takes 1.33 times as much carminic acid in F8 to approximate the performance of F6. Comparisons in other media confirmed these results.

The situation is not helped by an FDA specification for "carmine" that requires a minimum of 50% carminic acid. Without this restriction, new types of carminic colorants, which can only be manufactured at lower concentrations, could be introduced to the marketplace.

Finally, the analytical methods often do not take into account the fact that the absorption curve of carminic acid does not follow the Beer-Lambert law. It is linear only when the absorbance value is between 0.65 and 0.83.

WATER-SOLUBLE CARMINES

One of the most important features of the insoluble carmine pigment is that it can be converted to a water-soluble dye by treatment with alkali (sodium hydroxide, potassium hydroxide, or ammonium hydroxide). The process is simple. Mixing the carmine pigment with diluted alkali to a pH of 9.5–10.5 results in a bluish-red dye solution. Propylene glycol or glycerine may then be added if desired. If the mixture contains only carmine pigment, water, and alkali, it may be spray-dried to produce a water-soluble powdered dye with up to 50% carminic acid.

Food applications for solubilized carmine are typical of water-soluble dyes, with the caveat that they will precipitate in acid media. They are used extensively around the world in sausage meats, dairy products, ice creams, etc.

The property of these colorants to precipitate at low pH is often used to advantage by the food scientist. In many cases, especially when proteins are present, the food becomes dyed before the precipitate forms. In many fruit-based beverages it is desirable to color the fruit pulp rather than just the liquid phase. Solubilized carmines can do this quite predictably.

CONCLUSION

The taste of a food is not the only factor in its acceptance. Its aroma and color must be perfect. When a red food color is needed, there is always a carminic acid colorant available that can do the job.

REFERENCES

Committee of Natural Colorants of Adex, Exporter Association, Lima, Peru.

Federal Food, Drug and Cosmetics Act, 1960 (amended).

Food Chemicals Codex, Committee on Codex Specifications, National Academy Press, 1980.

Lloyd, A. G. 1979 Extraction and Chemistry of Cochineal. *Food Chem.* (July).

Schweppe, H. 1992. *Handbuch der Naturfarbstoffe.* Ecomed, Landsberg/Lech, Austria, 1992.

2

The Betalains

Joachim H. von Elbe and I. L. Goldman

University of Wisconsin–Madison
Madison, Wisconsin

INTRODUCTION

Plants containing betalains have colors similar to plants containing the more well-known anthocyanins, yet in nature these two pigment classes are mutually exclusive. Although there are numerous other sources of betalains suitable as food colorants, in the United States, the use is restricted to the red beet. Another potential source is the pigment from amaranth (*Amaranthus tricolor*), a widely distributed but underutilized vegetable. More than 50 species of amaranth are known to exist. Leaf extracts of amaranth containing the pigment amaranthine have found use as food colorants in different countries throughout the world. In the utilization of any pigment from a plant material as a food colorant, pigment concentration is of interest. Beet cultivars commonly used as vegetables contain a pigment concentration between 100 and 300 mg/100 g fresh weight. In recent years a red beet breeding program, selecting for high pigment, has made it possible to achieve four to five times higher pigment concentrations compared to the pigment content of commercial cultivars. The purpose of this review is to

TABLE 2.1 Average Betacyanin Content of Commercial
Beet Cultivars

Cultivar	Fresh weight (mg/100 g)	Dry weight (mg/g)	Solids (%)
Mono-King			
Burgundy	135	9.3	14.5
Detroit Red	137	10.5	13.1
Fire Chief	124	8.4	14.6
Ruby Queen	117	8.2	14.2
Pacemaker	125	9.3	13.4

Source: Adapted from von Elbe, 1977.

discuss the recent advances made in increasing the pigment content of
beets and the chemistry of betalains, with an emphasis on those reactions
that impact the application of beet pigment concentrate as a food colorant.

PIGMENT CONCENTRATION

Pigment concentration is of interest in the utilization of any pigment as a
food colorant. In general, colors added to foods posses high tinctorial
strength, and thus only small amounts are needed to achieve the desired result. Any food colorant used in the United States must meet the requirements stated under the Color Additive Amendment (FFD&C Act Sec. 721).
In the case of beet colorants, the amendment allows the use of beets as vegetable juice, juice concentrate, or powder. In the last two cases, the only technology permitted is the removal of water. Therefore, there is little that can
be done to achieve greater pigment concentration through processing
technologies. Certainly technologies that would result in increased pigment
concentration (more purified pigment preparation) are fermentation and
or enzyme treatments followed by modern separation techniques.

Average betacyanin concentrations of five commercial cultivars both on
a fresh and dry weight basis are presented in Table 2.1. On a dry weight
basis, the pigment content varies as much as 20%. In addition to variation of pigment content between cultivars, horticultural practices under
which beets are grown are shown. One important factor affecting the pigment concentration is the size of the root. This is illustrated within two cultivars in Table 2.2. These data demonstrate that to optimize pigment concentration in a colorant, the root size must be controlled. It also suggests
that as the beet matures there is an increase in cell size without further
increase in the pigment content. The present commercial cultivars yield

TABLE 2.2   Influence of Beet Root Size on Betacyanin Content

Variety	Diameter (in.)	Fresh weight (mg/100 g)	Dry weight (mg/g)	Solids (%)
Mono-King	2–3	117	8.1	12.4
Burgundy	3–4	113	8.8	12.9
	>4	85	6.8	12.5
W279 × W300C	2–3	170	11.6	14.7
	3–4	139	11.1	12.5
	>4	61	5.6	10.8

Source: Adapted from von Elbe, 1977.

a beet juice concentrate (65% solids) containing 0.4% pigment. The selection W279×W300C is an illustration that cultivars can be developed that will yield more pigmented beet roots.

BREEDING FOR ENHANCED BETALAIN PIGMENT CONCENTRATION IN TABLE BEET

The presence of dominant alleles at two closely linked loci (R and Y) is the condition for the production of betalain pigment in the beet plant (Keller, 1936). Wolyn and Gabelman (1989) demonstrated that three alleles at the *R* locus determine the ratio of betacyanin to betaxanthin in the beet root and shoot. Despite the importance of these simply inherited genes in pigment synthesis, pigment concentration is amenable to the type of selection practiced for other quantitative traits (Watson and Gabelman, 1984; Wolyn and Gabelman, 1990).

In 1978, a recurrent selection program was initiated by J. F. Watson II and W. H. Gabelman at the University of Wisconsin–Madison to increase pigment levels and decrease sugar (total dissolved solids, TDS) levels in table beet. In order to gain knowledge about the response of beet to simultaneous selection for pigment and solids, divergent selection for both high and low solids was practiced in conjunction with selection for high pigment. The two populations undergoing recurrent selection were named HPHS (high-pigment, high-solids) and HPLS (high-pigment, low-solids). Although table beet is a biennial plant, using a vernalization chamber following harvest in the fall and a winter greenhouse for pollination it is possible to compress the life cycle into a single year. Thus, one generation of selection is possible on an annual basis. Results from the first three cycles of selection demonstrated that pigment synthesis was highly responsive to selection (Wolyn and Gabelman, 1990). Pigment levels increased an average of 45% in three cycles. Simultaneous selection for low solids (HPLS) was ineffective for the first three cycles. Selection for high solids (HPHS) increased the relative mean values for solids, although not significantly.

Subsequent selection for high pigment and both high and low solids was practiced in these populations until cycle 8, at which time solids was no longer used as a selection criterion. Following cycle 8, selection continued for high pigment in both populations. Goldman et al. (1996) evaluated the response of HPHS and HPLS to eight cycles of recurrent half-sib family recurrent selection for pigment and solids concentration. Average gains per cycle were 13.15 and 4.06 mg and 7.59 and 2.61 mg per 100 g fresh weight for betacyanins and betaxanthins in HPHS and HPLS, respectively. These gains are more conservative than those realized when cycles are compared in different years as they are grown and selected. The experiment reported by Goldman et al. (1996) contained all cycles grown in a similar environment, thereby minimizing yearly fluctuations in pigment concentration.

Goldman et al. (1996) also reported that total pigment increased approximately 200% in both populations and that selection for low TDS was ineffective, while only a mild response was detected from selection for high TDS. Because betalain pigments are formed following glycosylation of cyclodopa and betalamic acid, sugar molecules are associated with pigment biosynthesis. These workers suggested that simultaneous selection for high pigment and low TDS may therefore be metabolically incompatible.

Beginning with cycle 9, selection was practiced for high total pigment concentration in the populations designated HPHS and HPLS. In the last two cycles (years 1997 and 1998), selected families with high pigment concentration from HPHS were introduced into the HPLS population via hybridization. Results from this selection (Fig. 2.1) demonstrate continued gains in pigment concentration in both populations. In each generation of selection, the F_1 hybrid cultivar Big Red was grown in replicated plots alongside both HPHS and HPLS populations. The HPHS population exhibited only ca. 125% of the total betalain pigment concentration in Big Red in 1990 but has increased to nearly 300% of Big Red in 1998 (Fig. 2.1A). Similar trends were noted for the HPLS population despite a slower rate of pigment gain from 1990 to 1996. Steeper gains in 1997 and 1998 may have been due to the introduction of selected families from HPHS into the HPLS population. In Fig. 2.1B, HPHS and HPLS are compared to Big Red for total betalain pigment concentration. Steep increases in pigment concentration for both HPHS and HPLS were noted from 1996 to 1998.

Since total pigment concentration increases across generations, it is reasonable to question whether such increases are due to gains in betacyanin, betaxanthin, or both pigments. Data presented in Fig. 2.1C suggest that much of the gain in total pigment concentration in HPHS and HPLS has

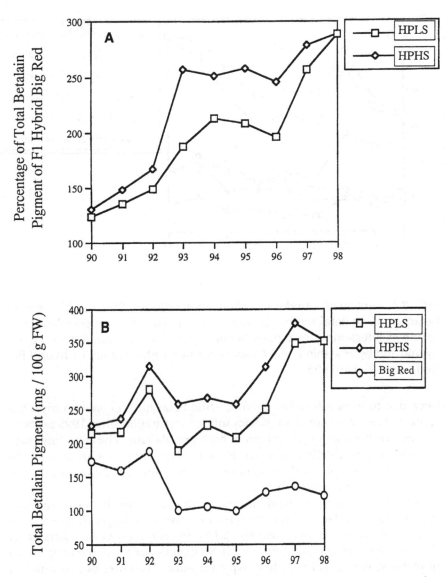

FIG. 2.1 (A) Response of two table beet populations to recurrent selection for enhanced betalain pigment concentration from 1990 to 1998. Pigment concentration is expressed as a percentage of the F1 Hybrid cultivar Big Red, which was grown each year as a control variety alongside populations undergoing selection. (B) Response of two table beet populations to recurrent selection for enhanced betalain pigment concentration and the control variety F1 hybrid Big Red grown in each year alongside populations

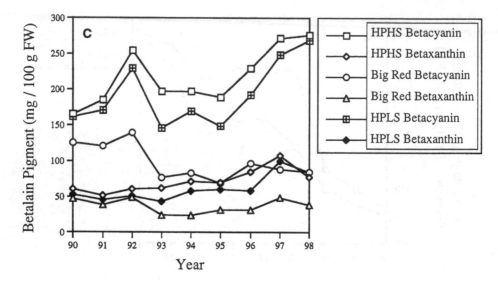

FIG. 2.1 (continued)   under selection from 1990 to 1998. (C) Changes in betacyanin and betaxanthin concentrations in two table beet populations to recurrent selection for enhanced betalain pigment concentration and betacyanin and betaxanthin concentrations for the control variety F1 hybrid Big Red from 1990 to 1998.

been due to increases in betacyanin, while betaxanthin concentration has remained relatively constant. Betacyanin concentration since 1995 account for most of the total pigment gain in both populations. These findings suggest that when selection pressure for total pigment concentration is applied, gains were primarily determined by the gain in betacyanin concentration in these two populations.

Selection for betalain pigment concentration in table beet continues to show progress. In fact, gains from selection have been larger in recent generations than in previous cycles (Fig. 2.1A). Selection response is dictated by three parameters: the amount of additive genetic variance, the proportion of plants selected, and the phenotypic variance. The latter two parameters are to a large extent under the control of the breeder, while the first parameter is determined at least in part by the biology of the pigment biosynthetic machinery in table beet. At this point, there does not appear to be a selection-response plateau in either HPHS or HPLS populations. If gains from selection were to continue in these populations, increased efficiency in pigment extraction might be expected, thus further enhancing the value of beet pigment as a natural colorant. Recent releases from this program

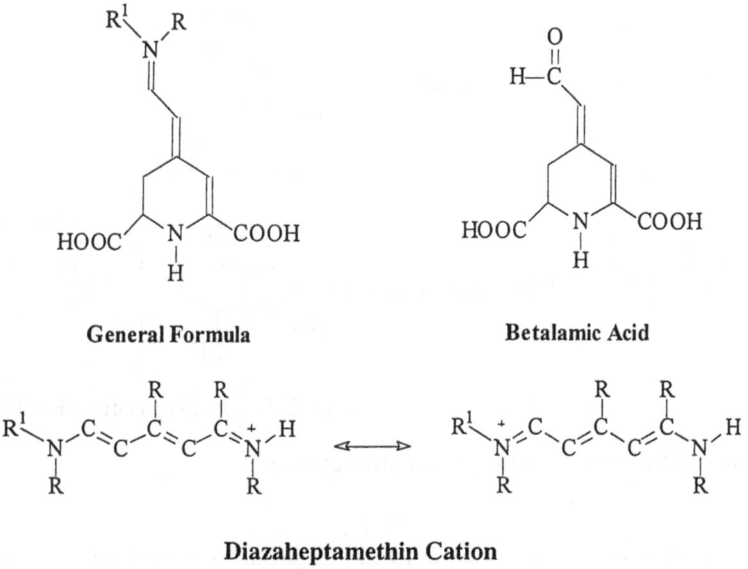

General Formula

Betalamic Acid

Diazaheptamethin Cation

FIG. 2.2 General formulas of betalains.

have made it possible to produce beet juice concentrates containing no less than 0.8% pigment (Seneca Foods Corp., 1998).

STRUCTURE

Like all natural pigments, the betalains are affected by a number of environmental factors. The structure of betalains must be considered to better understand the changes that do occur under food-processing conditions. The general structure for betalains is shown in Fig. 2.2. Betalains are a group of pigments containing betacyanins (red) and betaxanthins (yellow). They are water soluble and exist as internal salt (zwitterion) in the vacuoles of plant cells. The general formula of betalains represents the condensation of a primary or secondary amine with betalamic acid, and its structure can be described as a 1,2.4,7,7-pentasubstituted 1,7- diazaheptamethin system. When R' does not extend the conjugation of the system, the compound exhibits a maximum light absorption at about 480 nm, characteristic of yellow betaxanthins. If the conjugation is extended the light absorption shifts to 540 nm, characteristic of betacyanins.

Fig. 2.3 shows the structure of betanin and amaranthine. The difference between these two pigment structures is the glycosidic linkage. In betanin

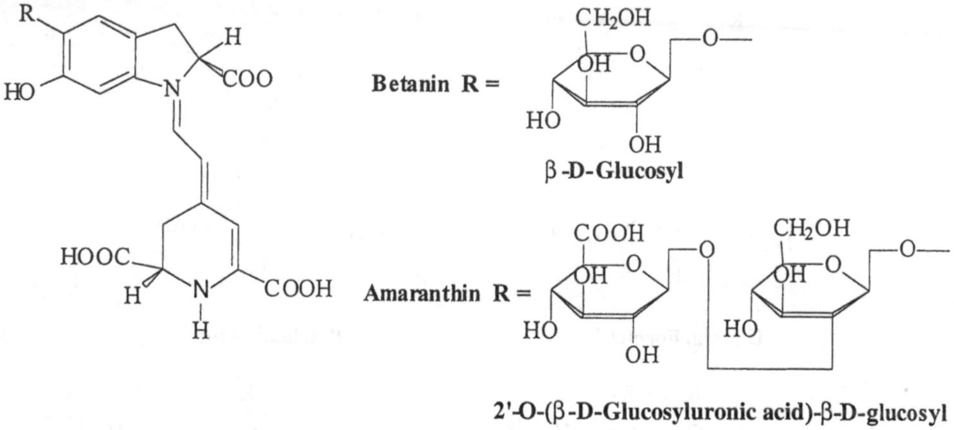

FIG. 2.3   Structure of betanin and amaranthin.

the R group is a glucose molecule, while in amaranth the basic structure is extended by a glucuronic acid molecule linked to the 2 position of glucose. Compared to betanin, this difference has little impact on the pigment's stability toward various environmental factors (Huang and von Elbe, 1986). Betacyanins are optically active because of the chiral carbons C-2 and C-15. Isomerization will occur under acidic conditions and/or the application of heat giving rise to the iso compounds (Fig. 2.4).

In the red table beet (*Beta vulgaris*), betanin and isobetanin are the major betacyanins and account for approximately 75% of all red pigments. The remainder is made up of betanidin and prebetanin (R = glucose-6-sulfate); (Fig. 4) and their C-15 isomers. The impact of heat on the isomerization of betacyanins in beets is illustrated in Fig. 2.5. Fresh beets had an isobetanin content of 3.9%. Blanching whole beets at 100C to an internal temperature of 66C increased the isobetanin content to 10.2%, and canning sliced beets, processed under agitation at 126°C for 10 minutes, further increased the isobetanin content to 28.1% (von Elbe et al., 1981). Twenty to 30% of the total pigments in all commercial beet cultivars are betaxanthins. In the red beet the betaxanthins are vulgaxanthin I and II. Their structures are shown in Fig. 2.6.

PROPERTIES

The hue (color) of betalains is unaffected at the pH of most foods. In the range of pH 3.5–7.0, the visible spectra of betalain solutions are identical and exhibit a maximum light absorption at 537 to 538 nm, therefore no

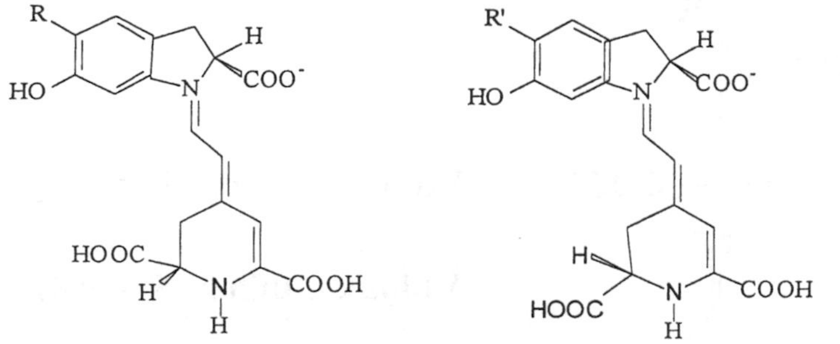

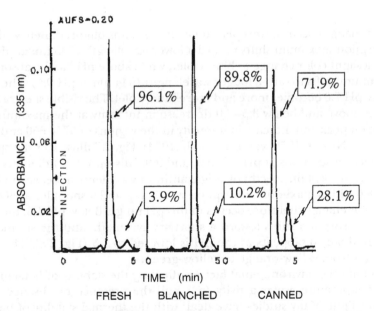

Betanidin, Isobetanidin   R, R' = - OH

Betanin, Isobetanin   R,R' = - Glucose

Amaranthin, Isoamaranthin   R,R' = 2'- Glucuronic acid- Glucose

FIG. 2.4   Structure of betacyanins.

FIG. 2.5   HPLC chromatograms of betanin/isobetanin as found in fresh, blanched, and canned beets. (From von Elbe et al., 1981.)

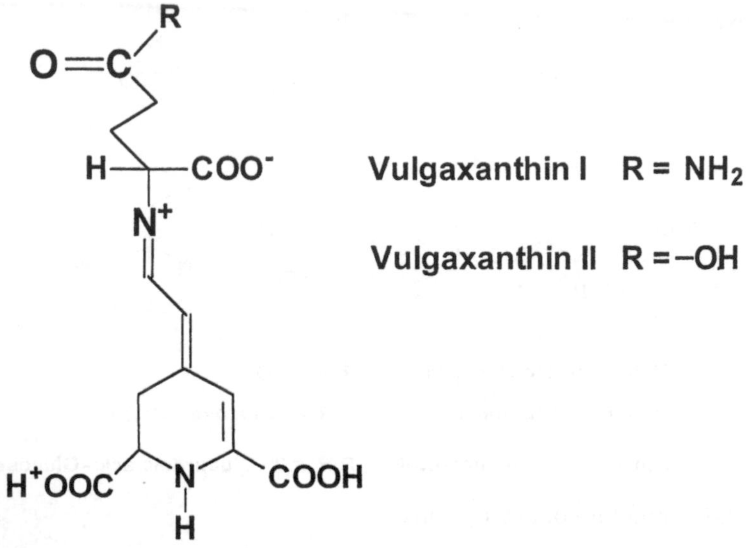

FIG. 2.6   Structure of vulgaxanthins I and II.

Vulgaxanthin I   R = NH$_2$

Vulgaxanthin II   R =−OH

color changes occur in this pH range and remains blue-red. Below pH 3.5, absorption maximum shifts toward a lower wavelength (535 nm at pH 2.0) with a slight color change to blue-violet, while above pH 7.0 the absorption maximum shifts toward a longer wavelength (544 nm at pH 9.0). The color above pH 7.0 becomes more and more blue-violet. The visible spectra at pH values above and below 3.5–7.0 decrease in intensity at the maximum absorption peak and increase in intensity in the regions of 575–650 and 400–450 nm (Nilsson,1970; von Elbe et al., 1974). Fig. 2.7 illustrates the spectra for betanin solutions at pH 2.0, 5.0, and 9.0. Nilsson (1970) demonstrated that the absorption spectra of vulgaxanthin I was identical between pH 4.5 and 8.0 with a maximum absorption at 476 nm and a shoulder at 460 nm. No color change therefore occurs in this pH range. Below pH 4.5, the maximum absorption shifts toward a shorter wavelength, and the shoulder at 460 gradually disappears as the pH is lowered to 1.5 (Fig. 2.8). The color changes from yellow-orange to yellow-green.

Among the environmental factors affecting the stability of betalains are pH, temperature, water activity (a$_w$), and the presence or absence of air and/or light. Many studies have dealt with the thermal stability of betanin (von Elbe et al., 1974; Saguy, 1979; Pasch and von Elbe, 1979; Sapers and Horstein, 1979; Saguy et al., 1979; Huang and von Elbe, 1985). The degradation reaction of betanin in solution has been shown to be at least partially reversible, and both the degradation and regeneration reactions are pH

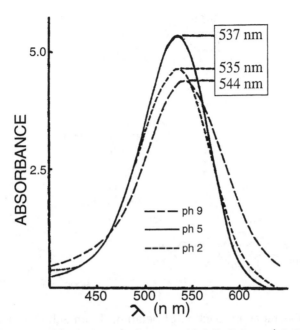

FIG. 2.7 Spectra of betanin solutions at pH 2.0, 5.0, and 9.0. (From von Elbe, 1975.)

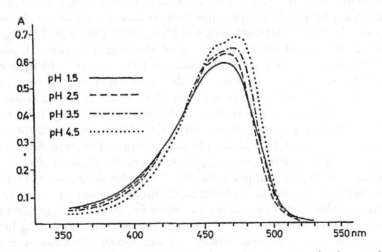

FIG. 2.8 Effect of pH (1.5–4.5) on the absorption spectra of vulgaxanthin I. (From Nilsson, 1970.)

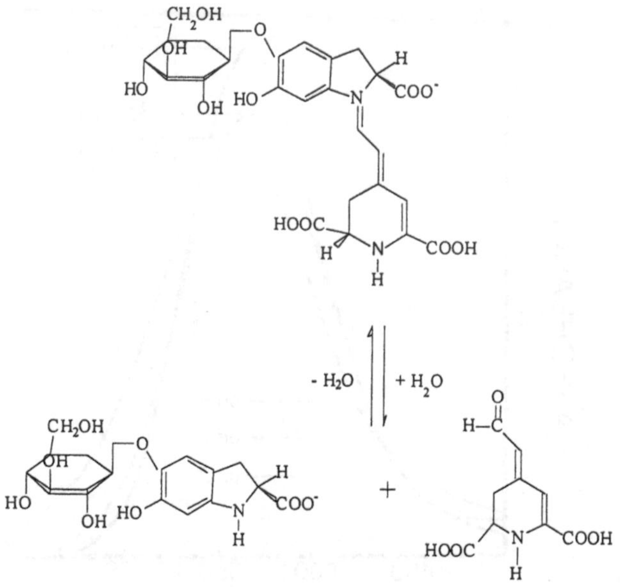

FIG. 2.9 Degradation reaction of betanin. (From Schwartz and von Elbe, 1983.)

dependent (von Elbe et al., 1981; Bilyk and Howard, 1982). Betanin in solution upon heating hydrolyzes into betalamic acid (BA) and cyclodopa-5-O-glycoside (CDG) (Schwartz and von Elbe, 1983) (Fig. 2.9). The regeneration of betanin, which is the reverse of the hydrolysis process, involves a Schiff base condensation of the amine of CDG with the aldehyde of BA (von Elbe et al., 1981). Betanin rapidly forms when solutions of BA and CDG are combined (Schwartz and von Elbe, 1983). It is well known that Schiff base reactions are reversible and temperature and pH dependent (Capon, 1972). It is therefore not surprising that a number of the above-mentioned studies indicated the pH dependence of betanin degradation.

Data in Fig. 2.10 demonstrate regeneration of the pigment when betanin in solution is degraded at pH 5.0 at 90°C. Samples were analyzed immediately after heating and after holding for 20 hours at room temperature. The 20-hour holding period was chosen to allow for maximum regeneration of betanin. The small amount of isobetanin that was formed during heating through the isomerization of betanin was considered as betanin. As betanin degrades (solid line), betalamic acid (BA) and cyclodopa-5-O-glycoside (CDG) accumulate up to 20 minutes, after which BA degrades while CDG accumulates at a lower rate. The increase in betanin after the 20-hour hold-

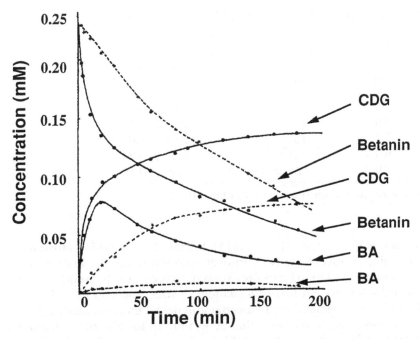

FIG. 2.10 Concentration changes of betanin, betalamic acid, and cyclodopa-5-O-glycoside before (———) and after (-----) regeneration. (From Huang and von Elbe, 1985.)

ing period is the result of regeneration (dashed line) from BA and CDG. The regeneration process is demonstrated by the lower concentration of both BA and CDG (dashed lines) when compared to the concentration (solid line) before regeneration. The partial regeneration was limited by the heat sensitivity of BA, whose concentration after regeneration is virtually zero. Therefore complete regeneration is not possible. In the absence of oxygen, BA will undergo an aldol condensation, thereby making unavailable the aldehyde group required in the regeneration of betanin from BA and CDG. It is interesting to note that a manufacturer of canned red table beets never quality grades for color immediately after sterilization, rather quality grading is done several hours after processing in order for the color to "set." The reason for this practice was certainly not understood, but it is obviously to take advantage of the reversibility of the degradation reaction of the betacyanin pigments after heating.

The ability of atmospheric oxygen to accelerate color loss in red beet pigment–containing products has been recognized for some time. Early

TABLE 2.3 Effect of Oxygen and pH on Half-Life Values
of Betanin in Aqueous Solutions at 90°C

pH	Half-life value of betanin (min)	
	Nitrogen	Oxygen
3.0	56 ± 6	11.3 ± 0.7
4.0	115 ± 1	23.3 ± 1.5
5.0	106 ± 8	22.8 ± 1.0
6.0	41 ± 4	12.6 ± 0.8
7.0	4.8 ± 0.8	3.6 ± 0.3

Source: Adapted from Huang and von Elbe, 1987.

studies by Vilece et al. (1955) and Habib and Brown (1956) attributed the darkening or loss of red color in beet products to the presence of oxygen. In solutions containing a molar excess of oxygen over betanin, the betanin loss follows apparent first-order kinetics (Attoe and von Elbe, 1982). The degradation deviates from first-order kinetics when the molar oxygen concentration is reduced to near that of betanin. In the absence of oxygen, the stability of betanin is increased. Molecular oxygen has been implicated as the active agent in oxidative degradation of betanin. Active oxygen species such as singlet oxygen or superoxide are not involved (Attoe and von Elbe, 1984; Attoe and von Elbe, 1985a). Table 2.3 shows half-life values for betanin in aqueous solution at 90°C between pH 3.0 and 7.0 in the presence and absence of oxygen. The greatest stability of the pigment is observed between pH 4.0 and 5.0. It is also in this range that maximum regeneration occurs.

Oxidation of betalains is accelerated in the presence of light. The presence of antioxidants, such as ascorbic acid and isoascorbic acid, improves the stability (Bilyk et al., 1981; Attoe and von Elbe, 1984; Han et al., 1998). The data by Han illustrate the antioxidant effectiveness of ascorbic and isoascorbic acid (Table 2.4). When 66.2% of the pigment remained after heating and ascorbic acid or isoascorbic acid were present, the pigment regenerated 98%. In contrast, when the pigment was degraded to 15%, regeneration resulted only 45% of the original pigment. The difference in these two results can be attributed to the instability of betalamic acid. Because copper and iron cations catalyze oxidation by ascorbic acid by molecular oxygen, they detract from the effectiveness of ascorbic acid as a protector of betalains. The presence of a chelator (EDTA or citric acid) greatly improves the effectiveness of ascorbic acid as a stabilizer of betalains (Pasch

TABLE 2.4   Regeneration of Pigment in Beet Juice in the Presence of
Ascorbic or Isoascorbic Acid

| | % pigment retention after regeneration[a] | | | |
| | When 66.2% remained after thermal degradation | | When 15.0% remained after thermal degradation | |
Additive (40 mM)	pH 3.8	pH 6.2	pH 3.8	pH 6.2
Control	85.4	65.0	39.6	20.4
Ascorbic acid	98.3	66.5	45.7	25.6
Isoascorbic acid	98.3	68.4	45.7	27.2

[a]Based on the absorbance at 535 nm.
Source: Adapted from Han et al., 1998.

and von Elbe, 1979). Several phenolic antioxidants, including butylated hydroxyanisole, butylated hydroxytoluene, catechin, quercetin, nordihydroguaiaric acid, chlorogenic acid, and α-tocopherol, inhibit free-radical chain autoxidation. Since free radical oxidation does not seem to be involved in betalain oxidation, these antioxidants are, not surprisingly, ineffective stabilizers of betalains. Similarly, sulfur-containing antioxidants such as sodium sulfite and sodium metabisulfite are not only ineffective stabilizers, they also hasten loss of color. Sodium thiosulfite, a poor oxygen scavenger, has no effect on betanin stability. Thioproprionic acid and cysteine are also ineffective as stabilizers. These observations confirm that betanin does not degrade by free-radical mechanism (Attoe and von Elbe, 1985b).

The degradation of betalains requires water. Thus, when water is unavailable or limited, betanin is very stable. This can be illustrated in a number of ways. Pasch and von Elbe (1977) reported a half-life value for betanin at 75°C as a function of water activity (a_w). The half-life was increased from 33 minutes at a_w of 1.0 to 124 minutes at a_w of 0.37. The protective effect is further illustrated by storing a betanin-containing product under water-limiting conditions (Fig. 2.11). When a dry gelatin dessert colored with betacyanin was packaged and stored at 25°C and 95% RH, the pigment degraded rapidly during the 6-month storage period, while almost 100% retention was retained by either storing the samples under dry conditions (in a disiccator over $CaCl_2$) or preventing moisture uptake through packaging in aluminum pouches.

Piatelli et al. (1965) showed that the betaxanthin indicaxanthin could be derived from the condensation of the betacyanin, betanin, and an excess of proline in the presence of 0.6 N ammonium hydroxide under vacuum.

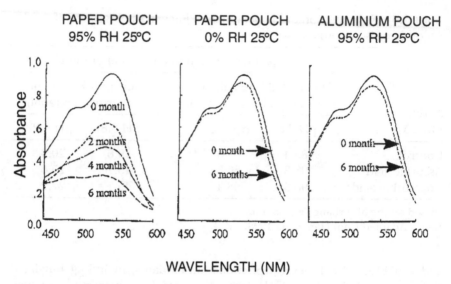

WAVELENGTH (NM)

FIG. 2.11 Spectra of extracts of betalain-colored gelatin stored for 6 months. (From von Elbe, 1977.)

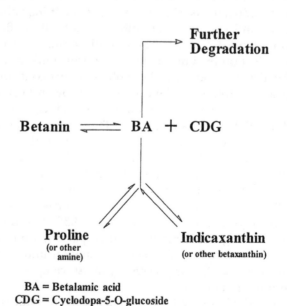

FIG. 2.12 Formation of indicaxanthin from betanin in excess of proline. (From von Elbe and Schwartz, 1996.)

TABLE 2.5 Percent Transformation of Base Exchange
Reaction (Betanin to Indicaxanthin)

Proline (M)	Proline/Betanin ratio	Transformation (%)
0.1	500	94.8
0.05	250	79.3
0.02	100	45.9
0.01	50	28.2

Source: J. H. von Elbe, unpublished data.

TABLE 2.6 Transformation of Betanin to Betaxanthin

Amino acid (0.01)	Relative light absorption	Maximum wavelength (nm)
Alanine	0.12 ± 0.01	472
Glutamic acid	0.028 ± 0.005	476
Glutamine	0.041 ± 0.004	474
Glycine	0.17 ± 0.01	474
Leucine	0.086 ± 0.005	474
Lysine	0.25 ± 0.01	469
Proline	0.64 ± 0.02	485
Serine	0.18 ± 0.01	475

Source: J. H. von Elbe, unpublished data.

This was the first conclusive evidence of the structural relationship between betacyanins and betaxanthins. It was further demonstrated that formation of betaxanthin from betanin involved condensation of a betanin hydrolysis product, betalamic acid, and an amino acid. Formation of indicaxanthin from betanin in excess of proline is illustrated in Fig. 2.12. The completion of the reaction is dependent on the relative concentration of proline (Table 2.5). At a proline:betanin ratio of 500:1, approximately 95% of the betanin was transformed to indicaxanthin, while at a ratio of 50:1 only 29% was transformed. Other amino acids will react similarly. Some relative data involving several amino acids is given in Table 2.6. These data only demonstrate the transformation of betanin to a betaxanthin with different amino acids. The maximum light absorption, depending on the amino acid, is between 469 and 485 nm. The greater light absorption of some con-

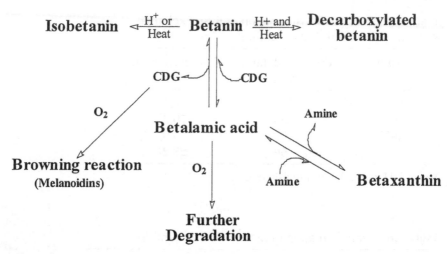

FIG. 2.13 Degradation of betanin under acid and/or heat. CDG = Cyclodopa-5-O-glucoside. (From von Elbe and Schwartz, 1996.)

densation products seems to be related to the water solubility of the amino acid. A good amino acid source for the reaction would be a bovine casein hydrolysate. Bovine casein contains 23.5% proline, 10.1% lysine, and 7% serine. This transformation reaction has not been found to generate yellow betaxanthins. Conversion of betacyanin to betaxanthin can occur in protein-rich foods. Betaxanthins are less stable than betacyanins under similar conditions (Saguy, 1979; Singer and von Elbe, 1980). The major reactions of betanin are summarized in Fig. 2.13.

REFERENCES

Attoe, E. L., and von Elbe, J. H. 1982. Degradation kinetics of betanin in solutions as influenced by oxygen. *J. Agr. Food Chem.* 30(4): 708–712.

Attoe, E. L., and von Elbe, J. H. 1984. Oxygen involvement in betanin degradation. *Z. Lebensm. Unters. Forsch.* 179: 232–236.

Attoe, E. L., and von Elbe, J. H. 1985a. Oxygen involvement in betanin degradation—Measurement of active oxygen species and oxydation reduction potential. *Food Chem.* 16: 49–67.

Attoe, E. L., and von Elbe, J. H. 1985b. Oxygen invovement in betanin degradation; Effect of antioxidants. *J. Food Sci.* 50(1): 106–110.

Bilyk, A., and Howard, M. 1982. Reversibility of thermal degradation of betacyanines under the influence of isoascorbic acid. *J. Agric. Food Chem.* 30: 906–909.

Bilyk, A., Kolodij, M. A., and Sapers, G. M. 1981. Stabilization of red beet pigments with isoascorbic acid. *J. Food Sci.* 46: 1616–1617.

Capon, B. 1972. Reaction of aldehydes and ketones and their derivatives. In *Organic Reaction Mechanisms*, B. Capon and C. W. Rees (Eds.), p. 401. Interscience Publishers, New York.

Goldman, I. L., Eagen, K. A., Breitbach, D. N., and Gabelman, W. H. 1996. Simultaneous selection is effective in increasing betalain pigment concentration but not total dissolved solids in red beet (*Beta vulgaris* L.). *J. Am. Soc. Hort Sci.* 121: 23–26.

Habib, A. T., and Brown, H. D. 1956. The effect of oxygen and hydrogen ion concentration in color changes of processed beets, strawberries and raspberries. *Am. Soc. Hort. Sci.* 68: 482–490.

Han. D., Kim, S. J., Kim, S. H., and Kim, D. M. 1998. Repeated regeneration of degraded red beet juice pigments in the presence of antioxidants. *J. Food Sci.* 63: 69–72.

Huang, A. S., and von Elbe, J. H. 1985. Kinetics of the degradation and regeneration of betanin. *J. Food Sci.* 50: 1115–1120, 1129.

Huang A. S., and. von Elbe, J. H. 1986. Stability of two betacyanin pigments—amaranthin and betanin. *J. Food Sci.* 51: 670–674.

Huang, A. S., and von Elbe, J. H. 1987. Effect of pH on the degradation and regeneration of betanin. *J. Food Sci.* 52: 1689–1693.

Keller, W. 1936. Inheritance of some major color types in beets. *J. Agric. Res.* 52: 27–38.

Nilsson, T. 1970. Studies into the pigments in beetroot (*Beta vulgaris* L. ssp. *vulgaris* var. *rubra* L.). *Lantbrukshoegsk. Ann.* 36: 179–219.

Pasch, J. H., and von Elbe, J. H. 1975. Betanin degradation as influenced by water activity. *J. Food Sci.* 40: 1145–1146.

Pasch, J. H., and von Elbe, J. H. 1979. Betanine stability in buffered solutions containing organic acids, metal ions, antioxidants or sequestrants. *J. Food Sci.* 44: 72–74, 81.

Piatelli, M., L. Minale and G. Prota. 1965. Pigments from Centospermae. III. Betanxanthines from *Beta vulgaris* L. *Phytochemistry* 4: 121–125.

Saguy, I. 1979. Thermostability of red beet pigments (betanin and vulgaxanthin I); influence of pH and temperature. *J. Food Sci.* 44: 1554–1555.

Saguy, I., Kopelman, I. J., and Mizrahi, S. 1979. Thermal kinetic degradation of betanin and betalamic acid. *J. Agric. Food Chem.* 26: 360–362.

Sapers, G. M., and Hornstein, J. S. 1979. Varietal differences in colorants properties and stability of red beet pigment. *J. Food Sci.* 44: 1245–1248.

Seneca Foods Corp. 1998. Beet juice concentrate, 0.8% betanin. Clyman, WI.

Schwartz, S. J., and von Elbe, J. H. 1983. Identification of betanin degradation products. *Z. Lebensm. Unters. Forsch.* 176: 448–453.

Singer, J. W., and von Elbe, J. H. 1980. Degradation rates of vulgaxanthin I. *J. Food Sci.* 45(3): 489–491.

Vilece, R. J., Fagerson, I. S., and Esselen, W. B. 1955. Darkening of food purees and concurrent changes in composition in headspace gas. *J. Agric. Chem.* 3: 433–435.

Von Elbe, J. H. 1975. Stability of betalains as food colorants. *Food Technol.* 29(5): 42–43, 46.

Von Elbe, J. H. 1977. The betalains. In *Current Aspects of Food Colorants*, T. E. Furia (Ed.), pp. 29–39. CRC Press, Cleveland, OH.

Von Elbe, J. H., Maing, I. Y. and Amundson, C. H. 1974. Color stability of betanin. *J. Food Sci.* 39: 334–337.

Von Elbe, J. H., and Schwartz, S. J. 1996. Colorants. In *Food Chemistry*, O. R. Fennema (Ed.), pp. 656–673. Marcel Dekker, New York.

Von Elbe, J. H., Schwartz, S. J. and Hildenbrand, B. E. 1981. Loss and regeneration of betacyanin pigments during processing of red beets. *J. Food Sci.* 46: 1713–1715.

Watson, J. F., and Gabelman, W. H. 1984. Genetic analysis of betacyanin, betaxanthin, and sucrose concentrations in roots of table beet. *J. Am. Soc. Hort. Sci.* 109: 386–391.

Wolyn, D. J., and Gabelman, W. H. 1989. Inheritance of root and petiole pigmentation in red table beet. *J. Hered.* 80: 33–38.

Wolyn, D. J., and Gabelman, W. H. 1990. Selection for betalain pigment concentrations and total dissolved solids in red table beets. *J. Am. Soc. Hort. Sci.* 115(1): 165–169.

3

Monascus

Richard E. Mudgett

University of Massachusetts*
Amherst, Massachusetts

INTRODUCTION

There is a general concern with respect to the safety of synthetic food colorants approved by the U.S. Food and Drug Administration (FDA). This
has resulted in the avoidance of "artificial" colorants by a growing number of health-conscious consumers and suggests a potential for developing new sources of natural food colorants from microbial organisms such
as *Monascus*. The genus *Monascus* produces several natural pigments, including the primary colors red and yellow and the secondary color orange.
These pigments are widely used in Asia as food colorants (Han, 1990;
Anonymous, 1999a).

   Monascus pigments are produced commercially in China, Japan, and Taiwan, primarily by a traditional method of food fermentation known as the
koji process. The process involves the inoculation of solid substrates, including a variety of natural agricultural products such as rice, wheat, soybeans, corn, and other grains. The advantages of solid substrate processes
include the ability of fungal and fungus-like organisms such as *Streptomyces*
to grow on solid surfaces at low moisture contents and water activities, low
shear environments, high productivity, and ease of product recovery by

*Current affiliation: Technical Liaisons, Amherst, Massachusetts

downstream processes such as countercurrent extraction (Mudgett, 1986). Major sources of information on such products and their production methods are available from authoritative publications by Steinkraus (1995) and Wang and Hesseltine (1979) and from a recent book by Ang et al. (1999) on traditional Asian foods.

The koji process may also be combined with a second-stage submerged culture process, known as a moromi, in which fermented solids are cultured in deep tanks to manufacture food products such as rice wine and soy sauce. *Monascus* pigments are used in Asia for many food applications such as coloring processed meats, seafood products, alcoholic and nonalcoholic beverages, dairy products, jellies and jams, and tomato ketchup. *Monascus* pigments are secondary metabolites of a polyketide nature derived from linear condensation of acetyl-CoA molecules and molecules of malonyl-CoA with subsequent modification of the condensate through a sequence of as-yet-unknown reactions (Miyake et al., 1984).

The major source of red pigments used in Asia as food colorants are strains of the organism *Monascus purpureus* Went by means of the Ang-kak, or red rice, process, also known as Anka in Taiwan. This process is known in Mainland China, the Philippines, and Thailand by names such as ankak, angquac, beni-koji, and aka-koji. The advantages of the process are that its raw materials are readily available, the yield is good, the color of the pigment produced is consistent and stable, the pigment is water-soluble and is easily mixed with other natural pigments and foods, and there is no evidence of carcinogenicity or toxicity. A number of countries have adopted natural pigments to replace coal-tar dyes because the latter have been implicated as carcinogens. Anka is said to have been first noted in the Yuan Dynasty (1271–1368 A.D.) and was introduced to Formosa by wine makers from China about 100 years ago. It is used to color foods such as fish, rice wine, red soybean cheese, pickled vegetables, and salted meats (Steinkraus, 1995).

Constant oxygen and carbon dioxide pressures maintained with a closed feedback control system, admitting oxygen as consumed and purging carbon dioxide as evolved during microbial respiration, significantly influenced growth and pigment production in solid-state fermentations of rice by a strain of *Monascus purpureus*. At a carbon dioxide pressure of 0.02 atm, red pigment yield increased from 130 to 204 mg/g of rice solids as the partial pressure of oxygen was increased from 0.05 to 0.50 atm. Conversely, increasing carbon dioxide pressures from 0.01 to 0.20 atm at a constant oxygen pressure of 0.21 atm decreased pigment production from 218 to 130 mg/g of rice solids, indicating a stimulatory effect of oxygen and an inhibitory effect of carbon dioxide, suggesting that pigment production by *Monascus* can be optimized by control of the gas environment (Han and Mudgett, 1991).

References cited in this chapter were obtained from published papers in available literature, personal communication with other researchers, *Food*

Science and Technology Abstracts (100), complete records from Agricola (77), abstracts from the Institute of Scientific Information's Web of Science (116), and information from websites located by Altavista (200). Other search engines had fewer citations, most of which were included in Altavista's websites. Many of the citations were in literature published in Asia and are not generally available in western libraries.

OVERVIEW OF THE GENUS *MONASCUS*

Taxonomic Considerations

Fungi are members of the division of the plant kingdom known as Thallophyta. They do not possess chlorophyll, nor are they differentiated into leaves, stems, or true roots, and they are widely distributed in nature, especially in the soil (Prescott and Dunn, 1959). The genus *Monascus*, a member of the class Ascomycetes, is descended from the Eumycetes, or true molds, the subclass Plectomycetes and the order Plectascales. However, the genus has also been considered as a member of the Fungi Imperfecti. The term "mold" is generally given to aerobic saprophytes that grow on organic matter or in solutions with the formation of expansive masses of mycelium, which may be thin and superficial in character or may occur as felted masses of tough and/or semi-gelatinous nature. The general distinction of molds as aerobic saprophytes is of some importance with regard to the safety of their general use in the production of traditional Asian foods and to their specific use in the production of pigments for use as food colorants or ingredients.

Structurally, molds consist principally of mycelia and spores. Of the four major classes of the Eumycetes, or true molds, the Ascomycetes possess septate mycelia and produce sexual spores exogenously in sacs. The mycelium is a collection, or aggregate, of hyphae, which are threadlike filaments of protoplasm. Hyphae are of two main types: (1) fertile hyphae, which are concerned with the production of reproductive cells, or fruiting bodies, the spores, and (2) vegetative hyphae, the function of which is to secure nutrients from their substrates. The septate hyphae of Ascomycetes contain crosswalls, or septa, which divide the mold into cells. Ascomycete cells are mononucleate. Sexual fusion is by gametangia or through somatic hyphae (Burnett, 1970) and colonies appear as blue-black or greenish molds or as powdery mildews. Ascomycetes are classified as members of the Eumycetes, or true molds, by the fact that sexual reproduction is totally absent among the Fungi Imperfecti. The life cycles of Ascomycetes and Plectomyces are described as haploid-dikaryotic. Ascospores are liberated by irregular rupture of both the asci and the cleistothecium. Gametes are represented by nuclei, which may come together by nuclear migration through the hyphae. Nuclei, which may eventually fuse, divide conjugately in the binucleate ascogenous hyphae, thus initiating a brief dikaryophase.

Growth of the mycelium is apical, i.e., by extension of the tip cells, and cell division is intercalary, i.e., cells may divide in any part of a hypha, or mycelial branch. This is possible because nuclei can migrate from one part of the mycelium to another. Young cells are filled with dense protoplasm, while older cells, prior to sporulation, contain many vacuoles and reserve food materials such as fat globules and/or glycogen. Cell walls are primarily composed of chitin. Ascospores of Ascomycetes are sexual spores and are produced characteristically in a special sac, or ascus. The number of mold species is large, but their classification is often difficult because of great differences in colonial appearance on different substrates. This will be of some importance in further discussion of the number of *Monascus* species.

Ainsworth et al. (1973) reported that the Monascaceae, one of the lower Ascomycetes, consists of a single genus *Monascus* van Tieghem, having a few species. More recent reports claim more than 20 species. An important insight in the systematic study of *Monascus* was provided by Carels and Sheperd (1977), who demonstrated that many of the reported differences in pigmentation were related to cultural conditions rather than to genotypic differences. This was supported later by Wong et al. (1981). Species concepts within *Monascus* were uncertain for many years before the revision by Hawksworth and Pitt (1983). They recognized three species on the basis of cultural and microscopic characteristics, namely *Monascus pilosus* strains Sato ex Hawksw. and Pitt sp. nov, *Monascus purpureus* Went, and *Monascus ruber* van Tieghem. On the other hand, Iizuka and Lin (1983) classified *Monascus* cultures into 12 species based on the formation of ascomata, length and shape of conidial chains, superficial appearance of colonies, and extent of pigment production. Biochemical tests were developed by Bridge and Hawksworth (1985) as an aid in identifying *Monascus* species. Enzymatic activities of nine *Monascus* strains were compared with API ZYM enzyme testing strips and other tests developed based on penicillin sensitivity. Consistent results were obtained between strains of the same species, confirming their taxonomy. Strains of *Monascus purpureus* had strong pectinase activity at pH 6 that was not evidenced in other species. Strains of *Monascus ruber* were the only strains exhibiting cellulase activity.

Nine species and 75 strains of *Monascus* are listed in the American Type Culture Collection (ATCC) catalogue (Jong and Edwards, 1991), including 21 strains of *Monascus purpureus* Went, the species most often cited in traditional Asian food fermentations for the production of red rice commonly used as a source of pigments for use as food colorants. In addition, 7 strains of *Monascus kaoliang* Iizuka et Manaki and 5 strains of *Monascus pilosus* Sato ex D. Hawksworth and Pitt used for the production of red rice are briefly described in terms of their country of origin, classification, and culture medium. The other 6 *Monascus* species appear to be isolates from various sources around the world, including the soil, spoiled prunes, milk curd, creosote-treated wood, earwax, shoes, canvas, and corn silage. Isolates were recently obtained from Turkish leather goods and shoes (Birbir et al., 1994). *Monascus filiformis*, originally discovered in 1819 by Rudolphi, is reported to have a life cycle involving three marine hosts in the southwest Atlantic Ocean, similar to the species in the shallow waters of the Argentine Sea. Species found in the North Sea differ from this schema by addition of intermediate planktonic invertebrates as intermediate hosts (Martorelli and Cremonte, 1998). Two new species have recently been isolated from surface sediments of the Shatt-al-Arab River in Iraq, namely *Monascus pallens* and *Monascus sanguineus* (Cannon et al., 1995).

Using ribosomal DNA sequencing, *Monascus purpureus*, although usually placed in a separate family, was grouped with Eupenicillium and the *Aspergillus* species of the Trichocomaceae (Berbee et al., 1995). Similarly, *Monascus purpureus* has been shown to be closely related to five species of *Aspergillus* based on nucelotide sequencing of the genes encoding the 18S-rRNA (Verweig et al., 1995). Perhaps more interestingly, four isozymes isolated from 15 *Monascus* species were analyzed by polyacrylamide gel electrophoresis (Nishikawa and Iizuka, 1993). Numerical analysis based on similarity values from the electrophoretic mobilities of the enzymes showed that these species could be divided into two subspecies, Groups I and II, and could be further divided into two subgroups. Strains of *Monascus ruber* and *Monascus kaoliang* were found to belong in Group I and strains of *Monascus pilosus* and *Monascus pubigherus* in Group II. Nearly 100% similarity was found between strains of *Monascus anka* and *Monascus rubiginosus*, strains of *Monascus anka* and *Monascus purpureus* Went, and among strains of *Monascus vitreus*, *Monascus fuliginosus*, and *Monascus serorubescens*. The authors concluded that electrophoretic analysis of endogenous enzymes may be useful in developing a new classification system for *Monascus* species, several strains of which studied in their work are known producers of pigments in the red rice process.

Nutritional and Growth Characteristics

The growth requirements of molds, yeasts, and bacteria, like those of plants, include sources of nitrogen, sulfur, phosphorus, hydrogen, potassium, sodium, magnesium, and other minerals such as trace elements. They may also require organic supplements such as vitamins. Unlike most plants, however, that derive carbon intermediates and energy from photosynthesis, molds also require a source of carbon and energy for catabolic and anabolic biochemical reactions in primary and secondary metabolism and an oxygen supply to oxidize the carbon and provide energy through oxidative phosphorylation. Species of *Monascus* used for traditional food fermentations are generally aerobic, as are most molds. Recommended media and temperatures for cultivation of *Monascus* species available from their culture collection are indicated in ATCC's *Catalogue of Filamentous Fungi* (Jong and

Edwards, 1991). These primarily involve variations of malt extract media, with or without yeast extract and/or peptone and glucose or sucrose as carbon and energy sources, and potato dextrose media, with glucose as the carbon energy source and temperatures from 24 to 30°C.

Nutritional and growth studies on *Monascus* have been mainly conducted in submerged culture fermentations. Lilly and Barnett (1962) reported that initial growth rates of *Monascus purpureus* on fructose, glucose, and invert sugar increased in that order. More mycelium was produced on media containing sucrose and one or both of its hydrolytic products than produced on control media containing only one of the two monosaccharides. McHan and Johnson (1970) showed that *Monascus purpureus* grew better in glucose-peptone-yeast extract broth than in any other complex medium used and that it did not grow nearly as well in Czapek-Dox mineral-salt-glucose broth. Zinc and a special combination of amino acids (glycine, leucine, and tryptophan) were identified as the important components of the richer medium responsible for the increased growth. Johnson and McHan (1975) also reported that in Czapek-Dox medium containing 5% D-glucose, zinc uptake by *Monascus purpureus* correlated with an increase in both phosphate and glucose uptake, increased growth, and increased production of carbon dioxide and ethyl alcohol. The growth responses of *Monascus purpureus* to the three amino acids and zinc were restricted to media containing carbohydrates or substrates related to the glycolytic pathway as the sole source of carbon and energy. McHan and Johnson (1979) also studied the effects of zinc on the utilization of nitrogen sources in which growth studies indicated that ammonium compounds were preferred over inorganic nitrogen sources and that nitrites were better than nitrates. Each of several amino acids, glycine, leucine, valine, aspartate, glutamate, arginine, and urea were found equal or superior to ammonium salts as a sole source of nitrogen. The addition of zinc to the media resulted in increased growth, regardless of the nitrogen source. The authors suggested that zinc has an apparent role in the interaction between amino acid and carbohydrate metabolism that must take place for optimum growth. Wong et al. (1981) investigated the effect of various glucose and ammonium nitrate concentrations on the growth of the mutant *Monascus purpureus* N11S and on its pigment synthesis. Mycelial growth increased with an increase in glucose and ammonium nitrate concentrations. Larger amounts of ammonium nitrate were required to give maximal growth in media containing higher concentrations of glucose. However, they noted that high ammonium ion concentrations may reduce nitrate reductase activity and nitrate assimilation.

More recently, Lin and Demain (1991) used *Monascus* strain TTWMB 6042 to investigate the regulation of pigment production based on nutrient composition. An initial medium containing 4% glucose, 0.3% ammonium nitrate, and inorganic salts as a control was supplemented with various inorganic compounds and with substitution of glucose by other carbon and energy sources. The formation of red pigment in these studies was monitored by optical density measurements at a wavelength of 500 nm. The carbon source and initial pH were found to be important. Pigment synthesis was strongly stimulated by monosodium glutamate (MSG) as the sole nitrogen source and was inhibited by high phosphate or magnesium sulfate concentrations. Based on these studies, they developed a chemically defined medium containing 5% maltose and 0.75 mM MSG. Similar studies were made by Chen and Johns (1993) based on HPLC analysis. Four pigments were detected in fungal extracts: the yellow pigments monascin and ankaflavin, the orange pigment rubropunctatin, and the red pigment monascorubramine. Growth and ankaflavin synthesis were favored at low pH, while production of the other pigments was relatively pH independent. The nitrogen source affected both growth and pigment synthesis, with ammonium salts and peptone giving higher fungal growth and pigment synthesis as compared with cultures on nitrates, in which ankaflavin was not detected. Further studies by Lin and Demain (1995) found that media employing ammonium nitrate as the sole nitrogen source gave low levels of pigment production by *Monascus* species and high levels of cell-bound pigments. They found that ammonium nitrate did not repress pigment synthase formation, enhance synthase decay, or serve as a nitrogen source for pigment production by resting cells. The high levels of cell-bound pigments did not exert a feedback effect on the further synthesis of pigments, suggesting that low pigment production was due to the poor ability of ammonium nitrate to donate nitrogen in the Schiff-based reaction converting orange pigments to red ones.

Recent studies have also shown that *Monascus purpureus* DSM 1379 oxidizes fatty acids to methyl ketones (Peters et al., 1993) and influences the order of monosaccharide metabolism in a two-phase growth cycle. The formation of methyl ketones from fatty acids was described as a known detoxifying mechanism that occurs during the lag phase of growth. Fermentations of fatty acid and saccharide mixtures led to two-phase growth kinetics, i.e, fatty acid oxidation took place prior to saccharide metabolism and fungal growth. This is described by the authors as a negative diauxic effect and discussed with reference to the regulation of fatty acid metabolism by methyl ketone and saccharride mixtures. Although it is known that *Monascus* strains produce ethanol in both submerged culture and solid substrate fermentations (Johnson and McHan, 1975; Endo, 1985; Matsumoto et al., 1987; Han and Mudgett, 1991), little information is available on nutritional effects on production of ethanol by *Monascus* and its role in pigment synthesis. Studies by Juzlova et al. (1994) show that pigment synthesis on 2% ethanol by *Monascus purpureus* CCM1852 was much higher than on maltose. Ammonium chloride and ammonium hydroxide favored synthesis of both yellow and red pigments, while yellow pigments were favored in cultures supplemented with peptone. A two-stage culture using maltose and ethanol, consecutively, was suggested to increase the efficiency of ethanol utilization for pigment production.

Monascus Pigments

Monascus pigments are of increasing interest in the West based on their traditional use as food colorants and their manufacture by solid substrate methods in the Far East. It was not until the early 1960s that systematic studies of *Monascus* resulted in isolation and structural characterization of the pigments monascin and monascorubrin from *Monascus purpureus* Went (Fielding et al., 1960, 1961; Haws and Holker, 1961; Inouye et al., 1962; Kumasaki et al., 1962). Monascin and rubropunctatin were isolated from *Monascus rubropunctatus* Sato (Fielding et al., 1961), monascin from *Monascus rubiginosus* Sato (Haws et al., 1959), and monascin and ankaflavin from *Monascus anka* Sato (Chen et al., 1969, 1971; Marchand and Whalley, 1973).

The structures of six water-insoluble *Monascus* pigments are shown in Fig. 3.1A (Lin and Demain, 1994). Some physical and chemical properties of these compounds are also shown in Table 3.1. The pyronoid oxygen atom of the orange pigments, monascorubin and rubropunctatin, is seen to be replaced by an amino nitrogen to form the red pigments monascorubramine and rubropunctamlne, respectively. A red water-soluble glutamate derivative is also shown in Fig. 3.1B (Lin and Demain, 1994). It is possible that only the monascorubrln-rubropunctatin pigments are produced biosynthetically and other pigments are formed by chemical transformations.

As seen in Figure 3.1, monascin bears the same structural relationship to ankaflavin as monascorubrin does to rubropunctatln. Rubropunctatin frequently occurs in admixture with its homolog, monascorubrin, from which it is separable with difficulty, a fact that has given rise to much confusion (Fielding et al., 1960; Chen et al., 1971). It has been observed that monascin is often associated with minor amounts of ankaflavin (Marchand and Whalley, 1973). On the other hand, monascorubrin can be selectively destroyed by hydrogen peroxide, since monascin is more stable to this reagent than monascorubrln (Kumasaki et al., 1962), suggesting this as a simple way to obtain pure monascin. Chen et al. (1971) found that the red pigment rubropunctatin could be obtained by the oxidation of monascin with dichlorocyanoquinone.

These *Monascus* pigments are readily soluble in ethanol and only slightly soluble in water (Su and Huang, 1980). Monascorubrin from *Monascus purpureus* Went (Kumasaki et al., 1962) is soluble in ether, methanol, ethanol, benzene, chloroform, acetic acid, and acetone and insoluble in water and petroleum ether. The ethanol solution of this compound becomes red in alkali and yellow in acid. Su and Huang (1980) showed that the pigment could provide color hues ranging from reddish-orange at pH 3–4, to red at pH 5–6, and to purplish-red at pH 7–9. Dominant wavelengths of the pig-

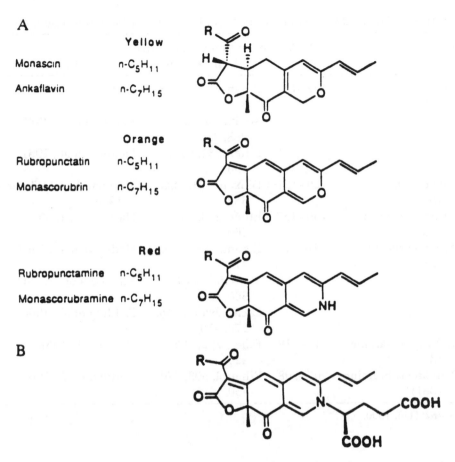

A

Yellow

Monascin   n-C_5H_{11}

Ankaflavin   n-C_7H_{15}

Orange

Rubropunctatin n-C_5H_{11}

Monascorubrin   n-C_7H_{15}

Red

Rubropunctamine   n-C_5H_{11}

Monascorubramine   n-C_7H_{15}

B

FIG. 3.1 Structures of *Monascus* pigments. (From Lin and Demain, 1994.)

ment were between 618–700 nm. Little information was found in the literature reviewed on pigment stability. A red pigment produced by *Monascus anka* faded in solution during prolonged exposure to strong light, like the carotenes, and was more photostable in 70% ethanol than in water. The pigment was heat stable in 70% ethanol up to 1000°C and was stable under neutral and alkaline conditions. Since *Monascus* pigments are insoluble or sparingly soluble in water, various methods to increase pigment solubility have been investigated. Yoshimura et al. (1975) suggested that pigments may be bound by protein-like compounds that make them apparently soluble. These results were confirmed (Broder and Koehler, 1980; Lin and Iizuka, 1982). The molecular weights of two protein-like substances were determined by gel column chromatography to be 5,000 and 35,000 daltons.

TABLE 3.1 Selected Physical and Chemical Properties of *Monascus* Pigments

Pigment (MW)	M.P. (°C)	Absorption wavelengths [Solvent: λ (nm)]	Ref.
Monascin (358)	142–145	Ethanol: 232, 238, 392	Chen et al., 1971
		Methanol: 225, 288, 385	Enouye et al., 1962
		Ether: 231, 287, 390	Fielding et al., 1961
Ankaflavin (386)	120–121	Dioxane: 212, 228, 382	Marchand and Whalley, 1973
Rubropunctatin (354)	156–157	Ether: 218, 246, 460	Haws et al., 1959
Monascorubrin (382)	141–143	Ethanol: 286, 416, 470	Hadfield et al., 1967
		Methanol: 470, 500, 556	Kumasaki et al., 1962
		Cyclohexane: 246, 278, 410	Fielding et al., 1960
Rubropunctamine (335)	217–218	Ether: 252, 305, 426	Haws et al., 1959
Monascorubramine (381)	207–208	Ethanol: 303, 370, 450	Hadfeld et al., 1967

Source: Han, 1990.

It was observed that the amount of *Monascus* pigment found in aqueous media depended on the concentration of water-soluble protein or amino acids in solution.

The orange pigments., monascorubrin and rubropunctatin, readily react with compounds containing amino groups via a ring-opening with a Schiff rearrangement to form water-soluble compounds (Francis, 1987). *Monascus* pigments have been reacted with amino sugars, polyamino acids, amino alcohols, ethanol, chitosan, and hexamine (Moll and Farr, 1978), proteins and peptides (Yamaguchi et al., 1973; Kawabata and Sato, 1976), bovine serum albumin (Broder and Koehler, 1980), casein (Nakagawa et al., 1976a), gluten (Tsunenaga and Once, 1975), amino acids (Broder and Koehler, 1980), RNA (Nakagawa et al., 1976b; Toyo Brewing Co., 1979), nucleic acids (Toyo Jozo Company, 1978), sugar–amino acid browning reaction products (Yoshimura et al., 1975), and aminoacetic and aminobenzoic acids (Wong and Koehler, 1983).

Sweeny et al. (1981) prepared *N*-glucosyl derivatives of the pigments rubropunctamine and monascorubramine and demonstrated a significant photoprotective effect by 1,4,6-trihydroxynaphthalene, suggesting that this compound provided necessary structural elements for molecular juxtaposition between the planar aromatic moieties of both partners and the formation of three hydrogen bonds involving unpaired electrons of the nitrogen and the two carbonyls of *N*-glucosylmonascorubramine as proton acceptors. Therefore it seems that hydrophobic and hydrophilic interactions between *Monascus* pigments and other components in foods may contribute to increases in the solubility and stability of *Monascus* pigments. Wong and Koehler (1983) modified *Monascus* pigments by reacting them with aminoacetic acid and found that water solubility, thermal stability, and photostability of modified pigments were significantly higher than those of unmodified pigments.

More recently, yellow pigments were found to result from reaction of the *Monascus* pigments with amines, followed by reduction of the ring carbonyl moiety to a hydroxyl group. Yellow pigments made from amino acid esters, dipeptide ester, and sugar alcohol amines were also found to be effective at levels imparting no objectionable taste in foods (St. Martin et al., 1991). Lin and Demain (1991) and Blanc et al. (1994) found that a chemically defined medium containing glutamic acid as sole nitrogen source resulted in optimal pigment production (either free or complexed) by *Monascus purpureus* and *Monascus ruber*. Structural data were obtained for monascorubrin and rubropunctatin complexed with side chains of capryl (c7) and caproyl (c5) linked to glutamic acid by amino groups where nitrogen replaced the pyronoid oxygen were obtained by IR, UV, NMR, and MS analysis. Hajjaj et al. (1997) found that *Monascus ruber* produced water-soluble red pigments in submerged culture with a chemically defined medium containing glucose as sole carbon source and monosodium glutamate (MSG) as sole nitrogen source. New molecules with polyketide structures, *N*-glucosylrubropunctamin and *N*-glucosylmonascorubramin, were isolated and studied by HPLC, NMR, Dionex, and mass spectometry methods. UV, polarographic, and thin-layer voltammetric methods showed an electron-donor complex. Extraction of the pigments with *N*-butanol also stabilized the pigments against daylight for several months. Watanabe et al. (1997) also developed a capillary electrophoretic method to analyze yellow pigments for use as food additives. Pigments extracted with micellar electrokinetic chromatography (MEKC) were found to contain xanthomonasin A. Sato et al. (1997) found eight major pigments, characterized as alanine or aspartate derivartives of monascorubrin and rubropunctatin, from various commercial *Monascus* pigments. The derivatives were found to have an azaphilone structure. Further analysis revealed that the isolated pigments included D and L forms of the amino acids. Hajjaj et al. (1998) also developed sampling and extraction methods to obtain pigments from *Monascus ruber* ATCC 96218 and showed that dropping the mycelia into liquid nitrogen or 60% methanol at −40°C were equally effective in terms of pigment recovery and stability. The best results for both recovery and stability were obtained with boiling buffered ethanol. Finally, Teng and Feldheim (1998) separated purplish pigments from *Monaascus anka* cultivated on rice into two fractions: the "orange" pigments, rubopuncatatin and monascorubrin, and the yellow pigments, monascin and ankaflavin. Treatment of the orange pigments with aqueous ammonia solutions gave the red nitrogen analogs rubropunctatamine and monascorubramine. These pigments were separated by HPLC or TLC mathods and identified by an HPLC-DAD (HPLC diode-array detector) method to determine their UV-VIS spectra.

Pigment Biosynthesis

Biosynthesis of fungal secondary metabolites, such as *Monascus* pigments, may be grouped in several metabolic classes according to the pathway from which they originate, as shown in Fig. 3.2 (Turner, 1871), from primary metabolites and appears to involve fatty acids. The polyketide pathway is almost exclusively involved in the production of secondary metabolites of the filamentous fungi, including the *Monascus* pigments, and to a much lesser extent in secondary metabolites of bacteria and higher plants. Polyketides are formed by the condensation of one molecule of acetyl-CoA with at least three molecules of malonyl-CoA. Fig. 3.3 shows the proposed biosynthetic pathway of *Monascus* pigments. The three-ring structure with one side chain is derived from the polyketide pathway, and, following modification steps, each of two side chains are derived from fatty acid synthesis and the one-carbon pool, respectively. The final steps of the pathway are not yet completely known.

Pathways of secondary metabolism do not usually function continuously throughout the fungal life cycle and become active only after the growth rate slows down. Borrow et al. (1961) identified three general phases in the production of most secondary metabolites. In the balanced growth phase, the rate of uptake and utilization of nutrients are maximal, and the fungus grows exponentially. Secondary metabolites are rarely produced in this phase. As some nutrient becomes limiting, the growth rate slows down and the storage phase begins. In this phase cell division ceases, but the dry weight may still increase due to the accumulation of storage products such as lipids and polysaccharides. It is in this phase that the synthesis of secondary metabolites usually begins. Eventually the fungus enters a maintenance phase, in which the dry weight becomes constant, the production of secondary metabolites slows down, and cell autolysis sets in. There is considerable experimental evidence to support the proposed stages of secondary metabolite formation, but variations may exist depending on the

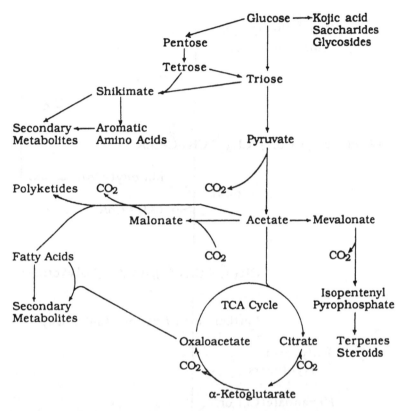

FIG. 3.2 Formation of secondary metabolites from intermediates of primary metabolism. (From Han, 1990.)

particular metabolite and species (Garraway and Evans, 1984) and culture conditions (Grootwassink and Gaucher, 1980).

More recently, Lin and Demain (1993) developed a resting cell system to determine the biosynthetic process of water-soluble red pigments by *Monascus* sp. strain TTWMB 6093. The system contained glucose, glycine, zinc sulfate, and magnesium sulfate in pH 7.0 buffered medium containing cyclohexamide to prevent protein syntheis. The effects of amino acids, carbon source, pH, various minerals, and inhibitors of polyketide synthesis were then investigated. Results showed that pigments were formed by polyketide synthesis and methylation. Inhibition of pigment formation by carbonyl reagents and stimulation by vitamin B_6 showed that endogenous conversion of water-insoluble orange pigments to water-soluble red pigments involved an enzymic Schiff base reaction. High concentrations of magnesium and phosphorus had a negative effect and several trace minerals a positive effect on pigment biosynthesis. Lin and Demain (1994) also

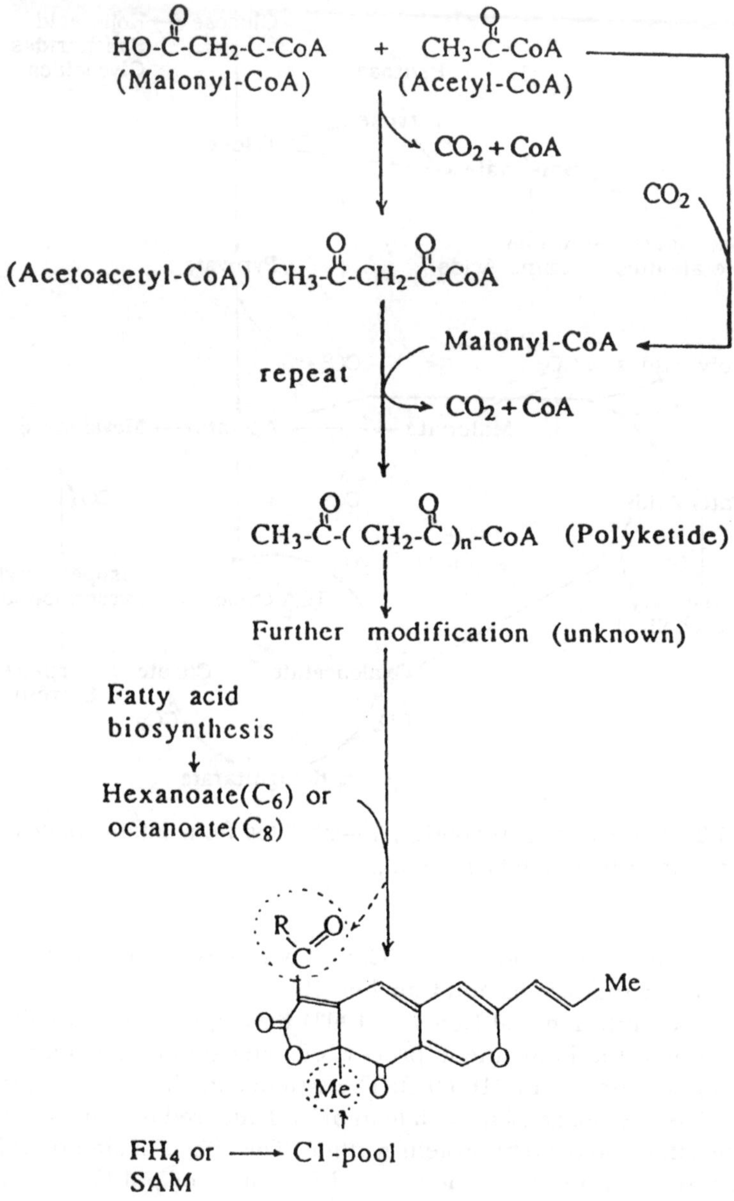

$$\underset{\text{(Malonyl-CoA)}}{\text{HO-}\overset{\text{O}}{\overset{\|}{\text{C}}}\text{-CH}_2\text{-C-CoA}} \quad + \quad \underset{\text{(Acetyl-CoA)}}{\text{CH}_3\text{-}\overset{\text{O}}{\overset{\|}{\text{C}}}\text{-CoA}}$$

$CO_2 + CoA$

CO_2

$$\text{(Acetoacetyl-CoA)} \quad \text{CH}_3\text{-}\overset{\text{O}}{\overset{\|}{\text{C}}}\text{-CH}_2\text{-}\overset{\text{O}}{\overset{\|}{\text{C}}}\text{-CoA}$$

Malonyl-CoA

repeat

$CO_2 + CoA$

$$\text{CH}_3\text{-}\overset{\text{O}}{\overset{\|}{\text{C}}}\text{-}(\text{ CH}_2\text{-}\overset{\text{O}}{\overset{\|}{\text{C}}}\text{)}_n\text{-CoA} \quad \text{(Polyketide)}$$

Further modification (unknown)

Fatty acid
biosynthesis
↓
Hexanoate(C_6) or
octanoate(C_8)

R O
C

Me

Me

FH$_4$ or ⟶ C1-pool
SAM

FIG. 3.3 Proposed biosynthetic pathway of *Monascus* pigments. (From Han, 1990.)

reported on the influence of amino acids on the synthesis of red pigments and found that leucine, valine, lysine, and methionine had negative effects on pigment formation with leucine having the strongest negative effect. They concluded that this was not due to feedback regulation by leucine, but rather to the enhanced decay of one or more pigment synthases, suggesting that enhanced decay was not simply due to de novo synthesis of a leucine-induced protease.

Glucose and maltose were the most effective carbon sources for pigment production by *Monascus purpureus* 192 F in a study of carbon source effects on ethanol and pigment production by *Monascus purpureus* 192 F in pH-controlled submerged cultures (Chen and Johns, 1994). Five pigments were detected in fungal extracts by HPLC analysis. These included the yellow pigments, monascin and ankaflavin, orange pigments, rubropunctatin and monascorubrin, and the red pigment monascorubramine, the major product in all extracts. Fed-batch cultures with low glucose concentrations resulted in minimal ethanol and high monascorubramine production and favored production of monascorubrin, while ankaflavin was favored at high glucose concentrations. Santerre et al. (1995) developed a novel method for evaluating pigment production during fermentation of *Monascus ruber*. This involved reflectance measurements at 10 degrees off normal under an artificial illuminant.

Hong et al. (1995) investigated the chain elongation process by which the polyketide carbon skeleton is assembled and the possibility of using precursors as starter units for synthesizing *Monascus* pigments. Crotonate and sorbate at low concentrations enhanced pigment synthesis without significantly increasing mycelial dry weight. Sorbate and its ethyl ester gave about twice that of crotonate and its ethyl ester. Cinnamate and vinylacrylate were also examined as possible precursors. Cinnamate cultures were dark red but gave lower pigment levels. When its ethyl ester was added, production of the red pigment increased significantly. Pastrana et al. (1995) investigated the production of *Monascus ruber* red pigment in synthetic media with glucose as carbon and monosodium glutamate as the sole nitrogen source. At high glucose and monosodium glutamate concentrations, high ethanol and carbon dioxide levels were associated with large pellet sizes, while for smaller pellet sizes, lower ethanol production and increased pigment production prevailed. Growth limitation was overcome by aeration with carbon dioxide–air mixtures. In all cases, maximum specific rates of pigment synthesis were attained before maximum specific growth rates were achieved.

Finally, Shin et al. (1998) explored the effects of co-culturing *Monascus* sp. J101 on sucrose with strains of *Aspergillus oryzae, Saccharomyces cerevisiae, Bacillus cereus, Aspergillus oryzae*; broth filtrates of *Saccharomyces cerevisiae* cultures resulted in morphological changes and in 2-fold and 30- to 40-fold increases in levels of biomass and pigment synthesis, respectively. The authors suggested that enzymes produced by *Saccharomyces cerevesiae* and *Aspergillus oryzae*, such as amylase, chitinase, and protease, were the effectors of morphological changes through degradation of cell wall constituents. But lysozyme, amylase, protease, and chitinase from species of *Bacillus*, *Staphylococcus*, and *Streptomyces* were not effective.

Antimicrobial Activity

Wang and Hesseltine (1979) commented on the absence of mycotoxins in traditional fermented foods, citing investigation of 73 industrial strains of *Aspergillus* by samples of soy sauce, miso and wheat, rice and soybean tempehs, and samples of *Monascus* red koji collected from various producers in Japan. They point out that if an aflatoxin-producing strain such as *Aspergillus flavus* were accidentally used to make koji for soy sauce, the soy sauce would obviously contain the toxin. Parenthetically, *Monascus* is not reported to cause disease in humans by Dubos and Hirsch (1965) and several other medical references, although serious diseases related to fungal aflatoxins are well documented.

There are, however, a number of reports in the recent literature on bacteriostatic compounds from *Monascus* species. Nozaki et al. (1991) studied a secondary metabolite of *Monascus anka*, ankalactone, which they reported as bacteriostatic. Similarly, Blanc et al. (1995a) have reported the presence of monascidin A in submerged cultures of *Monascus pupureus* and *Monascus anka*, which they identified as citrinin, a toxic fungal metabolite. They concluded that *Monascus* red pigments used as food colorants must be free of citrinin and that monascidin A, an antibiotic, is not appropriate as an antimicrobial food additive. Blanc et al. (1995b) also found that addition of cerulenine and ethionine, specific inhibitors of citrinin and pigment synthesis, inhibited both growth and secondary metabolite synthesis in submerged cultures on ethanol, as did the use of methionine or urea as nitrogen sources.

Martinkova et al. (1995) found antibiotic activity in *Monascus pupureus* extracts containing the orange pigments monascorubrin and rubropunctatin against various strains of bacteria, yeast, and fungi, depending on the composition of the culture medium and method of culture. They also reported embryotoxicity and teratogenicity of the extracts. When amino acids, peptides, or proteins were available during cultivation on the solid substrates rice, wheat, or barley or in submerged cultures with an organic nitrogen source, the bioactive compounds were converted to inactive complexes. The orange pigments monascorubramine and rubropunctamine retained some biological activity, which the authors believe is due to their detoxification by binding with cellular amino groups.

Blanc et al. (1995c) investigated the production of citrinin using various media and culture conditions. Maximum production was obtained from *Monascus ruber* in submerged culture. In a further study, Pastrana et al. (1996) found a metabolic pattern in which citrinin was produced during the lag phase of growth, but not during the growth phase. No tropophase-idiophase transition was detected, suggesting that biomass and pigment formation occurred simultaneously. Citrinin found in pigments made by the traditional koji process and imported from China was traced to production of this compound during storage of the substrate, corn, under warm and humid conditions prior to fermentation (M. Mandt, personal communication).

Food Spoilage

Ascomycetes are not infrequent contaminants in foods and are classified as members of the Eumycetes, or true fungi, by Frazier (1967). Colonies of *Monascus purpureus* found in dairy products, for example, are thin and spreading and reddish or purple in color. The fact that they are not infrequent contaminants is indicated by the fact that Frazier omitted genera that only grow occasionally on specific foods. *Monascus* is one of the mold genera most often found on meats, poultry, and seafood products (Jay, 1986). While it may occur, the genus is not one of the most often found in fresh and refrigerated meats.

Interestingly, Jay cites a species designated as *Monascus bisporus*, also known as *Xeromyces bisporus*, which was able to germinate in 120 days at a water activity of 0.605. Higher moisture levels were required for sexual and asexual sporulation. Other citations from the recent literature indicate that *Monascus* species can grow at extremely low water activities and moisture contents. For example, *Monascus bisporus* was found in spoiled prunes and in licorice, confirming the organism's ability to grow at extremely low water activities (Jong and Edwards, 1991). Similarly, a strain of *Monascus* was isolated from shoes and other leather goods and is considered a serious problem by the leather industry.

Immunological methods for the detection of molds in foods have been developed recently at Purdue. A double sandwich enzyme-linked immunoabsorbent assay (ELISA) method for the detection of *Botyris* and *Monascus* species was developed by Cousin et al. (1990) and is described in detail by Cousin (1990). Sensitivity of the method for detecting *Monascus* species in foods was 2–4 ng/mL. *Monascus* species were not detected in any of the commercial foods examined. More recently, an enzyme-linked assay was developed at Purdue (Yong, 1997) to detect both food spoilage and toxin-producing fungi. Antibodies from *Fusarium moniliforme* detected antigens from all *Fusarium* species and two *Monascus* species. In a survey of 49 commercial food products, a mixture of antibodies from *Aspergillus parasiticus* and *Fusarium moniliforme* detected fungal antigens in 45 samples.

APPLICATIONS OF *MONASCUS* PIGMENTS

Ang-kak—The Red Rice Process

A flow diagram for a tray-based red rice process used in mainland China is shown in Fig. 3.4 (Anonymous, 1999a). After moistening and sterilizing the rice, a fungal culture incubated at 25–35°C for 7 days is seeded in beds of rice packed in trays with closely controlled temperatures and humidities and fermented for 2–4 weeks. The rice particles are then sifted, dried, and packaged either as a food product or as a powder after milling. Sterilization is designed to meet the standards used for laboratory cultures. *Monascus purpureus* pigments produced by the process are mainly monascorubrin (red) and monascin (yellow). The amount of pigments produced is very small compared with the weight of the unfermented solids, so the rice does not lose its nutritive value (Anonymous, 1999b). Red rice and its extracted pigments are also being used as an alternative method to nitrite curing salts—suspected carcinogens—for coloring meats.

The following is a condensed summary from a colorful description of the ang-kak, or red rice, process, the major source of red pigments used as colorants by the food and beverage industries in Mainland China (Xu and Bao, 1999).

> Ang-kak was first mentioned in the literature of the Song Dynasty (960–1279 A.D.). The traditional techniques for making red rice were recorded by Li Shizhen in a monograph of Chinese medicine and in Song Yinxing's "Tian Gong Kai Wu". Traditional producing areas of ang-kak were centered in South China in the Fujian, Zhejiangh, Jiangsu, Jiangxi and Taiwan Provinces. Gutian, in the Fujian Province is cited as the most "famous" red rice production center. Red rice, known by names such as ang-kak, ankak, anka, angquac, beni-jiuqu, aga-jiuqu in various countries of the Orient, is produced in the fermentation industry of China. Extracts are used to prepare red rice wines and foods such as sufu, or Dou-fu-ru in Chinese, a fermented soybean curd product, fish sauce, fish paste, and red soybean curd, a cheese-like product used as a flavor compound. Large quantities of hydrolytic enzymes, such as α-amylase, glucoamylase, protease, and lipase, are produced by *Monascus* spp. which break down rice constituents during growth, with penetration of mycelia into the rice kernels. Pigments produced on rice by *Monascus purpureus* and *Monascus anka* are used in China as household and industrial food colorants. The main pigments produced by *Monascus* spp. are monascorubin and rubropunctatin, monascin and ankaflavin, rubropunctamine and monascorubramine. In addition to its value as a good colorant, ang-kak

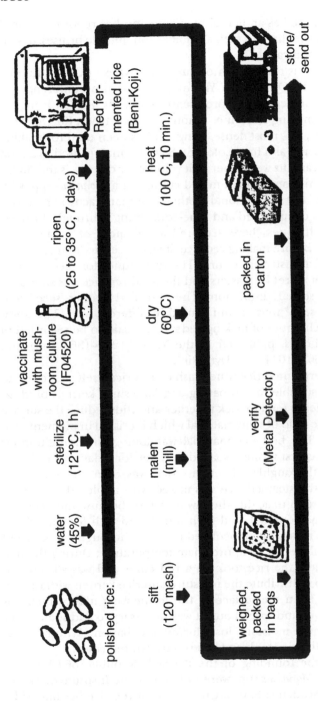

FIG. 3.4 Flow diagram for tray-based red rice process used in Mainland China. (From Anonymous, 1999a.)

has been used as an herbal medicine and has reportedly cured ailments and diseases such as indigestion, muscle bruises, dysentery and anthrax.

Red rice is classified into four major types in China: Ku Qu, Qing Qu, Se Qu and Wu Yi Hong Qu, based on differences in their raw materials and production methods. Ku Qu is mainly used for making rice wine. Among the four types, Ku Qu is the "heaviest", i.e., most dense. Qing Qu has been used for making rice wine and as a food colorant. Se Qu is mainly used as a food colorant, and its weight per unit volume, or density, is the "lightest". Wu Yi Hong Qu is a mixed culture of a *Monascus* spp. with *Aspergillus niger*. Traditional methods for manufacturing red rice were very complicated and time-consuming. Through the ages, these techniques "have acquired an element of mystery and wonder". Before making red rice, it was necessary to make "seed red rice" as a starter culture. The techniques for seed red rice were kept secret by masters and the seed red rice was supposed to have sexual distinction. The so-called "father seed" and "mother seed" were manufactured in different ways. The traditional techniques of making red rice are said to be recorded in ancient books, published in the Yuan (1271–1368 A.D.) and Ming (1368–1644 A.D.) Dynasties.

The fermented rice wine mash or red rice itself was found to be most suitable for preparing ang-kak, since kernels of glutinous varieties tend to stick together and thus reduce the surface to volume ratio of solid material which is critical in pigment production. The best raw material was long-grain non-glutinous rice. It was first washed, soaked in water for a day or more, and drained thoroughly. The moist rice was then cooked. Upon cooling, the steamed rice was mixed with a diluted vinegar or alum solution to acidify the raw material (because the *Monascus* spp. are acidophilic), and then inoculated with fermented red rice wine mash. The inoculated rice was thoroughly mixed and then incubated at an appropriate temperature. During the first several days, the rice took on a red color and was stirred and shaken to redistribute the moisture and rice kernels with respect to depth from the surface of the fermenting mass, which was turned over and spread out. It was necessary to add some water to replenish moisture lost during incubation. Within about 2 weeks, the rice took on a deep purplish red color.

After the founding of the People's Republic of China, 11 strains of *Monascus* spp. were isolated by the Institute of Microbiology, Academia Sinica. Cultivation of red rice became widespread and traditional operating procedures were standardized.

Large-scale production of red rice with "through-flow systems", thick-layer beds aerated during cultivation, were more recently developed in Zhejiang Province. Besides their use as saccharifying agents and use in rice wine brewing, red rice has been mainly used as a food colorant and as an herbal medicine. Large-scale production of food colorants by extracting red rice has developed more recently and most products are exported to foreign countries. In China, red rice and its extracts are considered as safe food additives and herbal medicines of significant value in the treatment of disease.

Although less romantic in terms of its historical development, a more detailed description of the process is provided in a chapter on traditional Asian foods (Wang and Hesseltine, 1979) as indicated in the following excerpts:

Unlike other mold-modified products, which are usually used as a flavoring agent or as a protein source, ang-kak or red rice is primarily a color agent. It was originated in China and is a product made by fermenting rice with strains of *Monascus purpureus*. Because of its red color, ang-kak is used for coloring various foods and as a color additive for manufacturing fermented foods such as sufu, red wine, fish sauce, and fish paste in the Orient. . . .

Only those strains that produce a dark-red growth throughout the rice kernels, at low enough moisture levels to allow the individual grains to remain separate from one another, are suitable for the fermentation. *Monascus purpureus* NRRL 2897 maintained in the AR Culture Collection was isolated from an ang-kak sample bought in the Philippines market and has demonstrated the ability to carry out the fermentation successfully. Palo et al. (1960) studied various conditions of the fermentation and found that the optimum temperature for pigment formation is about 27°C. Growth will occur at as low as 20°C and as high as 37°C, but at these extremes poor pigmentation results. The mold will produce the pigment over a wide range of pH values from 3–7.5. Lin (1973) isolated a strain of *Monascus* sp. F-2 from kaoliang koji and found that it produced large amounts of pigment in submerged culture with rice as a sole carbon source. Treatment and successive isolation (Lin and Suen, 1973) greatly improved the yield of pigment production by *Monascus* sp. F-2. Two hyperpigment-producing strains, R-1 and R-2, were isolated which produced five times more red pigment than the original parent strain. . . .

All varieties of rice are suitable except the glutinous ones, which are unsatisfactory, because the rice becomes gluey and

COLUMN: RP-8 MICROCOLUMN 2.1 mm I.D.×10 cm
MOBILE PHASE:ACETONITRILE-WATER (37:63)
FLOW RATE: 120μl/min
UV WAVELENGTH:214nm

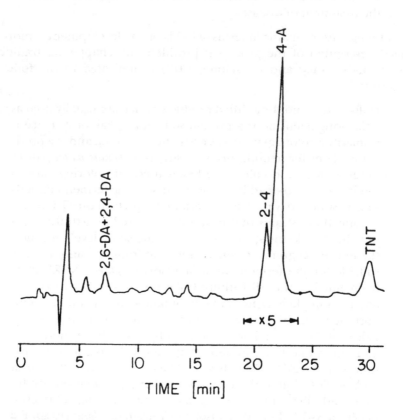

FIGURE 17.   Micro HPLC-UV chromatogram of urine sample containing metabolites of TNT. (From Yinon, J. and Hwang, D.-G., *Biomed. Chromatogr.*, 1, 123, 1986. With permission).

column. Hydrogen was used as carrier gas. On-column injection was done at 40°C followed by a 2-min hold and then a temperature programming at 7.5°C/min up to 205°C. As the XAD-4 extraction was done in the field, water samples had not to be brought to the laboratory. Lowest detection level of TNT was 0.1 μg/l.

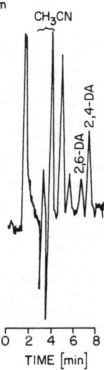

COLUMN: RP-8 MICROCOLUMN 2.1 mm I.D. × 10 cm
MOBILE PHASE: METHANOL-ACETONITRILE-WATER (20:18:62)
FLOW RATE: 120 μl/min
UV WAVELENGTH: 214 nm

CH₃CN

2,6-DA

2,4-DA

TIME [min]

FIGURE 18.   Micro HPLC-UV chromatogram of urine sample showing the separation of 2,4-DA and 2,6-DA. (From Yinon, J. and Hwang, D.-G., *Biomed. Chromatogr.*, 1, 123, 1986. With permission).

2. *High-Performance Liquid Chromatography (HPLC)*

Kaplan and Kaplan[153] developed a method of concentration and detection of trace quantities of TNT and its biotransformation products in water. Concentration was done by passing solutions through a C-18 SEP-PAK cartridge with μOndapak packing on octadecylsilane treated silica bonded phase. After washing the cartridge with methanol and water, 300 ml solutions were drawn through it at about 10 ml/min by suction. Compounds collected on the cartridge were then eluted with 5 ml methanol. Analysis was done by HPLC using a reversed-phase 30 cm × 3.9 mm I.D. C₁₈μOndapak column. Mobile phase was methanol-water using a methanol gradient from 40 to 78% at a flow rate of 2.5 ml/min. A UV detector was used at 230 nm. Compounds tested included

products consisting of fermented fish and rice colored with ang-kak are called buron dalag, burong hipon, or burong bungus, depending on the fish used in the mixture.

In Germany, batches of sausages were prepared with 1000–4000 ppm of *Monascus purpureus* extract for comparison with 0–72 ppm of nitrite salts. Color stability was better with these extracts than with nitrites (Fink-Gremmels et al., 1991). From Thailand, a mutant of *Monascus araneosus* produced no pigments but a substantial amount (28 g/L) of L-malic acid on 10% glucose (Wongleung et al., 1993). In France, pigments of *Monascus ruber* were extracted and used as colorants for sausage and pate. The color of these products was stable for 3 months at 4°C (Fabre et al., 1993). In Belgium, a species of *Monascus* was used to produce monascin as a food additive (Vandamme, 1993), and again in Germany, *Monascus purpureus* was used to convert short-chain fatty acids to methyl ketone mixtures to obtain flavors resembling those of *Penicillium roqueforti* (Kranz et al., 1992). In The Czech Republic, the effects of cochineal, *Monascus* pigments and betalaine were evaluated as colorants for sausage mixtures, frankfurters, and meat products containing some soy protein. Amounts of pigment compensating for meat product brightness caused by soy protein were established (Pipek et al., 1996). *Monascus* pigments have been recommended as ideal substitutes for nitrate-curing salts by the Institute of Meat Research in Kulmbach, Germany, and as food colorants for a large number of food products at concentrations shown in Table 3.2 (Anonymous, 1999e). The above citations clearly show that research and development of *Monascus* pigments as food colorants is centered in Asia and in Western Europe.

Medical Applications

The number of reports citing *Monascus* pigments used as herbal medicines and dietary supplements is very large, as seen from internet websites through Altavista and other search engines. These include the suppression of tumor promotion, regulation of immunoglobulin production, lowering of serum lipids in hyperlipidemia, and reduction of aminocetaphen-induced liver toxicity by antioxidase action. The National Institutes of Health (NIH) has established a National Center for Complementary and Alternative Medicine (NCCAM) that conducts and supports basic and applied biomedical research and disseminates information on dietary supplements, or "neutraceuticals," to practitioners and the general public. The agency cautions users not to seek therapies described on manufacturer or distributor web pages without consulting a licensed health care provider. Neutraceuticals are marketed in the United States under the Dietary Supplement and Health Education Act of 1994 (DSHEA). Accordingly, hard scientific data supporting health claims based on research in the United States are not as available for neutraceuticals as they are for traditional pharma-

TABLE 3.2 Recommended Dosages for Use of Red Rice Powders as Colorants of Various Food Products (% w/w)

Sausages, ham	0.005–0.025
Ice cream	0.02–0.10
Tomato paste	0.5–2.0
Curry	0.005
Fruit juices	0.002–0.005
Meats	0.10–0.20
Sauces	0.02–0.10
Jams	0.02–3.00
Sweets	0.001–0.005
Wines	0.2–1.0

Source: Anonymous, 1996.

ceuticals, since they are not regulated as drugs. Consumers are also urged to note that rigid quality controls are not required for nutraceuticals and that the potency and purity of such products may be questionable, since they are imported from countries of origin that may not have manufacturing methods meeting ISO standards. Most of these are products used as traditional herbal medicines in the Far East and are becoming very popular in the West as "dietary supplements." Health claims are not permitted in the advertising and marketing of these products, although belief in health benefits appears to be the basis for their popularity.

Cholestin, or monascolin K, distributed by the Pharmanex Company and obtained from red rice imported from China as a neutraceutical, contains a natural inhibitor of the rate-limiting step in cholesterol synthesis mediated by HMG-CoA (3-hydroxy-3-methyl glutaryl coenzyme) reductase (Harkness, 1998). Cholestin is classified as a cardiovascular agent for treatment of hyperlipemia based on inhibition of HMG-CoA reductase and is claimed to reduce total cholesterol, low-density lipoprotein (LDL) and triglyceride levels, while increasing high-density lipoprotein (HDL) levels. The U.S. Food and Drug Administration (FDA) ruled that Cholestin was an unapproved drug, not a dietary supplement, and banned its sale. The Pharmanex Company then brought a suit against FDA. The court made a temporary ruling in favor of the company by lifting the FDA ban. FDA had argued that the product is not a true dietary supplement, because the company altered the traditional fermentation process to produce high levels of the drug lovastatin, the same drug used as the prescription-only statin drug Mevacor produced by Merck. However, the researcher who developed Mevacor said the two products were not the same because lovastatin is an isolated pure product, while Cholestin contains natural HMG-CoA reductase inhibitors, including the natural metabolite in Cholestin, which FDA called lovastatin. Clinical reports on Cholestin were provided by its manufacturer from China (Wang et al., 1997; Li et al., 1998; Qin et al., 1999). These were generally confirmed by a preliminary study at UCLA showing reductions in total cholesterol up to 18% and significant improvements in HDL/LDL ratios for some 83 individuals with initial cholesterol levels of about 255 (Heber et al., 1999). Similar clinical results have been reported for other commercially available dietary supplements, such as Cholestene, Choles-Control, and Monastin or Monacolin K.

The statins—lovastatin or mevalonin, mevastatin (originally known as compactin), pravastatin, and simvastatin—are HMG-CoA reductase inhibitors produced by strains of *Penicillium, Aspergillus,* and *Monascus.* A chemically defined medium supporting growth and providing cells converting compactin to pravastatin has been reported (Peng et al., 1997, 1999; Yashphe, 1997; Peng and Demain, 1998) based on a new hydroxylase found in a species of *Actinomadura.* The statins are also claimed to reduce coronary atherosclosis (Kritchevsky et al., 1998). There is some doubt about the linkage between reduced cholesterol levels and atherosclerosis because available research has not considered other biochemical mechanisms that might be responsible for decreases in cholesterol levels.

Several reports were found on the inhibition of tumor promotion in mice and rats. Oral administarion of *Monascus* pigments used as food colorants in Asia suppressed tumor promotion by 12-O-tetradecanolyphorbo-13-acetate (TPA) in mice following initiation by 7,12-dimethylbenz[a]anthracene (Yasukawa et al., 1994, 1996). It was suggested that inhibition of tumor promotion resulted from antinflammatory activity. Antioxidant activity of *Monascus anka* pigments has also been reported to reduce experimentally induced liver injury to rats by galactosamine (Aniya et al., 1999) or acetaminophen (Aniya et al., 1998). This is attributed to antioxidant activity of mold extracts as free radical scavengers and to inhibition of acetaminophen (AAP) metabolism. *Monascus* pigments have also been shown to inhibit IgG or IgM production at concentrations as low as 1 mM, suggesting that natural pigments may have a role in regulating immunoglobulin activity (Kuramoto et al., 1996).

Nontraditional Products

Monascus species have been used to produce many valuable products such as enzymes, coenzymes, monascolins (antihypertensive compounds), and flocculants, in addition to the pigments widely used as food colorants in Asia. Iizuka and Mineki (1977) isolated two forms of glucoamylase from *Monascus kaoliang.* In a subsequent study (Iizuka and Mineki, 1978), they examined substrate specificities of these enzymes. Imanaka et al. (1972) investigated a-galactosidase production from a *Monascus* species isolated from the soil. When glucose and galactose were supplied simultaneously, the mold utilized glucose before using galactose, indicating diauxic cell growth. Optimal processing conditions were determined for -galactosidase production in multistage continuous culture, using a kinetic model of enzyme production (Imanaka et al.,1973). They found that galactose uptake was constitutive and had a relatively short induction period. Yang et al. (1985) investigated a-glucosidase from *Monascus ruber* and found it was irreversibly inactivated by conduritol B epoxide and inhibition of first order with respect to time and inhibitor concentration. Acid protease was purified and its properties studied from extracts of *Monascus kaoliang* (Tsai et al., 1978) and *Monascus* sp. No. 3403 (Yasuda et al., 1984).

Monascus anka IFO 4478 produced coenzyme Q_{10} in the solid state (Dainippon Ink and Chemicals, Inc., 1982a). *Monascus purpureus* M-023 also produced coenzyme Q_{10} in submerged culture (Dainippon Ink and Chemicals, Inc., 1982b). A nuclease inhibitor was also obtained from *Monascus purpureus* Went 908 that was cultured on bran medium, purified, and its properties studied (Saruno et al., 1981). The mode of action for this inhibitor was noncompetitive and its molecular weight estimated to be about 2500 by Sephadex G-75 gel filtration activity using the SHR rat model (Tarui et al., 1987). Monascolin, a potent inhibitor of cholesterol biosynthesis, has been isolated from cultures of *Monascus ruber* and its mutants. Dihydromonacolin L and Monacolin X have been produced by a mutant of *Monascus ruber*. Their structures were determined by a combination of physical methods (Endo et al., 1985a). The structures of Monascolin J and Monascolin L were also determined by mass spectrometry, NMR, IR, and UV analysis (Endo et al., 1985b). Monacolin K was found to be formed from a chain of 11 acetate residues contributing 22 carbon atoms (Endo et al., 1985c). Monacolin M was also produced by cultivating *Monascus ruber* No. 1005 in submerged culture at 25°C for 10 days (Endo et al., 1987).

Monascus anka was also found to produce substances that flocculate *Saccharomyces cerevesiae* (Nakamura et al., 1976). Flocculates in the cultures were readily precipitated by the addition of acetone. These flocculants also also found to precipitate a number of other microorganisms. *Monascus purpureus* NRRL 2897 was also cultivated on cotton seed flour to examine the nutritional quality of protein recovered from the fermented product (Plating and Cherry, 1979).

More recently, an albino mutant of *Monascus araneosus* did not produce any pigments, but did produce L-malic acid at 28 g/L (Lumyong and Tomita, 1993). α-Galactosidase was also produced on extracts of sugarcane and soybean wastes. An extract prepared from 5% soybean and 7% sugarcane wastes gave the highest concentration and was superior to a chemically defined medium used in submerged cultures (Wongleung et al., 1993). A ribonuclease was obtained from *Monascus* sp No. 3403 and purified by column chromatography and its molecular weight determined as between

2600 and 3000 daltons (Yasuda et al., 1995b). Glucoamylase was also obtained from cultures of *Monascus rubiginosus* (Dong et al., 1996). Long-chain fatty acids have also been obtained from mutant strains of *Monascus purpureus* (Juzlova et al., 1996a). Some 39 fatty acids were identified, including 22 unsaturated, 14 monoenoic, and 2 dienoic, in addition to some 8% α-linolenic acid.

Secondary metabolites, including *Monascus* pigments and antihypertensive compounds and a new compound, ankalactone, are discussed in terms of their production methods, isolation procedures, and biological activities of such compounds (Juzlova et al., 1996b). A mutant of *Monascus purpureus*, DSM 1379, was used to study pigment and lipid synthesis (Rasheva et al., 1997). The mutant lost its ability to produce pigments and produced a high amount of lipids consisting of 88% triglycerides containing 51% of saturated fatty acids. Eighty volatile compounds were identified from samples of rice fermented by *Monascus purpureus* and in cultivation media following fermentation (Patakova-Juzlova et al., 1998). These compounds included alcohols, aldehydes, ketones, esters, and terpenoids. The compounds were identified by gas chromatography and mass spectrometry following distillation and extraction with dichloromethane.

PRODUCTION OF *MONASCUS* PIGMENTS

Submerged Culture Methods

Although traditional food processes in Asia have predominantly employed solid substrate fermentation methods, for which there has been relatively little research in the West, cultivation of *Monascus* species in submerged cultures depends on much the same cultural considerations as those in the solid-state in terms of process parameters. Submerged culture methods offer a distinct advantage over solid-state methods for research studies on nutrition and growth. The medium in which microbial metabolism takes place is accessible for media manipulation and for direct measurement and control of parameters in the liquid phase such as substrate consumption, secondary metabolite formation, pH, temperature, oxygen consumption, etc. In solid-state fermentations, the liquid phase is distributed in liquid films on solid surfaces and is not directly accessible for measurement and control, except through the external gas environment.

Condition of the inoculum and initiating the fermentation are much more important for mycelial cultures than for bacteria. This is because mycelial cells are discrete entities that age during apical mycelial growth. That is, growing mycelia have an average age that increases more rapidly with time than unicellular bacteria, which form new cells by binary fission. Fungal mycelia have a longer biochemical history and are therefore more sensitive to process mismanagement. Process control is therefore most critical during the early stages of the process.

For *Monascus* pigment production, one of three inoculation methods are generally used. These include spore suspensions (Lin, 1973), mycelial suspensions (Lin and Suen, 1973; Yoshimura et al., 1975; Bau and Wong, 1979; Broder and Koehler, 1980; Su and Huang, 1980; Wong et al., 1981; Wong and Koehler, 1981a,b), and mycelial homogenates (Carels and Shepherd, 1977, 1978). The effects of inoculum source, size, and preparation method on growth and pigment formation are not well documented. Maximum pigment yields are generally obtained at an initial medium pH between 5.5 and 6.5. Optimum temperatures for pigment production are reported at 30 (Su and Huang, 1980), 35 (Wong and Bau, 1977; Bau and Wong, 1979; Broder and Koehler, 1980; Wong and Koehler, 1981a,b; Wong et al., 1981), 33 (Miyake et al., 1982), 32 (Lin and Suen, 1973; Yoshimura et al., 1975; Su and Huang, 1980: Riken Vitamin Co., Ltd., 1982), 28 (Broder and Koehler, 1980), and 25°C (Carels and Shepherd, 1977, 1978, 1979).

Yoshimura et al. (1975) cultivated a *Monascus* strain in 1-L glass fermentors. The medium was prepared by replacing glucose in a yeast-malt-glucose medium with other carbon sources. Among these, ethyl alcohol gave the highest pigment yield. Suitable nitrogen sources for pigment production included polypeptone, yeast extract, nitrate, casamino acids, monosodium glutamate, and L-aspartate. Carels and Shepherd (1977) investigated the effect of various nitrogen sources on *Monascus* pigment production in shake flask culture. When the nitrogen source was yeast extract or nitrate, pigments were red; with ammonium nitrate, pigments were orange. Broder and Koehler (1980) investigated pigment production by *Monascus purpureus* NRRL 2897 utilizing various carbon and nitrogen sources. Maltose yielded pigments with the longest dominant wavelength (darkest red); glucose and fructose yielded lighter red pigments. Yeast extract was a superior nitrogen source for red pigment production. Wong et al. (1981) investigated effects of the glucoseammonium nitrate ratio in a synthetic medium on mycelial growth and pigmentation by *Monascus purpureus*. The concentration of ammonium nitrate for maximum pigment production was much lower than that for maximum growth, even at high glucose concentrations. More red pigments were formed in media with high levels of ammonium nitrate at low glucose concentrations.

Monascus species have been cultivated in media with maltitol (Yamanaka et al., 1975), ethanol (Yoshimura et al., 1975; Yamanaka et al., 1975; Johnson and McHan, 1975; Ajinomoto Co., 1979), rice powder (Lin and Suen, 1973; Su and Huang, 1980; Miyake et al., 1982), and sweet potato starch (Zhang et al., 1982) as carbon and energy sources. Pigment production was stimulated by the addition of a nonionic surfactant and by controlling the supply of oxygen (Ajinomoto Co., 1977).

60   MUDGETT

Agitation rates significantly affect fungal morphology in submerged culture. Mycelia grow in pulp, pellet, and intermediate forms (Takahashi et al., 1965; Yoshimura et al., 1975) that depend on media composition and may be quite sensitive to mycelial shear effects in stirred-tank bioreactors. Mycelia may grow into homogeneous and filamentous suspensions ("pulp-type" growth), globose colonies ("pellet-type" growth), or intermediate forms between pulp and pellet forms (Takahashi et al., 1965). Yoshimura et al. (1975) investigated relationships between morphology and pigment formation by a *Monascus* sp. At 300 rpm, mycelia were produced in the pulp form with low pigment formation. At 600–800 rpm, mycelial growth and pigment production were high and mycelial forms were intermediate between pulp and pellet forms. At 1200 rpm, pellets were formed and pigment production decreased significantly. This apparently resulted from impeller shear effects at this high agitation rate. A mycelial inoculum that is not homogenized generally yields pellet formation, which also reduces oxygen transfer due to mass transfer limitation. Pellet formation in *Aspergillus* and *Penicillium* has been related to adhesion, absence or presence of trace metals or chelating agents, alkaline pH, high carbon-to-nitrogen ratios, and inorganic and complex organic nitrogen sources.

Lin and Suen (1973) treated ascospores of *Monascus* sp. F-2 with *N*-methyl-*N*-nitro-*N*-nitrosoguanidine (NTG) and isolated an overproducing mutant. *Monascus purpureus* was also mutated by UV irradiation (Wong and Koehler, 1981a) and by ionizing radiation (Wong and Bau, 1977) to select overproducing mutants. Such mutants have also shown antibiotic activity (Wong and Bau, 1977; Wong and Koehler, 1981b), which may be of concern for clearance as a food colorant in the United States (Francis, 1987).

Porous resin spheres (Nagasawa et al., 1987) and a polymeric resin (Evans and Wang, 1984) were used to adsorb pigments. Adding a sterilized nonionic adsorbent resin directly to a growing submerged culture did not enhance pigment production. However, when resins were used to create an immobilized *Monascus* culture (Evans and Wang, 1984), pigment production was much higher, suggesting that mycelial support may offer an explanation for enhanced pigment production as in solid-substrate fermentations. Adsorption of *Monascus* pigments on a porous resin, Amberlite XAD-7, prevented pigment binding with protein in submerged cultures.

Demain et al. (1990) devised a medium containing maltose, monosodium glutamate, phosphate, and magnesium sulfate at pH 5.5 and found that specific pigment production was 3.8 times higher than on the original medium containing glucose, ammonium nitrate, phosphate, and magnesium sulfate at an initial pH of 4.6. Changes in the medium also had a favorable effect on excretion of the pigment produced. Lin et al. (1992) found that adding glutamate, glycine, or leucine produced water-soluble red pigments different from monascorubrin and rubropunctatin that corresponded to the exogenous amino acid. Lin and Demain (1994) also found that leucine, valine, lysine, or methionine used as sole nitrogen sources had strong negative effects on water-soluble red *Monascus* pigments. The negative effect of leucine was attributed to the decay of pigment synthase caused by de novo synthesis of a leucine-induced protease. Further studies on the regulation of pigment synthesis include those of Lin and Demain (1991, 1993).

Fabre et al. (1993) studied the properties of red pigments from *Monascus ruber* in submerged culture using ethanol and glutamate as carbon and nitrogen sources. After extraction, the pigments added to sausage and pate remained stable for 3–4 months at 4°C. Sensory tests showed that *Monascus* pigments could be used to replace traditional food additives such as nitrite salts or cochineal. Pastrana and Goma (1995) evaluated two stoichiometric models in glucose and ethanol batch cultures of *Monascus ruber* to estimate process variables. The first model provided a basis for joint prediction of biomass and monosodium glutamate values and the second for joint estimation of biomass and malate values. A mass balance model was also used for biomass and monosodium glutamate estimation. Studies of *Monascus* pigment production in submerged cultures were also made (Santerre et al., 1995; Pastrana et al., 1995; Hamdi et al., 1997) in cooperation with the Institut des Sciences Appliquees at Toulouse, France (INSA), where a major symposium on *Monascus* culture and its applications was held in July 1998.

Hamdi et al. (1996) found it necessary to increase the oxygen supply in order to control ethanol production in the growth phase of *Monascus purpureus*. Han (1990) also found that significant amounts of ethanol were produced during submerged culture of *Monascus purpureus*, suggesting that anaerobic metabolism was caused by oxygen limitation and/or the Crabtree effect and that ethanol yields may be significantly affected by aeration and agitation rates. It is often necessary to use oxygen enrichment at high rates of growth and pigment production in submerged cultures to maintain aerobic conditions and to avoid adverse shear effects at high impeller agitation rates. Aeration at normal atmospheric pressures is not sufficient to provide the necessary oxygen transfer rates in submerged cultures at stringent oxygen demands (Mudgett, 1980). Yields and productivities in submerged culture fermentations from 1973 to 1990 are shown in Table 3.3 (Han, 1990). These will be compared with yields and productivities in the solid state in the section that follows.

Solid Substrate Fermentations

A traditional koji or solid substrate process used in Taiwan for producing *Monascus* pigments is shown In Figure 3.5 (Su and Huang, 1977). The process is used to make rice wines and tofuyu (a fermented soy protein) as well as a colorant for many traditional oriental food products. Solid substrate methods have been and are being used to produce a wide variety of tradi-

TABLE 3.3 *Monascus* Pigment Yields and Productivities in Submerged
Culture Fermentations, 1973–1990

Strain	Yield[a]	Productivity[b]	Ref.
Monascus anka Nakazawa et Sato	5	55	Lin and Iizuka, 1982
Monascus sp. No.2 AJ7744	36	215	Yoshimura et al., 1975
Monascus sp. F-2	24	250	Lin, 1973
Monascus sp. NRRL1993	60	389	Evans and Wang, 1984
Monascus kaoliang S-11	92	1282	Lin and Suen, 1973
Monascus anka V-204	156	1080	Su and Huang, 1980
Monascus purpureue ATCC 16365	14	282	Han, 1990

[a]OD_{500} units/mL of broth
[b]OD_{500} units/L/hr of broth
Source: Han, 1990.

tional and nontraditional food products around the world (Wang and Hesseltine, 1979; Mudgett, 1986; Steinkraus, 1995).

Solid-state fermentations (SSF) may be briefly described as those in which microbial growth and product formation occur in thin liquid films on surfaces of solid substrates such as rice, corn, millet, and soybeans, (Hesseltine, 1977a, b). The process is not commonly used by the food and pharmaceutical industries of the West because of its deep commitment to submerged culture methods. The method is said to have originated in China about 2500 years ago and was later brought to Japan by Buddhist missionaries. Traditional solid-state, or koji, processes are used in Japan to produce shoyu (soy sauce) and miso (soup flavorant) and a starter culture for sake (rice wine). The organisms used in such processes produce enzymes that break down proteins, lipids, and carbohydrates (Hesseltine, 1972). SSF are limited to fungi or fungal-like bacteria such as *Streptomyces*, which can grow at low moistures and water activities. This ability gives filamentous organisms a distinct competitive advantage over nonfilamentous yeasts and bacteria with respect to process contamination.

After the discovery of penicillin, submerged culture methods were developed in the West for the production of antibiotics and other microbial metabolites. Nevertheless, solid-state fermentations have a future in both advanced and developing countries due to their numerous advantages over submerged fermentations, such as higher productivity, simpler techniques, low capital investment, reduced energy requirements, low wastewater output, improved product recovery, and elimination of foaming. Process disadvantages include difficulty in controlling moisture and temperature and in monitoring process parameters, e.g., biomass production, substrate consumption, and product formation (Mudgett, 1986; Han, 1990).

Major process variables in the solid state may include substrate pretreatment, nutrient supplementation, particle size, moisture content, sterilization, temperature, pH, inoculum density, aeration, and agitation. These are discussed in depth in an earlier review of solid-state fermentations (Mudgett, 1986). Control of moisture content, and therefore water activity, is a critical parameter in the solid state and is governed by the sorption-desorption isotherm of the substrate, relative humidity (RH), aeration method, and depth of substrate. The moisture content of steamed rice is quite sensitive to the ambient RH based on its sorption-desorption characteristics, and that of saturated water vapor is also logarithmically dependent on temperature.

Heat evolved during fermentation is directly related to metabolic activities of the microorganism and depth of the substrate. Therefore, heat transfer is of considerable importance, since it is closely linked with moisture transfer. High temperatures may result in severe moisture loss if temperature and RH are not well controlled. These parameters are of special importance because they affect spore germination, growth, product formation, and sporulation (Lonsane et al., 1985). Loss of moisture can be compensated by supplying moisture using air humidified with a saturated salt solution (Bajracharya and Mudgett, 1980; Sato et al., 1982, 1983), by spraying with water (Narahara et al., 1982, 1984), or by incubation in a humidified chamber (Kundu et al., 1983). Temperatures within a bed of solid particles are dependent on depth from the surface.

Aeration is of major importance in the solid state and is implicated not only in gas transfer, as in submerged cultures, but also in heat and moisture transfer between the fermenting solids and the external gas environment. Aeration also involves the coupled processes of oxygen consumption and carbon dioxide evolution in microbial respiration (Nyiri et al., 1975). Aeration rates depend on oxygen requirements for growth and product synthesis, heat evolution, thickness of the substrate layer, and the degree to which carbon dioxide and other volatile metabolites need to be removed (Lonsane et al., 1985).

Fermentations conducted without agitation gave much lower production of ochratoxin A (Lindenfelser and Ciegler, 1975), suggesting inadequate aeration. Agitation and mixing of the fermenting solids provides homogeneity, enhances heat and mass transfer and gas exchange, prevents aggregate formation, and distributes spore inocula more effectively (Lonsane et al., 1985). Moderate agitation rates and mixing of substrate particles have enhanced secondary metabolite production (Hesseltine, 1972; Lindenfelser and Ciegler, 1975).

Temperature and pH of the fermenting solids are significant variables in the solid state and are specific for the microbial-substrate system to be cultivated, as in submerged culture fermentations. Both are more difficult to control in the solid state, since direct measurements in the liquid phase are impractical because of their intimate association with the solids at low moistures. The rate of heat generation in the bed at high levels of biological activity may also induce thermal gradients within the bed, which lead to heat transfer limitations and suboptimal conditions for biomass or product formation (Mudgett, 1986). Good buffering capacity of substrates may help to eliminate the need for pH control during fermentation. This advantage is exploited by the initial adjustment of pH during moistening by using buffers at the desired pH level (Lonsane et al., 1985).

Nutritional factors may limit growth and product formation in the solid state due to either lack of a required nutrient or the inhibitory effect of a compound present in the natural substrate. Although most traditional food fermentations do not require nutritional supplementation, it may be necessary in nontraditional fermentations for secondary metabolite production requiring specific nutrients not available from the substrate. An important indicator of nutritional regulation of growth in solid-state fermentations is the C:N ratio. Optimal C:N ratios vary in a wide range from 10 to 100 in solid substrate fermentations, but the availability of carbon and nitrogen may be more important than their ratio. The nitrogen source is also important. Ammonium chloride, ammonium sulfate, and urea have resulted in less ATP expenditure and to a lower pH, while sodium nitrate led to a higher ATP expenditure and, because of ammonia evolution, to a higher pH (Moo-Young et al., 1983). It is generally necessary to optimize inoculum density since excessively low densities give insufficient biomass formation and permit overgrowth by undesirable organisms, while excessively high densities may produce too much biomass and deplete the solid substrate of nutrients necessary for product formation.

A novel liquid-solid system (Lee et al., 1995) produced red and yellow pigments with a strain of *Monascus* that gave biomass concentration of 37.5 g/L and optical densities of 145 units at OD\dil(500) and OD\dil(400) for the red and yellow pigments, respectively, in a 170-hour fermentation. The system used a 400 g/L gelatinized starch cake as its carbon-energy source and a liquid phase supplemented with all of the other nutrients except the starch. It was not possible to determine the productivity and yields of biomass and pigment formation from the data presented.

Relatively little information on the cultivation of *Monascus* pigments in the solid state is available in the American literature. However, research reports from the Orient are available on the internet from China, Indonesia, Japan, Korea, Singapore, and Thailand and European reports from Bulgaria, Czech Republic, England, France, Germany, the Netherlands, and Spain. These indicate a broad global interest in *Monascus* pigment production. Pigment yields and productivities in the solid state are shown in Table 3.4 (Han, 1990). His results were about 10% higher than those reported by

TABLE 3.4 *Monascus* Pigment Yields and Productivities in Solid Substrate Fermentations, 1973–1990

Strain	Yield[a]	Productivity[b]	Ref.
Monascus anka Nakazawa et Sato	174	1040	Lin and Iizuka, 1982
Monascus anka (mass production)	240	Not reported	Lin, 1973
Monascus anka	1690	8800	Lin and Iizuka, 1982
Monascus anka Nakazawa et Sato UN202-13[c]	666	3470	Hiroi et al., 1979
Monascus kaoliang R-1087[c]	5433	37730	Lin and Iizuka, 1982
Monascus purpureus ATTC 16365	6016	41780	Han, 1990

[a]OD_{500} units/ g dry fermented solids.
[b]OD_{500} units/kg dry fermented solids/hr.
[c]Mutant strains.

Lin and Iizuka (1982) for *Monascus kaoliang* and much higher than earlier studies in the solid state shown in the table. Pigment yields and productivities in the solid state to date are clearly higher than those of submerged culture fermentations, as seen by values previously shown for submerged cultures in Table 3.3.

Effects of Oxygen and Carbon Dioxide

Oxygen and carbon dioxide partial pressures and dissolved concentrations in gas-liquid equilibrium have been shown to significantly affect secondary metabolite production for both submerged culture and solid substrate fermentations, suggesting that such processes may be optimized by maintaining the gas environment under constant partial pressures (Mudgett, 1980).

Studies were made in a packed bed fermentor at constant oxygen and carbon dioxide partial pressures with *Monascus purpureus* ATCC 16365 (Han, 1990). When oxygen pressures were varied from 0.05 to 0.50 atm at a constant carbon dioxide pressure of 0.02 atm, biomass yields were essentially constant, while red and yellow pigment yields in ODU/g of initial dry solids (IDS) increased monotonically as oxygen pressures were increased. Conversely, when carbon dioxide pressures were varied from 0.01 to 0.20 atm at a constant oxygen pressure of 0.21 atm, biomass yields were again relatively constant, while yields of the yellow and red pigments decreased monotonically. These results indicated stringent oxygen demand rates at a constant low carbon dioxide pressure and a strong inhibitory effect by carbon dioxide at normal atmospheric oxygen pressures. At final moisture contents of about 50% (wet basis), oxygen uptake and carbon dioxide evolution rates varied from 100 to 150 mmol/L/hr at respiratory quotients close to unity. Maximum pigment yield and productivity were some 6000 ODU/g IDS and 650 ODU/kg IDS/hr equivalent to 650 ODU/L/hr in the liquid phase (Han and Mudgett, 1991).

Previous studies showed stimulation of lignase activity by oxygen and inhibition by carbon dioxide in birch lignin degradation by *Phanerochaete chrysosporium* (Mudgett and Paradis, 1985). Similarly, amylase productivity was significantly stimulated by oxygen at pressures above normal atmospheric and severely inhibited by carbon dioxide at pressures above 0.01 atm in *Aspergillus oryzae* cultured on rice (Bajracharya and Mudgett, 1980). Stimulation of cellulase and pectinase activities was observed at both oxygen and carbon dioxide pressures above atmospheric in plant protein recovery (Bajracharya and Mudgett, 1979; Mudgett and Bajracharya, 1979). Progress of *Aspergillus oryzae* growth on rice in the solid state was studied by electron micrograph. These show the surface characteristics of the rice solids prior to inoculation, mycelial development in early to middle stages of growth, and extensive sporulation in the final stages of cultivation, by seen in SEM miccrographs shown in Fig. 3.6 (Mudgett et al., 1982). Deep mycelial penetration of the rice kernels is shown by TEM micrographs of growth during the stages of cultivation, as seen in Fig. 3.7 (Mudgett et al., 1982).

Solid-state fermentations are seen as gas-liquid-solid systems in which the aqueous phase is intimately associated with solid surfaces at varying moisture contents that are directly in contact with the external gas environment. This general concept suggested that the gas environment in SSF might be involved in the regulations of biomass and product formation. The role of oxygen in regulating microbial metabolism is not entirely clear but is seen as a possible switch from one metabolic pathway to another in addition to its role as a nutrient. It may also be toxic at partial pressures above atmospheric for some organisms and at pressures below atmospheric for microaerophilic organisms. The role of carbon dioxide is even less clear. It is a product of microbial metabolism and provides a means of dissimilating carbon in substrate oxidation to obtain metabolic intermediates. Kirtzman et al. (1977) reported that carbon dioxide inhibited the tricarboxylic acid cycle in *Sclerotium rolfsii* and that the organism overcame this inhibition by using the glyoxylate pathway. High concentrations of dissolved carbon dioxide were found to inhibit the specific growth rate and penicillin production rate (Ho and Smith, 1986a). Edwards and Ho (1987) also suggested that high dissolved carbon dioxide pressures caused *Penicillium chrysogenum* to lose control of its plasticizing effect, resulting in severe morphological changes. High pressures stimulated amylase production and depressed microbial growth in submerged cultures of *Bacillus* (Zajic and Liu, 1969;

FIG. 3.6 Characteristics of surface growth of *Aspergillus oryzae* on rice solids. (From Mudgett et al., 1982.)

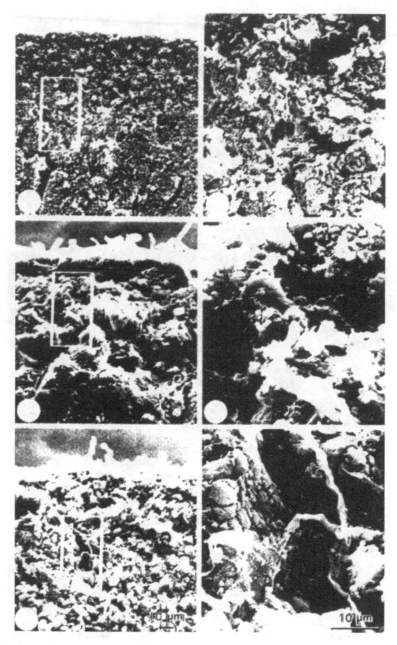

FIG. 3.7 Characteristics of interior growth by *Aspergillus oryzae* on rice. (From Mudgett et al., 1982.)

Gandhi and Kjaergaard, 1975). Ulmer et al. (1981) found that biomass production by *Chaetomium cellulolyticum* was not oxygen limited but required ventilation of carbon dioxide.

In aerobic processes, oxygen is the terminal acceptor in electron transport mediated by an oxidase system and may also be incorporated by an oxygenase. Respiration rates are generally independent of dissolved oxygen concentrations in equilibrium with the gas environment for a range of partial pressures critical for growth and product formation. Bench-scale SSF experiments carried out by Levonen-Munoz and Bone (1985) showed that at least 50% oxygen was necessary to achieve preferential lignin removal. However, oxygen toxicity has been reported for *Pseudomonas, Escherichia,* and *Staphylococcus* species at partial pressures above 1.0 atm (Wiseman et al., 1966) and *Penicillium* species at pressures greater than 1.5 atm (Pirt, 1967). While the specific mechanisms of control in microbial metabolism are not clear, oxygen appears to have a significant regulatory role. Amylase production by *Asperglllus* on rice was much higher in oxygen-enriched gas environments, as noted above. While the role of carbon dioxide in microbial metabolism is less clear, it seems likely that observed inhibitory effects result from feedback regulation of primary and secondary metabolism and/ or as metabolic switches from one pathway to another. The role of oxygen may also involve such mechanisms. Oxygen is often the major rate-limiting nutrient in biooxidation of carbon and energy sources for biomass and product formation

These observations suggest that control of the gas phase may be used to direct microbial metabolism toward specific products based on regulation of microbial metabolism. Secondary metabolites are currently produced in uncontrolled gas environments for most industrial fermentations. Solid substrate processes have been shown to meet stringent oxygen demands at normal atmospheric pressures (M. Mudgett, unpublished). This is believed to be the result of high interfacial surface area–to–liquid phase volume ratios for thin liquid films on solid surfaces. For extremely high demands, excess oxygen would need to be supplied by oxygen enrichment. A closed system, such as the one shown in Fig. 3.8, could be used to study the effects of oxygen and carbon dioxide in both submerged and solid-state cultures and would provide a method for optimizing yield and productivity for increased profitability.

CONCLUSION

The potential for using *Monascus* pigments as food additives in the United States is unclear at this time, since the regulatory climate in the United States for the use of red rice and its derivatives is uncertain. The use of any natural colorant as a food additive requires premarket approval with convincing evidence that the product will cause no harm. Petitions for the use

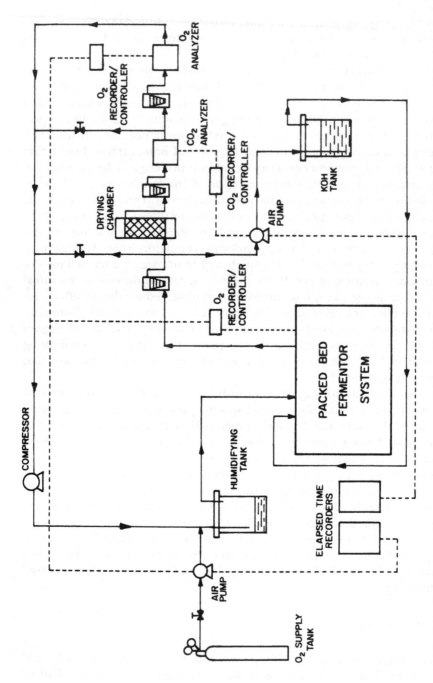

FIG. 3.8 Experimental gas control system used in packed bed fermentation studies. (From Han and Mudgett, 1991.)

of natural pigments as food colorants, whether exempt or certified, must include a full description, including safety data, product and process specifications, limits if any, and assurance of Good Manufacturing Practice and its approval as a color standard (Hallagan, 1999).

The situation is even more complicated by an unresolved conflict between FDA and the Pharmanex Company of Simi Valley, California, over the product Cholestin. FDA's ban on the sale of this product is on hold pending a court's final ruling. Cholestin is produced by the traditional koji red rice process and is imported from mainland China. It is marketed as a "neutraceutical" or "dietary supplement" under the Dietary Supplement and Health Insurance Act of 1994 (DSHEA) in accordance with current guidelines of the National Institutes of Health's National Center for Complementary and Alternative Medicine (NCCAM). FDA has designated the company's product a "pharmaceutical" based on the fact that one of Cholestin's constituents, lovastatin, is identical to a prescription-only drug, Mevacor, manufactured by Merck. One argument used by Pharmanex is that the researcher who developed Mevacor has said that the two products are not the same because lovastatin is an isolated pure product, while Cholestin contains natural HMG-CoA reductase inhibitors, including the natural metabolite the FDA equates with lovastatin. This seems to be a technicality that begs the question as to whether the product meets FDA standards for food and pharmaceutical products and whether the product is safe for unregulated over-the-counter, self-administered use by the general public. A final ruling by the court has not yet been made.

This case is important for both the food and pharmaceutical industries. Of direct interest to the pharmaceutical industry, a ruling favorable to Pharmanex would allow a less rigorous approval process for many products with publicly perceived health benefits as dietary supplements, without regulatory control of off-shore manufacturing processes employed. This could very well open the door to the displacement of more rigorously monitored pharmaceutical products by unregulated off-shore products. An exmple of this is seen by the large number of HMG-CoA reductase inhibitors derived from the red rice process by other companies, such as "Cholestene" and "Choles-Control," in addition to the growing number of dietary supplements from abroad which are freely marketed by companies in both the United States and Europe. Such products are sold in most supermarkets and pharmacies as "natural dietary supplements."

The situation in the European Economic Union (EEU) with respect to regulatory approval of *Monascus* pigments as food colorants appears to be pretty much the same as in the United States. EEU's Food Commission considers *Monascus* both a food colorant and a food additive subject to color standards, regulatory approval, and Good "Clinical" Practice (GCP). In Eastern Europe, however, red rice flour is used for coloring meats. Some 95% of the 500 tons of red rice imported each year from China by a German company goes mainly to Eastern Europe for meat processing. (M. Mandt, personal communication). However, red rice and its pigments are freely used in Asia, especially in China and Japan. *Monascus* pigments are included as natural food colorants in the seventh edition of related Japanese Standards (Henry, 1999).

There seems to be little economic incentive for American companies to engage FDA's regulatory approval process to manufacture red rice pigments as food colorants by solid substrate methods. The food and pharmaceutical industries have little experience with the method and are deeply committed to submerged culture methods, for which there is an extensive research database. The issue is further complicated by the fact that there are special requirements to have any material, natural or synthetic, accepted as a "food colorant."

There would need to be an economic incentive for a company to apply for FDA approval based on either method of cultivation. One approach might be to file an application to approve off-shore products based on clinical data provided by the manufacturer. There remains the problem of monitoring manufacturing processes for off-shore products in accordance with GMP standards Many clinical studies on *Monascus* pigment inhibitors on HMG-CoA reductase and their effects on cholesterol synthesis are reported in the Asian literature. There appears to be a catch-22 situation with respect to interrelationships between NCCAM, FDA, American industry, and the general public. It seems that the real question as to whether a product is safe for human consumption may depend on different standards, which, at this time, seem very unclear.

It seems paradoxical that products consumed as "dietary supplements" should have different approval requirements than those used as "food additives." This appears to result from NIH's creation of NCCAM, presumably based on a demand by some Americans, both young and old, for herbal medicines and dietary supplements, primarily from Asia, appealing to religious and philosophical connections with Eastern cultures and the search for a "fountain of youth." It seems even more paradoxical that Americans concerned about potentially harmful effects of secondhand smoke, herbicides, pesticides, food additives, microwaves, nuclear waste, global warming, automobile emissions, and a hole in the ozone layer are willing to take a chance on products and processes not subject to uniform regulatory standards.

REFERENCES

Ainsworth, G. C., Sparrow, F. K., and Sussman, A. S. (1973). *The Fungi: An Advanced Treatise*. Academic Press, New York.

Ajinimoto Co., 1977. Red pigment. Jpn. Pat. 52,027,720.

Ajinimoto Co., 1979. *Monascus* pigment production. *Jpn. Pat.* 79,145,269.

Ang, C. Y. W., Liu, K., and Huang, Y. (1999). *Asian Foods: Science and Technology*. Technomic Publishing Company, Lancaster, PA.

Aniya, Y., Yokomakura, T., Yonamne, M., Nagamine, T., and Nakamishe, T. (1998). Protective effects of the mold *Monascus anka* aginst aminoceta-induced liver toxicity in rats. *Jpn. J. Pharmacol.* 78: 79–82.

Aniya, Y., Yokomakura, T., Yonamine, M., Shimada, K., Nagamine, T., Shimabukuru, M., and Gibo, H. (1999). Screening of antioxidant action of various molds and protection of *Monascus anka* against experimentally induced liver injuries of rats. *Gen. Pharmacol.* 32: 225–231.

Anonymous. (1999a). MONASCUS RED, red fermented rice as alternative to nitrite-curing salt. *http://www.allok.comerot.htm.*

Anonymous (1999b). MONASCUS RED, red fermented rice for usage in meat products. *http://www.allok.com/efleisch.htm.*

Anonymous (1999c). Doosan Training and Technology Center. *http://www.dst.doosan.com/hot.htm.*

Anonymous (1999d). Yiwu Natural Pigment Industrial Corporation. *http://www.ccc.zjcninfo.net/company/A000000266/ejs.htm.*

Anonymous (1999e). *Monascus* for the health. *http://www.allok.com/epharma.htm.*

Bajracharya, R., and Mudgett, R. E. (1979). Solid substrate fermentation of alfalfa for enhanced protein recovery. *Biotechnol. Bioeng.* 21: 551–560.

Bajracharya, R., and Mudgett, R. E. (1980). Effects of controlled gas environments in solid substrate fermentations of rice. *Biotechnol. Bioeng.* 22: 2219–2235.

Bau, Y. S., and Wong, H. C. (1979). Zinc effects on growth, pigmentation and antibacterial activity of *Monascus purpureus. Plant Physiol.* 46: 63–67.

Berbee, M. L., Yoshimura, A., Sugiyama, J., and Taylor, J. W. (1995). Is penicillium monophyletic—an evaluation of phylogeny in the family Trichocomaceae from 18s, 5.8s and its ribosomal DNA-sequence data. *Mycologia* 87: 210–222.

Birbir, M. Ozyaral, O., Johansson, C., and Ilgaz, A. (1994). Mold strains isolated from unfinished and finished leather goods and shoes. *J. Am. Leather Chem. Assoc.* 89: 14–19.

Blanc, P. J., Loret, M. O., Santerre, A. L., Pareilleux, A., Prome, D., Laussac, J. P., and Goma, G. (1994). Pigments of *Monascus. J. Food Sci.* 59: 862–865.

Blanc, P. J., Laussac, J. P., Lebars, P., Loret, M. O., Pareilleux, A., Prome, D., Prome, J. C., Santerre, A. L., and Goma, G. (1995a). Characterization of monascidin A from *Monascus* as citrinin. *Int. J. Food Microbiol.* 27: 201–213.

Blanc, P. J., Loret, M. O., and Goma, G. (1995b). Production of citrinin by various species of *Monascus. Biotechnol. Lett.* 17: 291–294.

Blanc, P. J., Loret, M. O., and Goma, G. (1995c). Production of citrinin by various species of *Monascus. Biotechnol. Lett.* 17: 291–294.

Borrow, C. D., Jefferys, E. G., Kessek, H. J., Lloyd, E. C., and Nixon, L. X. (1961). The metabolism of *Gibberella fujikuroi* in stirred culture. *Can. J. Microbiol.* 7: 227–226.

Bridge, P. D., and Hawksworth, D. L. (1985). Biochemical tests as an aid to the identification of *Monascus* species. *Lett. Appl. Microbiol.* 1: 25–29.

Broder, C. U., and Koehler, P. E. (1980). Pigments produced by *Monascus purpureus* with regard to quality and quantity. *J. Food Sci.* 45: 567–569.

Burnett, J. H. (1970). *Fundamental of Mycology.* Edward Arnold Publishers, London.

Cannon, P. F., Abdullah, S. K., and Abbas, B. A. (1995). Two new species of *Monascus* from Iraq with a key to known species of the genus. *Mycol. Res.* 99: 659–662.

Carels, M., and Sheperd, D. (1977). The effect of different nitrogen sources on pigment production and sporulation of *Monascus* species in submerged shaken cultures. *Can. J. Microbiol.* 24: 1360–1372.

Carels, M., and Sheperd, D. (1978). The effect of pH and amino acids on conidiation and pigment production of *Monascus major* ATCC 16367 in submerged shaken culture. *Can. J. Microbiol.* 24: 1346–1357.

Carels, M., and Sheperd, D. (1979). The effect of changes in pH on phosphate and potassium uptake by *Monascus rubinosus* in submerged culture. *Can. J. Microbiol.* 25: 1484–1488.

Chen, M. H., and Johns, M. R. (1993). Effect of pH and nitrogen source on pigment production by *Monascus purpureus. Appl. Microbiol. Biotechnol.* 40: 132–138.

Chen, M. H., and Johns, M. R. (1994). Effect of carbon source on ethanol and pigment production by *Monascus purpureus. Enz. Microb. Technol.* 16: 584–590.

Chen, F. C., Manchard, P. S., and Whalley, W. B. (1969). The structure of monascin. *Chem. Commun.* 3: 130–131.

Chen, F. C., Manchard, P. S., and Whalley, W. B. (1971). The chemistry of fungi. Part LXIV. The structure of monascin: the relative sterochemistry of azaphilones. *J. Chem. Soc. C* 1971: 3577–3579.

Cousin, M. A. (1990). Development of an enzyme-linked immunosorbent assay for detection of molds in foods. *Dev. Ind. Microbiol.* 31: 157–163.

Cousin, M. A., Dufrenne, J., Rombouts, F. M., and Notermans, S. (1990). *Food Microbiol.* 7: 227–235.

Dainippon Ink and Chemicals, Inc. (1982a). Production of coenzyme Q_{10} by fermentation. *Jpn. Pat.* 82,08,789.

Dainippon Ink and Chemicals, Inc. (1982b). Production of coenzyme Q_{10}. Jpn. Pat. 82,08,790.

Demain, A. L., Kobayashi, T., Lin, T. F., Nochur, S., Romaniec, M. P. M., and Roberts, M. F. (1990). Biotechnology of enzymes and pigments. In *Microbiological Applications in Food Biotechnology*, B. H. Nga and Y. K. Lee (Eds.). Elsevier Applied Science, New York.

Detilly, G., Mou, D. G., and Cooney, C. L. (1983). Optimization and economics of antibiotic production. In *The Filamentous Fungi*, Vol. 4, J. E. Smith, D. R. Berry, and B. K. Kristiansen (Eds.). Edward Arnold Publishers, London.

Dong, Z. Y., Yang, S. J., and Zhang, S. Z. (1996). Effect of N-linked oligosaccharide on the conformation and properties of glucoamylase from *Monascus rubiginosus*. Enzyme Engineering XIII 799: 193–196.

Dubos, R. J., and Hirsch, J. G., Eds. (1965). *Bacterial and Mycotic Infections of Man*, 4th ed. J. B. Lippincott Company, Philadelphia.

Edwards, A. G., and Ho, C. S. (1987). Effects of carbon dioxide on *Penicillium chrysogenum*. *Biotechnol. Bioeng.* 32: 1–7.

Endo, A. (1985). History and recent trends concerning red koji fungi. *Hakko Kogyo* 43: 544–562.

Endo, A., Hasumi, K., Nakamura, T., Kunishima, M., and Masuda, M. (1985a). Dihydromonacolin L and monacolinX, new metabolites that inhibit cholesterol synthesis. *J. Antibiot.* 38: 321–327.

Endo, A., Hasumi, K., and Negishi, S. (1985b). Monascoli J and L, new inhibitors of cholesterol synthesis produced by *Monascus ruber*. *J. Antibiot.* 38: 420–422.

Endo, A., Negishi, Y., Iwashita, T. Mizukawa, K., and Hirama, M. (1985c). Biosynthesis of ML-236B (compactin) and monacolin K. *J. Antibiot.* 38: 444–448.

Endo, A., Komakawa, D., and Shimada, H. (1987). Manufacture of novel physiologically active monacolin M with *Monascus*. Jpn. Pat. 62,132,878.

Evans, P. J., and Wang, H. Y. (1984). Pigment production from immobilized *Monascus* species utilizing polymeric resin adsorption. *Appl. Environ. Microbiol.* 46: 1323–1326.

Fabre, C. E., Santerre, A. L., Loret, M. O., Baberian, R., Pareilleux, A., Goma, G., and Blanc, P. J. (1993). Production and food application of red pigments of *Monascus ruber*. *J. Food Sci.* 58: 1099–1102.

Fielding, B. C., Haws, E. J., Holker, J. S. E., Powell, A. D. G., Robertson, A., Stanway, D. N., and Whalley, W. B. (1960). Monascorubrin. *Tetrahed. Lett.* 1960(5): 24–27.

Fielding, B. C., Holker, J. S. E., Jones, D. F., Powell, A. D. G., Richmond, K. W., Robertson, A., and Whalley, W. B. (1961). The structure of monascin, *J. Chem. Soc.* 1961: 4579–4589.

Fink-Gremmels, J., Dresel, J., and Leistner, L. (1991). Use of *Monascus* extracts as an alternative to nitrite in meat products. *Federal Meat Research Institute*, Kulmbach, FRG.

Francis, F. J. (1987). Lesser known food colorants. *Food Technol.* 41: 62–68.

Frazier, W. C. (1967). In *Food Microbiology*, McGraw-Hill Book Company, New York.

Gandhi, A. P., and Kjaergaard, L. (1975). Effect of carbon dioxide on formation of a-amylases by *Bacillus subtilis* growing in continuous and batch cultures. *Biotechnol. Bioeng.* 17: 1109–1118.

Garraway, M. O., and Evans, R. C. (1984). *Fungal Nutrition and Physiology*. Wiley Interscience, New York.

Grootwassink, J. W. D., and Gaucher, G. M. (1980). De novo biosynthesis of secondary metabolism enzymes in homogeneous cultures of *Penicillium urticae. J. Bacteriol.* 141: 443–445.

Hadfield, J. R., Holker, J. S. E., and Stanway, D. N. (1967). The biosyntheis of fungal metabolites. Part II. The b-oxo-lactone equivalents in rubropunctatin and monascorubrin. *J. Chem Soc.* (C) 1967: 751–755.

Hajjaj, H., Klaebe, A., Loret, M. O., Tzedakis, T., Goma, G., and Blanc, P. J. (1997). Production and identification of N-glucosylrubropunctamine and N-glucosylmonascorubramine from *Monascus ruber* and occurrence of electron donor-acceptor complexes in these red pigments. *Appl. Env. Microbiol.* 63: 2671–2678.

Hajjaj, H., Blanc, P. J., Goma, G., and Francois, J. (1998). Sampling techniques and comparative extraction procedures for quantitative determination of intra- and extracellular metabolites in filamentous fungi. *Microbiol. Lett.* 164: 195–200.

Hallagan, J. (1999). Regulation of color additive exempts from certification in USA. IFT Basic Symposium, Chicago, July 24.

Hamdi, M., Blanc, P. J., and Goma, G. (1996). Effect of aeration conditions on the production of red pigments by *Monascus purpureus* on prickly pear juice. *Proc. Biochem.* 31: 543–547.

Hamdi, M., Blanc, P. J., Loret., M. O., and Goma, G. (1997). A new process for pigment production by submerged culture of *Monascus purpureus. Bioproc. Eng.* 17: 75–79.

Han, O. (1990). Optimization of *Monascus* pigment production in solid-state fermentation. Ph.D. thesis, University of Massachusetts, Amherst, MA.

Han, O., and Mudgett, R. E. (1991). Effects of oxygen and carbon dioxide partial pressures on *Monascus* growth and pigment production. *Biotechnol. Prog.* 8: 5–10.

Harkness, R. (1998). Cholestin has an unusual history. *Sun-Herald, Sept. 19*, Los Angeles, California.

Hawksworth, D. L., and Pitt, J. I. (1983). A new taxonomy for *Monascus* species based on cultural and microscopical characters. *Aust. J. Bot.* 31: 51–61.

Haws, E. J., and Holker, J. S. E. (1961). Further evidence for structure of rubopunctatin. *J. Chem. Soc.* 1961: 3820.

Haws, E. J., Holker, J. S. E., Lelley, A., Powell, A. D. J., and Robertson, A. (1959). The structure of rubropunctatin. *J. Chem. Soc.* 1959: 3598.

Heber, D., Yip, I., Ashley, J. M., Elashoff, D. A., Elashoff, R. M., and Go, V. W. (1999). Cholesterol-lowering effects of a proprietary Chinese red-yeast-rice dietary supplement. *Am. J. Clin. Nutr.* 69: 231–236.

Henry, B. S. (1999). Regulations in Europe and Japan. *IFT Basic Symposium,* Chicago, July 24.

Hesseltine, C. W. (1972). Solid-state fermentation. *Biotechnol. Bioeng.* 14: 517–532.

Hesseltine, C. W. (1977a). Solid state fermentation—Part 1. *Proc. Biochem.* 12: 24–27.

Hesseltine, C. W. (1977b). Solid state fermentation—Part 2. *Proc. Biochem.* 12: 28–32.

Hiroi, T., Shima, T., Suzuki, T., Tsukioka, M., and Kimura, S. (1978). Nutritional and physiological roles of monascorubrin in *Monascus anka,* a new strain to produce a natural pigment. *J. Ferment. Technol.* 56: 149–154.

Ho, C., and Smith, M. D. (1986a). Effect of dissolved carbon dioxide on penicillin fermentations: mycelial growth and penicillin production. *Biotechnol. Bioeng.* 28: 668–677.

Ho, C., and Smith, M. D. (1986b). Morphological alterations of *Penicillium chrysogenum* caused by carbon dioxide. *J. Gen. Microbiol.* 132: 149–154.

Hong, Y. J., Kim, J. G., Woo, H. C., and Kim, S. U. (1995). Effects of feeding intermedate and starter units on *Monascus* pigments production. *Agr. Chem. Biotechnol.* 38: 31–36.

Iizuka, H., and Lin, C. F. (1983). On the genus *Monascus* of Asia and its specific characteristics. *Proc. 6th Int. Fermentation Symp.* London, Canada.

Iizuka, H., and Mineki, S. (1977). Studies on the genus *Monascus* I. Purification and properties of two froms of glucoamylase from *Monascus kaoliang* nov. sp. F-1. *J. Gen. Appl. Microbiol.* 23: 217–230.

Iizuka, H., and Mineki, S. (1978). Studies on the genus *Monascus* II. Substrate specificity of two glucoamylases obtained from *Monascus kaoliang* nov. sp. F-1. *J. Gen. Appl. Microbiol.* 24: 185–192.

Imanaka, T., Kaieda, T., Sato, K., and Taguchi, H. (1972). Optimization of α-galactosidase production by mold in batch and continuous culture and a kinetic model for enzyme production. *J. Ferment. Technol.* 50: 633–646.

Imanaka, T., Kaieda, T., and Taguchi, H. (1973). Optimization of α-galactosidase production in multi-stage continuous culture of mold. *J. Ferment. Technol.* 51: 431–439.

Inouye, Y., Nakanishi, K., Nishikawa, H. Ohashi, M., Terahara, A., and Yamamura, S. (1962). The structure of monascoflavin. *Tetrahedron.* 18: 1195–1203.

Jay, J. M. (1986). *Modern Food Microbiology*, 3rd ed. Van Nostrand Reinhold Company, New York.

Johnson, G. T., and McHan, F. (1975). Some effects of zinc on the utilization of carbon sources by *Monascus purpureus. Mycologia* 67: 806–816.

Jong, S. C., and Edwards, M. J. (1991). *Catalogue of Filamentous Fungi*, 18th ed. *American Type Culture Collection*, Rockville, MD.

Jung, S. T., Kang, S. G., Hwang, K. T., Rhim, J. W., Park, Y. K., Park, H. J., Park, I. B., and Yun, S. W. (1996). Characteristics of red pepper paste using *Monascus anka* koji. IFT Annual Meeting (Abstract).

Juzlova, P., Martinkova, L., Lozinski, J., and Machek, F. (1994). Ethanol as substrate for pigment production by the fungus *Monascus purpureus. Enz. Microb. Technol.* 16: 996–1001.

Juzlova, P. Rezanka, T., Martiinkova, M., and Kren, V. (1996a). Long-chain fatty acids from *Monascus purpureus. Phytochemistry* 43: 151–153.

Juzlova, P., Martinkova, M., and Kren, V. (1996b). Secondary metabolites of the fungus *Monascus*: a review. *J. Indust. Microbiol.* 16: 163–170.

Kawabata, S., and Sato, K. (1976). Water-soluble *Monascus* pigment. *Jpn. Pat.* 76,099,519.

Kirtzman, G., Chen, I., and Henis, Y. (1977). Effect of carbon dioxide on growth and carbohydrate metabolism in *Sclerotium raffoni. J. Gen. Microbiol.* 100: 167–175.

Kranz, C., Panitz, C., and Kunz, B. (1992). Biotransformation of free fatty acids in mixtures to methyl ketones by *Monascus purpureus. Appl. Microbiol. Biotechnol.* 36: 436–439.

Kritchevsky, D., Li, C., Wei, W., and Wang, Y. (1998). Influence of *Monascus purpureus* fermented rice (red yeast rice) on experimental atherosclerosis in rats. *FASEB J.* 12: 1200.

Kumasaki, S., Nakanishi, K., Nishikawa, E., and Ohashi, M. (1962). Structure of monascorubrin. *Tetrahedron.* 18: 1171–1184.

Kundu, A. B., Ghosh, B. S., Ghosh, B. L., and Ghose, S. N. (1983). Role of water in the hydrolysis of cellulose in a solid state. *J. Ferment. Technol.* 61: 185–188.

Kuramoto, Y., Yamada, K., Tsuruta, O., and Sugano, M. (1996). Effect of natural food colorings on immunoglobulin production in vitro by rat spleen lymphocyes. *Biosci. Biotechnol. Biochem.* 44: 1712–1713.

Lee, Y. K., Chen, D. C., Chauvathatcharin, S., Seki, T., and Yoshida, T.

(1995). Production of *Monascus* pigments by a solid-liquid state culture method. *J. Ferm. Bioeng.* 79: 516–518.

Levonen-Munoz, E., and Bone, D. H. (1985). Effects of different gas environments on bench-scale solid-state fermentation of oat straw by white rot fungi. *Biotechnol. Bioeng.* 27: 382–387.

Li, C. L., Zhu, Y., Wang, Y. Y., Zhu, J. S., Chang, J., and Kritchevsky, D. (1998). *Monascus purpureus*—fermented red rice (red yeast rice): a natural food product the lowers blood cholesterol in animal models of hypercholesterolemia. *Nutr. Res.* 18: 71–81.

Lilly, V. G., and Barnett, H. L. (1962). The utilization of sucrose and its constituent sugars by *Monascus purpureus. Proc. W. Va. Acad. Sc.* 24: 27–32.

Lin, C. (1973). Isolation and cultural conditions of *Monascus* sp. For the production of pigment in a submerged culture. *J. Ferment. Technol.* 51: 407–414.

Lin, T. F., and Demain, A. L. (1991). Effect of nutrition of *Monascus* sp. on formation of red pigments. *Appl. Microbiol. Biotechnol.* 36: 70–75.

Lin, T., and Demain, A. L. (1993). Resting cell studies on formation of water-soluble red pigments by *Monascus* sp. *J. Ind. Microbiol.* 12: 361–367.

Lin, T., and Demain, A. L. (1994). Leucine interference in the production of water-soluble red *Monascus* pigments. *Arch. Microbiol.* 162: 114–119.

Lin, T. F., and Demain, A. L. (1995). Negative effect of ammonium nitrate as nitrogen source on the production of water-soluble red pigments by *Monascus* sp. *Appl. Microbiol. Biotechnol.* 43: 701–705.

Lin, T. F., Yakushiji, K., Buchi, G. H., and Demain, A. L. (1992). Formation of water-soluble red pigments by biological and semi-synthetic processes. *J. Indust. Microbiol.* 9: 173–179.

Lin, C., and Iizuka, H. (1982). Production of extracellular pigment by a mutant of *Monascus kaoliang* sp. Nov. *Appl. Environ. Microbiol.* 43: 671–676.

Lin, C., and Suen, S. J. (1973). Isolation of hyper-pigment-productive mutants of *Monascus* sp. F-2. *J. Ferment. Technol.* 51: 757–759.

Lindenfelser, L. A., and Ciegler, A. (1975). Solid-state fermentor for ochratoxin A production. *Appl. Microbiol.* 29: 323–327.

Lonsane, B. K., Ghyldyal, N. P., Budiatman, S., and Ramakrishna, S. V. (1985). Engineering aspects of solid-state fermentation. *Enz. Microb. Technol.* 7: 258–265.

Lumyong, S., and Tomita, F. (1993). L-Malic acid production by an albino strain of *Monascus araneosus. World J. Microbiol. Biotechnol.* 9: 383–384.

Marchand, P. S., and Whalley, W. B. (1973). Isolation and structure of anakflavin: a new pigment from *Monascus anka. Phytochemistry.* 12: 2531–2532.

Martinkova, L., Juslova, P., and Vesely, D. (1995). Biological activity of polyketide pigments produced by the fungus *Monascus. J. Appl. Bacteriol.* 79: 609–616.

Martorelli, S. R., and Cremonte, F. (1998). A proposed three-host history of *Monascus filiformis* (Rodolphi, 1819) (Digenea : Fellodistomidae) in the southwest Atlantic Ocean. *Can. J. Zool.* 76: 1198–1203.

Matsumoto, S., Shimazaki, S., and Ebine, H. (1987). Flavor formation by *Monascus* species during fermentation on cooked rice. Misouo Kukaku (Japan) 35: 262–264.

McHan, F., and Johnson, G. T. (1970). Zinc and amino acids: important components of medium promoting growth of *Monascus purpureus*. *Mycologia* 62: 1108–1131.

McHan, F., and Johnson, G. T. (1979). Some effects of zinc on the utilization of nitrogen sources by *Monascus purpureus*. *Mycologia* 71: 169–179.

Miyake, T., Ohno, S., and Sakai (1982). Pigment of *Monascus*. French Pat. 2,505,856.

Miyake, T., Ohno, S., and Sakai, S. (1984). Process for the production of *Monascus* pigment. U.S. Pat. 4,442,209.

Moll, H. R., and Farr, D. R. (1978). Preparation of a red coloring agent. *Swiss Pat.* 606,433.

Moo-Young, M., Moreira, A. R., and Tengerdy, R.P. (1983). Principle of solid-state fermentation. In *The Filamentous Fungi*, Vol. 4, J. E. Smith, D. R. Berry, and B. K. Kristiansen (Eds.). Edward Arnold Publishing Company, London.

Mudgett, R. E., and Bajracharya, R. (1979). Effects of controlled gas environments in microbial enhancement of plant protein recovery. *J. Food Biochem.* 3: 135–149.

Mudgett, R. E., and Paaradis A. J. (1985). Solid-state fermentation of natural birch lignin by *Phanerochaete chrysogenum*. *Enzyme Mirob. Technol.* 7: 150–154.

Mudgett, R. E., Nash, J., and Rufner, R. (1982). Controlled gas environments in solid substrate fermentations. In *Developments in Industrial Microbiology* 23: 397–405.

Mudgett, R. E. (1980). Controlled gas environments in industrial fermentations. *Enzyme Microb. Technol.* 2: 273–279.

Mudgett, R. E. (1986). Solid-state fermentations. In *Manual of Industrial Microbiology and Biotechnology*, A. L. Demain and N. A. Solomon (Eds.). American Society of Microbiology, Washington, DC.

Nagasawa, T., Yamata, H., Ohara, M., and Yoshida, T. (1987). Production of *Monascus* pigments. *Jpn. Pat.* 62,210,993.

Nakagawa, N., Watanabe, S., and Kobayashi, J. (1976a). Water-soluble *Monascus* pigment. *Jpn. Pat.* 76,091,938.

Nakagawa, N., Watanabe, S., and Kobayashi, J. (1976b). Water-soluble *Monascus* pigment. *Jpn. Pat.* 76,091,937.

Nakamura, J., Miyashiro, S., and Hirsoe, Y. (1976). Screening, isolation and some properties of microbial cell flocculants. *Agric. Biol. Chem.* 40: 377–383.

Narahara, H., Koyama, Y.,Yoshida, T., Pichangjura, T., Ueda, R., and Taguchi, H. (1982). Growth and enzyme production in a solid-state culture of *Aspergillus oryzae. J. Ferment. Technol.* 60: 311–3139.

Narahara , H., Koyama, Y., Yoshida, T., Atthasampunna, P., and Taguchi, H. (1984). Control of water content in a solid-state culture of *Aspergillus oryzae. J. Ferment. Technol.* 62: 453–459.

Narahara, H. (1977). Effect of water activity on the growth and the yield of *Aspergillus conidia. Hakko Koyaku Kaiishi* 55: 254–261.

National Institutes of Health (1999). Bringing together the best of healing. *http://nccam.nih.gov.*

Nishikawa, J., and Iizuka, H. (1993). Numerical taxonomy of *Monascus* species based on the electrophoretic analysis of intracellular enzymes. *J. Basic Microbiol.* 33: 331–342.

Nozaki, H., Date, S., Kondo, H., Kiyohara, H., Takaoka, D., Tada, T., and Nakayama, M. (1991). Spectral evidence and results of an X-ray diffraction analysis relating to the propoosed structure of ankalactone, a new bacteriostatic. *Agric. Biol. Chem.* 55: 899–900.

Nyiri, L. K., Toth, G. M., and Charles, M. (1975). On-line ,measurement of gas-exchange conditions in fermentation processes. *Biotechnol. Bioeng.* 18: 1663–1678.

Pastrana, L., and Goma, G. (1995). Estimation of bioprocess variables from *Monascus ruber* cultures by means of stoichiometric models. *Proc. Biochem.* 30: 607–613.

Pastrana, L., Blanc, P. J., Santerre, A. L., Loret, M. O., and Goma, G. (1995). Production of red pigments by *Monascus ruber* in synthetic media with a strictly controlled nitrogen source. *Proc. Biochem.* 30: 333–341.

Pastrana, L., Loret, M. O., Blanc, P. J., and Goma, G. (1996). Production of citrinin by *Monascus ruber* submerged culture in chemically defined media. *Acta Biotechnol.* 16: 315–319.

Patekova-Juslova, P., Rezanka, T., and Viden, I. (1998). Identification of volatile metabolites from rice fermented by the fungus *Monascus purpureus* (ang-kak). *Folia Microbiol.* 43: 407–410.

Peng, I., and Demain, A. L. (1998). A new hydroxylase system in *Actinomadura* sp. cells converting compactin to pravastatin. *J. Indus. Microb. Biotechnol.* 20: 373–375.

Peng, Y., Walker, E. A., Davis, J. C., and Demain, A. L. (1999). A chemically defined medium supporting growth and providing cells converting compactin to pravastatin. *J. Ind. Microbiol. Biotechnol.* 22: 78–79.

Peng, Y., Yashphe, J., and Demain, A. L. (1997). Biotransformation of compactin to provastatin by *Acinomadura* sp. 2966. *J. Antibiot.* 50.

Peters, N., Panitz, C., and Kunz, B. (1993). The influence of carbohydrate dissimilation on the fatty acid metabolism of *Monascus purpureus*. *Appl. Microbiol. Biotechnol.* 39: 582–589.

Pipek, P., Baco, B., and Bebarovs, V. (1996). Colorants for sausages. *Potravinarske Vedy* 14: 405–414.

Pirt, S. J. (1967). A kinetic study of the mode of growth surface colonies of bacteria and fungi. *J. Gen. Microbiol.* 47: 181–197.

Plating, S. J., and Cherry, J. P. (1979). Protein and amino acid composition of cottonseed flour fermented with selected filamentous fungi. *J. Food Sci.* 44: 1178–1182.

Prescott, S. C., and Dunn, C. G. (1959). *Industrial Microbiology*, 3rd ed. McGraw-Hill Book Company, New York.

Qin, S. U., Zhang, W. Q., Zhao, M. L., Dong, Z. N., Li, Y. C., Xu, X. M., Fang, X., Fu, L., Zhu, J. S., and Chang, J. (1999). Elderly patients with primary hyperlipidemia benefited from treatment with *Monascus purpureus* rice preparation: a placebo-controlled double-blind clinical trial. *Circulation* 99: 89.

Rasheva, T., Kujumdzieva, A., and Hallett, J. N. (1997). Lipid production by *Monascus purpureus* albino strain. *J. Biotechnol.* 56: 217–224.

Riken Vitamin Co., Ltd. (1982). Production of *Monascus* pigments. Jpn. Pat. 82,079,888.

Santerre, A. L., Qyeinnec, I., Destruhaut, Pourcile, J. B., and Blanc, P. J. (1995). Online evaluation of fermentation broth color. *Biotechnol. Techniques* 8: 319–324.

Saruno, R., Setoyama, T., Nakashima, C., Kato F., and Murata, A., 1981. Purification and some properties of nuclease inhibitor from *Monascus purpureus*. *Agric. Biol. Chem.* 45: 133–139.

Sato, K., Goda, Y., Sakamoto, S. S., Shibata, H., Maitani, T., and Yamada, T. (1997). Identification of major pigments containing D-amino acid units in commercial *Monascus* pigments. *Chem. Pharm. Bull.* 45: 227–229.

Sato, K., Nagatani, M., and Sato, S. (1982). A method of supplying moisture to the medium in a solid-state culture with forced aeration. *J. Ferment. Technol.* 60: 607–610.

Sato, K., Nagatani, M., Nakamura, K., and Sato, S. (1983). Oxygen uptake in a solid-state culture with forced aeration. *J. Ferment. Technol.* 61: 623–629.

Shin, C. S., Kim, H. J., Kim, M. J., and Ju, N. Y. (1998). *Biotechnol. Bioeng.* 59: 576–581.

St. Martin, E. J., Kurek, P. R., Schumacher, E. F., and Rorbach, R. P. (1991). Reduced Monascus pigment derivatives as yellow food colorants. U.S. Pat. 5,013,564.

Steinkraus, K., Ed. (1995). *Handbook of Indigenous Fermented Foods.* 2nd ed. Marcel Dekker, New York.

Su, Y. C., and Huang, W. H. (1977). Chinese red-rice anka. In *Handbook of Indigenous Fermented Foods,* Steinkraus, K. (Ed.), Marcel Dekker, New York.

Su, Y. C., and Huang, J. H. (1980). Fermentative production of anka pigments. *Proc. Natl. Sci. Council* 4: 4201–215.

Sweeny, J. G., Estada-Valdes, M. C., Iacobucci, G. A., Sato, H., and Sakamura, S. (1981). Photoprotection of the red pigments of *Monascus anka* in aqueous media by 1,4,6-trihydroxynaphthalene. *J. Agric. Food Chem.* 29: 1189–1193.

Takahashi, J., Hidaka, H., and Yamada, K. (1965). Effect of mycelial forms on citric acid formation in submerged mold culture. *Agric. Biol. Chem.* 29: 331–336.

Tarui, S., Tanabe, N., and Yoshida, A. (1987). An anti-hypertensive fraction and its manufacture from *Monascus.* Jpn. Pat. 62,298,598.

Teng, S. S., and Feldheim, W. (1998). Analysis of anka pigments by liquid chromatography with diode array detection and tandem mass spectrometry. *Chromatographia* 47: 529–536.

Toyo Brewing Company (1979). Coloring of meat products. *Jpn. Pat.* 54,01,374.

Toyo Jozo Company (1978). Food coloring. *Jpn. Pat.* 52,006,003.

Tsai, M., Hseu, T. H., and Shen, Y. S. (1978). Purification and characterization of an acid protease from *Monascus kaoliang. Int. J. Peptide Protein Res.* 12: 293–302.

Tsunenaga, T., and Once, A. (1975). Food coloring agents from *Monascus* cultures. *Jpn. Pat.* 75,036,519.

Turner, W. B. (1971). *Fungal Metabolites.* Academic Press, New York.

Ulmer, D. C., Tengerdy, R. P., and Murphy, V. G. (1981. Solid state fermentation of manure fibers. *Biotechnol. Bioeng. Symp.* Series II: 449–641.

Vandamme, E. J. (1993). Production of vitamins and related biofactors via microorganisms. *Agro Food Ind. Hi-tech* 4: 29–31.

Verweig, P. E., Meis, J. F. G. M., Vandenhurk, P. Zoll, L. Samson, R. A., and Melchers, W. J. G. (1995). Phylogenetic relationships of 5 species of *Aspergillus* and related taxa as deduced by comparison of sequences of small subunit ribosomal rna. *J. Med Vet. Mycol.* 33: 185–192.

Wang, H. L., and Hesseltine, C. W. (1979). Mold-modified foods. In *Microbial Technology,* Vol. II, H. J. Peppler and D. Perlman (Eds.). Academic Press, New York.

Wang, H. L., and Hesseltine, C. W. (1982). Oriental foods. In *Prescott and Dunn's Industrial Microbiology,* 4th ed., G. Reed (Ed.). Avi, Westport, CT.

Wang, J. X., Lu, Z. I., Chi, J. M., Wang, W. H., Su, M. Z., Kou, W. R., Yu, P. L., Zhu, J. S., and Chang, J. (1997). Multicenter clinical trial of the serum-lowering effects of a *Monascus purpureus* (red yeast) rice preparation from traditional koji. *Current Ther. Res. Clin. Exp.* 58: 964–978.

Watanabe, T., Yamamoto, A., Nagai, S., and Terabe, S. (1997). Separation and determination of *Monascus* yellow pigments for food by micellar electrokinetic chromatography. *Anal. Sci.* 13: 571–575.

Wiseman, G. M., Violago, F. C., Roberts, C., and Penn, I. (1966). Effect of hyperbaric oxygen on aerobic bacteria. *Can. J. Microbiol.* 12: 521.

Wong, H. C., and Bau, Y. S. (1977). Pigmentation and antibacterial activity of fast neutron and x-ray induced strains of *Monascus purpureus* Went. *Plant Physiol.* 60: 578–581.

Wong, H. C., and Koehler, P. E. (1981a). Mutant for *Monascus* pigment production. *J. Food Sci.* 46: 956–957.

Wong, H. C., and Koehler, P. E. (1981b). Production and isolation of an antibiotic from *Monascus purpureus* and its relationship to pigment production. *J. Food Sci.* 46: 589–692.

Wong, H. C., Lin, Y. C., and Koehler, P. E. (1981). Regulation of growth and pigmentation of *Monascus purpureus* by carbon and nitrogen concentrations. *Mycologia* 73: 649–654.

Wong, H. C., and Koehler, P. E. (1983). Production of red water-soluble *Monascus* pigments. *J. Food Sci.* 48: 1200–1203.

Wongleung, Y. L., Fong, W. F., and Lam, W. L. (1993). Production of α-galactosidase by *Monascus* grown on soybean and sugarcane wastes. *World J. Microbiol. Biotechnol.* 9: 529–533.

Xu, R. Y., and Bao, F. T. (1999). Grandiose survey of Chinese alcoholic drinks and beverages. *http://wxuli.edu.cn/wine/u2-3.htm.*

Yamaguchi, Y., Ito, H., Watanabe, S., Yoshhida, T., and Kumatsu, A. (1973). Water-soluble *Monascus* pigment. Jpn. Pat. 73,765,906.

Yamanaka, S., Oida, A., and Mitzugi, H. (1975). Red pigments produced by *Monascus.* Jpn. Pat. 74,075,793.

Yang, S., Ge, S., Zeng, Y., and Zhang, S. (1985). Inactivation of α-glucosidase by the active site-directed inhibitor conduritol B expoxide. *Biochim. Biophys. Acta* 823: 236–240.

Yashphe, J., David, J., Peng, Y., Song, H. B. and Demain, A. L. (1997). New organisms which convert compactin to provastatin. *Acinomycetol.* 11: 20–25.

Yasuda, M., Seishi, K., and Miyahira, M. (1984). Purification and properties of acid protease from *Monascus* sp. No. 3403. *Agric. Biol. Chem.* 48: 1637–1639.

Yasuda, M., Matsumoto, T., Sakaguchi, M., and Kobamoto, N. (1993). Changes in chemical components of tofu prepared by *Monascus* fungus during fermentation. *J. Jpn. Soc. Food Sci. Technol.* 40: 331–338.

Yasuda, M., Matsumoto, T., Sakaguchi, M., and Kinjyo, S. (1995a). Production of tofuyo using the combination of red and yellow kojis. *J. Jpn. Soc. Food Sci. Technol.* 42: 38–43.

Yasuda. M., Ikehara, K., Tawata, S., Kobamoto, N., and Toyama, S. (1995b). Purification and properties of a ribonuclease from a species of the genus *Monascus. Biosci. Biotechnol. Biochem.* 59: 327–328.

Yasuda, M., Kinjyo, S., and Miki, E. (1996). Changes in breaking characteristics, creep behavior and microstructure of tofuyu during fermentation. *J. Jpn. Soc. Food Sci. Technol.* 43: 322–327.

Yasukawa, K., Takahashi, M., Natori, S., Kawai, K., Yamazaki, M., Takeuchi, M., and Takido, M. (1994). Azaphilones inhibit tumor promotion by 12-O-tetradecanoylphorbol-13-acetate in two-stage carcinogenesis in mice. *Oncology* 51: 108–112.

Yasukawa, K., Takahashi, M, Yamanuchi, S. and Takido, S. (1996). Inhibitory effect of oral administration of *Monascus* pigment on tumor promotion in two-stage carcinogenesis in mouse skin. *Oncology* 53: 247–249.

Yong, R. K. (1997). Immunological detection of molds, including mycotoxin producers, in foods. Ph.D. thesis, Purdue University, Lafayette, IN.

Yoshimura, M.,Yamanaka, S., Mitzugi, K., and Hirose, Y. (1975). Production of *Monascus* pigment in a submerged culture. *Agric. Biol. Chem.* 39: 1789–1795.

Zajic, J. E. and Liu, F. S. (1969). Effect of carbon dioxide upon α-amylase, protease and spore formation in *Bacillus subtilis. Dev. Ind. Microbiol.* 11: 350–356.

Zhang, M., Chang, H. and Lin, C., (1982). Studies on the red pigment production of *Monascus. Shipin Yu Faxiao Gongye* (China) 1984: 1–7.

4
Gardenia Yellow

Yoshiaki Kato
San-Ei Gen F.F.I., Inc.
Osaka, Japan

INTRODUCTION

Gardenia yellow is a yellowish food colorant and a member of the carotenoid family. Its principal pigments are crocetin derivatives. It is very soluble in water compared to other carotenoids. Gardenia yellow is obtained by extraction with water or ethanol from the fruit of *Gardenia jasminoides* Ellis (*Shan-zhi-i* in Chinese) or *Gardenia augusta* Merrill var. *grandiflora* Hort. (*Shui-zhi-i* in Chinese) (Rubiaceae). Both fruits have been used in Japan and China as herbal drugs for their antiphlogistic, diuretic, and cholagogic effects. They have also been used as a yellow dye and to treat contusions. Furthermore, they are important crude drugs in Kanpo (traditional Asian drugs) prescriptions, Shishikankyo-to, Ourengedoku-to, Inchinko-to, Shishishi-to, etc., and they have sedative, antipyretic, diuretic, choleretic, and anti-inflammatory effects. They were listed in *The Japanese Pharmacopoeia* as crude drugs and in the list of existing food additives. On the other hand, both fruits have been used as food colorants for Japanese traditional foods such as *ohan* (yellow rice colored with gardenia fruit) and *kuri-kanroni* (yellow chestnuts colored with gardenia fruit and soaked in syrup).

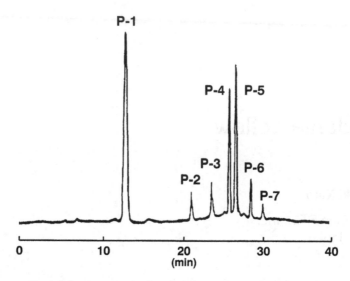

FIG. 4.1 HPLC of pigments isolated from gardenia. Conditions: Nucleosil 5C18 column with a 4.6 mm diameter and 250 mm height. Flow rate, 1.0 mL/min mobil phase and time program, 50% methanol (0–15 min), 50–100% methanol (liner gradient, 15–30 min), 100% methanol (30–40 min).

Many workers have described the chemical constituents of gardenia fruits in pharmacological and biosynthetic studies, but their food chemical properties have not been reported. Recently, gardenia yellow has been widely used to give a yellowish color to food products such as noodles, pasta, candy, beverages, and pickled products in Japan (Shimizu and Nakamura, 1993; Tanimura et al., 1993). For this reason, it would be helpful for food manufactures to understand the food chemical properties of this pigment.

CONSTITUENTS OF GARDENIA YELLOW

Karrer and Miki (1929) reported the main component of yellow pigments to be crocin. Recently, seven crocetin glycosides crocetin have been isolated from *Gardenia jasminoides* and characterized by UV, LC/MS, and NMR (Ichi et al., 1995a). Figure 4.1 shows a HPLC chromatogram of yellow pigments extracted from fully ripened gardenia fruits. The main components were crocetin-digentiobioside ester (crocin: P-1), crocetin-monogentiobioside-monoglucoside ester (P-2), crocetin-monogentiobioside (P-3 and 4), crocetin-diglucoside ester(P-5), and crocetin-monoglucoside ester (P-6 and P-7) (Fig. 4.2). Compounds P-4 and P-7 were *cis* isomers. Crocetin was hardly present. The composition (%) of these compounds in fully matured fruit was also investigated (Table 4.1). Furthermore, Pfister et al. (1996) reported that 13Z-crocin was isolated from *Gardenia jasminoides* and characterized by

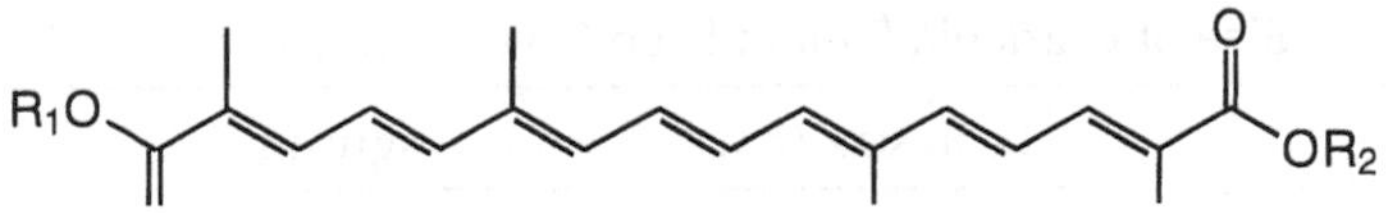

Symbol	R_1	R_2
P-1	β-D-gentiobiosyl	β-D-gentiobiosyl
P-2	β-D-gentiobiosyl	β-D-glucosyl
P-3	β-D-gentiobiosyl	H
P-4	β-D-gentiobiosyl	H
P-5	β-D-glucosyl	β-D-glucosyl
P-6	β-D-glucosyl	H
P-7	β-D-glucosyl	H

FIG. 4.2 Chemical structures of pigments isolated from gardenia yellow.

TABLE 4.1 Composition of Crocetin Derivatives from Gardenia Fruit

Symbol	Compound	Composition (%)
P-1	Crocetin-digentiobioside ester (*trans*): crocin	68.34
P-2	Crocetin-monogentiobioside-monoglucoside ester (*trans*)	4.50
P-3	Crocetin-monogentiobioside ester (*trans*)	2.45
P-4	Crocetin-monogentiobioside ester (*cis*)	5.33
P-5	Crocetin-diglucoside ester (*trans*)	15.33
P-6	Crocetin-monoglucoside ester (*trans*)	2.54
P-7	Crocetin-monoglucoside ester (*cis*)	0.88

UV, LC/MS, and NMR . Pfander and Rychener (1982) reported that the main components of yellow pigments obtained from stigma of saffron are crocetin-digentiobioside ester (crocin), crocetin-monogentiobiose-mono-glucoside ester, crocetin-diglucoside ester, crocetin-monogentiobioside ester, and crocetin-monoglucoside ester. Gardenia fruits and saffron stigma produced almost the same yellow pigments as crocetin derivatives, but the composition ratio of gardenia yellow pigments was different from saffron.

YELLOW PIGMENTS AT DIFFERENT STAGES OF GROWTH

Gardenia plants produce white blossoms with six petals at end of April and fruits soon thereafter. Table 4.2 shows the changes in morphological char-

TABLE 4.2 Size of Gardenia Fruit at Each Stage of Growth

Stage	Date of harvest	Fruit size (cm) Length	Width	Fruit weight (g) Fresh	Dry	Color of fruit
S-1	27 May	2.0–3.0	0.6–1.3	1.60	0.38	White-gray
S-2	26 June	3.4–4.5	1.3–1.6	3.79	0.76	Pale green
S-3	24 July	3.9–4.7	1.2–1.9	5.98	1.08	Pale yellow
S-4	26 August	4.7–6.7	1.5–2.2	5.56	1.32	Yellow
S-5	25 Sept.	4.0–6.0	1.6–2.1	3.56	1.18	Orange
S-6	27 Oct.	4.0–6.0	1.5–2.0	3.82	1.37	Reddish orange
S-7	26 Nov.	3.9–5.7	1.5–2.0	4.05	1.42	Red

acteristics during the development of gardenia fruits (Ichi et al., 1995a). One month after blossoming, fruits were small and elliptical in shape with six longitudinal wings and corollas typical of gardenia fruits.

Sarcocarps grow rapidly and thicken in a period of 3 months (S-3) after blossoming, so that fruits become large in size. As the fruit became larger the fresh weight increased, but no accumulation of yellow pigments on sarcocarps was seen at this stage. Multiplication of fruits stops 3 months after blossoming, and the sarcocarps start turning yellow. Rapid accumulation of yellow pigments begins 5 months after blossoming.

Table 4.3 shows changes in total carotenoid content in association with growth together with that of gardenia fruit fresh and dry weight. As described above, carotenoid pigments were rarely recognized in fruits (S-1 and S-2) at the initial stage of growth, and the carotenoid content in fruits 3 months after blossoming (S-3) was as low as 0.23 mg/g dry weight. Even

TABLE 4.3 Contents of Crocetin Derivatives in the Fruit at Each Stage of Growth

Stage no.	Crocetin derivative composition (%) P-1	P-2	P-3	P-4	P-5	P-6	P-7	Total carotenoid (mg/g dry weight)
S-3	28.00	ND	15.60	18.90	31.70	2.90	1.12	0.05
S-4	60.84	3.20	1.79	14.55	11.30	3.51	0.56	0.18
S-5	70.74	3.29	1.52	8.36	8.80	3.20	0.54	0.81
S-6	69.26	7.11	1.25	6.76	9.46	2.89	0.34	1.76
S-7	71.59	5.84	2.02	4.19	10.78	2.93	0.44	4.49

ND: Not detected.

after the end of vegetative growth and thickening, the fruit (S-4) showed a carotenoid content of 0.45 mg/g dry weight. In the ripening stage after the increase in dry weight ends, however, the carotenoid pigment content increases rapidly to 4.49 mg/g dry weight in completely ripe fruits 7 months after blossoming (S-7). This result indicates that very little carotenoid is formed in gardenia fruits in the vegetative stage and biosynthesis progresses actively in the ripening process, suggesting a rise in the activity of enzyme groups participating in biosynthesis.

To investigate the phenomenon of crocin biosynthesis, the change in gardenia yellow pigment components at each growth stage was traced by calculating five composition ratios in crocin and four kinds of crocetin derivatives (Ichi et al., 1995a). The composition ratio is expressed by the ratio (%) of each peak area to total peak areas of five components measured with HPLC (Table 4.3). The crocin composition ratio in fruits (S-3) was as low as 28% at the stage of continuous thickening, and the ratio occupied by peak components (P-3, P-4, and P-5) other than crocin was high. As soon as the ripening stage starts (S-3), the carotenoid composition changed drastically to show an overwhelming presence of crocin, as high as 61%. In association with further advancement of the ripening stage with the crocin composition ratio reaching 70%, the composition ratios of three components—P-3, P-4, and P-5, the main components at the stage of growth—decreased.

Based on the change of the carotenoid composition with time, it was presumed that some crocin derivatives such as P-3, P-4, and P-5 are direct intermediate precursors of crocin biosynthesis.

FOOD CHEMICAL PROPERTIES OF GARDENIA YELLOW

Color Tone and Absorption Spectra

Gardenia yellow pigments are used to color noodles, beverage, pickles, etc. a brilliant yellow. Since these gardenia yellow pigments are a mixture of five kinds of crocetin derivatives, color tone and absorption spectra of these five kinds of pigment components were measured for comparison (Ichi et al., 1995b) (Table 4.4). The color tones of the yellow pigments P-1 through P-5 were nearly same both visually and numerically. However, the color tone of crocetin was different from these pigments. The two maximum wavelengths of P-1 (crocin) in the visible area were highest at 459 and 435 nm, and those from P-2 to P-5 gradually decreased. Crocetin at 442 and 420 nm was the lowest. Apparently the color tone varies with the sugar linkage.

Heat and Light Stability of Pigments

The stability of commonly used food-coloring substances when exposed to the heat of cooking is an important consideration. Therefore, the heat resistance of G-P, G-P–containing crocetin derivatives, and crocetin was

TABLE 4.4 Color Tone of Crocetin Derivatives from Gardenia Fruit

	G-P	Crocin	P-2	P-3	P-4	P-5	P-6	P-7	Crocetin
L	96.77	97.26	97.45	97.55	97.47	97.44	97.41	97.55	98.72
a	−20.19	−20.70	−21.30	−20.85	−21.18	−21.23	−20.59	−21.22	−22.04
b	58.43	59.15	58.97	58.77	58.52	58.44	56.18	54.85	50.95
$\tan^{-1}(b/a)$	109.06	109.28	109.86	109.53	109.90	110.00	110.13	111.15	113.39
Chroma	61.82	61.72	62.70	62.36	62.23	62.18	59.83	58.81	55.51
λ_{max} (nm)	441	444	459	457	453	453	450	447	447
	—	436	434	431	431	428	424	424	420
				320			314		

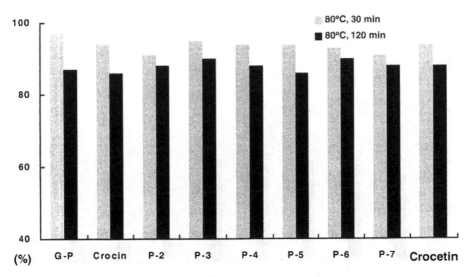

FIG. 4.3   Heat stability of crocetin derivatives isolated from gardenia yellow.

examined (Ichi et al., 1995b). Test materials were dissolved in a McLlvaine buffer solution (pH 7.0). They then underwent heat treatment at 80°C for periods of 30 and 120 minutes, at which time the percentage of remaining color was determined (Fig. 4.3).

G-P and all the crocetin derivatives subjected to heat treatment for 30 minutes showed a color loss of less than 5%. The same materials subjected to heat treatment for 120 minutes had color retention exceeding 85%. Spectral patterns and color tone changes were not significant.These results show that all constituents have a relatively high resistance to heat. The addition of various sugars to crocetin does not affect heat resistance.

The light resistance of commonly used food-coloring substances during distribution and use is an important consideration. Therefore the light resistance of G-P, G-P–containing crocetin derivatives, and crocetin was examined (Koda et al., 1992b) under high-strength white light irradiation (Fig. 4.4). Gentiobiose in the crocetin nuclear material for ester-combined crocin, P-2, P-3, and P-4 light resistance showed a marked disparity when compared to other constituents (P-5 through P-7 and crocetin). The former constituents had a 40% residual color ratio, while the latter constituents had approximately a 10% residual color ratio following 4-hour irradiation exposure. This is a high degree of color fading. The results also show a wide range of light-related stability for different crocetin derivatives. This seems to be dependent more on derivative type than on the number of ester-combined sugars. The light resistance characteristics of the G-P used in these tests was slightly inferior as compared to crocin. G-P contains crocetin

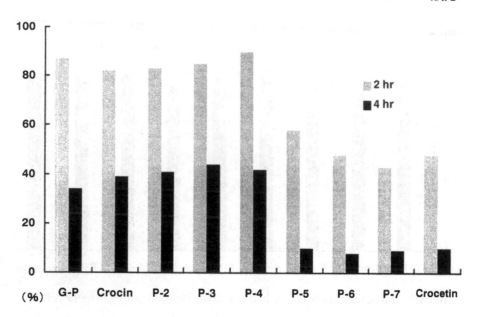

FIG. 4.4 Light stability of crocetin derivatives isolated from gardenia yellow. (Apparatus: Fade-Ometer.)

derivatives with highly light-sensitive constituents. The origin of this is thought to be the roughly 20% of components with no gentiobiose. This is based upon the fact that crocin with a 30% pigment composition (components with no gentiobiose were approximately 40%) was used. During the gardenia yellow pigment manufacturing process, when high polarity–low extraction ratio gentiobiose-containing components were used, the resulting products had weak red coloring as well as low light resistance.

A diffraction grating spectroscope was used to check the wavelength-dependent color fading and light resistance of crocetin derivatives and crocetin (Fig. 4.5). P-3 was used as a representative specimen for crocin and P-2 through P-4 samples containing gentiobiose components. P-5 was used as a representative specimen for crocetin and P-5 through P-7 samples. The seven varieties of crocetin and crocetin derivatives were measured and tested in the same way as the gardenia yellow pigment marketed by Yoshizumi (1974). All materials were found to have excellent stability when exposed to visible light wavelengths greater then 450 nm. At shorter light wavelengths, significant color fading occurred. This was especially true in the ultraviolet light range of 250–300 nm. Light wavelengths of 275 (±10 nm) showed maximum effect.

Identical light resistance tests detected different light resistances for short wavelength light sensitivity for components containing gentiobiose (crocetin and P-2 through P-4) as well other components. The former com-

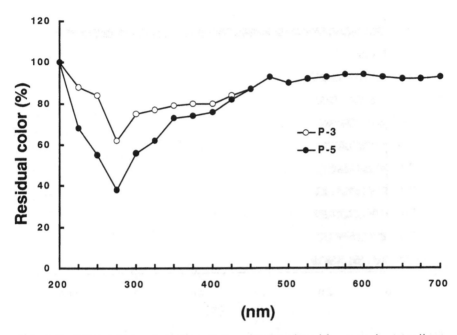

FIG. 4.5   Effect of wavelength on P-3 and P-5 isolated from gardenia yellow.

ponents all had light sensitivity characteristics similar to P-3. The latter components all had light sensitivity characteristics similar to P-5.

Antioxidant Effects of Gardenia Yellow Pigments

It is known that carotenoid pigments have a high oxidation resistance. Recently much attention has been given to the use of β-carotene pigments and similar materials. Crocetin derivatives and carotenoid pigments with polyene composition can also be expected to display oxidation resistance. Linoleic acid oxidation-prevention effectiveness for β-carotene and bixin containing a carotenoid with a 25-count carbon equivalent (the principal component of annato pigment) together with the oxidation resistance of the seven types of crocetin derivatives have been studied (Ichi et al., 1995b). Figure 4.6 shows the degree of linoleic acid oxidation at the time of test material addition. In the control group (no addition), linoleic acid oxidation was 100%. Relative oxidation values at test material addition were then determined. β-Carotene showed an oxidation ratio of only 9.0%, indicating an extremely high oxidation resistance. The bixin oxidation ratio was 22.3%. There was no relationship between oxidation for crocetin and crocetin derivative additives including various sugar and ester combinations of crocetin including *cis* and *trans* types. No matter what the constituents, the degree of oxidation was in the 20–24% range. This indicates a certain

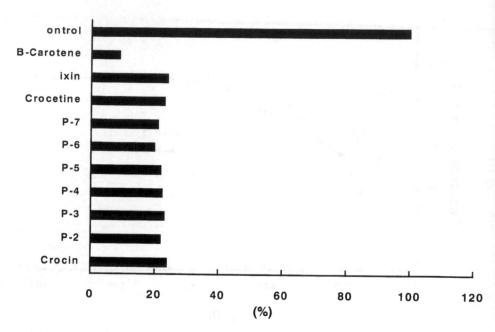

FIG. 4.6 Antioxidative activity of crocetin derivatives isolated from gardenia yellow.

resistance to oxidation. These results suggest that gardenia yellow pigment used as a food coloring additive not only improves the appearance of the food but also acts to prevent oxidation.

REFERENCES

Ichi, T., Higashimura, Y., Katayama, T., Koda, T., Shimizu, T., and Tada, M. (1995a). *Nippon Shokuhin Kagaku Kogaku Kaishi* 42: 776–783.

Ichi, T., Higashimura, Y., Katayama, T., Koda, T., Shimizu, T., and Tada, M. (1995b). *Nippon Shokuhin Kagaku Kogaku Kaishi* 42: 784–789.

Karrer, P., and Miki, K. (1929). *Helv. Chim. Acta* 12: 985.

Pfander, H., and Rychener, M. (1982). *J. Chromatogr.* 234: 443–447.

Pfister, S., Meyer, P., Steck, A., and Pfander, H. (1996). *J. Agric. Food. Chem.* 44: 2612–2615.

Shimizu, T., and Nakamura, M. (1993). *Outline of Natural Food Colors*, pp. 49–57. Korin, Tokyo.

Tanimura, A., Katayama, S., Endow, H., Kurokawa, K., and Yoshizumi, T. (1993). *Handbook of Natural Food Colors*, pp. 212–231. Korin, Tokyo.

Yoshizumi, T. (1974). *New Food Ind.* 16: 41.

5
Paprika

Carol L. Locey and James A. Guzinski

KALSEC, Inc.
Kalamazoo, Michigan

INTRODUCTION

Paprika is a rich red powder used for its flavor and color in such foods
as Hungarian goulash, chicken paprikash, and French salad dressing. Its
source is the dried sweet pepper, the fruit of *Capsicum annuum* L., of the
Solanacea family. In the 1950s technology was developed that allowed the
extraction and concentration of the pigments responsible for the beautiful
red-orange color. Today, oleoresins of paprika are one of the most economically significant sources of natural food color produced from a botanical source. The food coloring ability of paprika is a result of a mixture of
carotenes and xanthophylls present as free pigments or mono- or diesters
of fatty acids. Other components of an extract include oil, waxes, lecithin,
sludges, capsaicinoids (pungent compounds), and flavor compounds (Govindarajan and Sathyanarajana, 1991).

The worldwide requirement for paprika is estimated at approximately
50,000 tons. New Mexico, the largest growing area in the United States, produced 10,880 dry tons in 1997 and 6800 in 1998 (USDA, 1998). An accurate
determination of the world demand for paprika oleoresin is difficult. In
1993, Hornero-Mendez estimated it at 1600 tons. Total usage in the United

TABLE 5.1   Imports of Paprika Oleoresin into the
United States

Year	kg	Value (US$)
1994	48,869	$2,300,000
1997	262,291	$8,506,967
1998	178,417	$4,411,977

Source: USDA, 1999.

States is not available, but import statistics for paprika oleoresin are published by USDA (Table 5.1).

RAW MATERIAL

Capsicum peppers are one of the oldest cultivated crops having been grown by Native Americans 5,000–7,000 years ago (Heiser, 1976). It is generally agreed that these indigenous peppers were highly pungent. Columbus and post-Columbian traders collected the small hot peppers and introduced them to Europe. The milder varieties that evolved from the selective breeding in Spain and Hungary became known as paprika. Today, important growing regions for paprika are India, the United States, Mexico, Spain, and Africa (ASTA, 1999).

Five domesticated species of Capsicum are recognized: C. annuum L., C. baccatum L., C. chinense Jacq, C. frutescens L., and C. pubescens R&P (Bosland, 1996). Only two, C. annuum L. and C. frutescens L., are of economic importance. In addition, approximately 20 wild species growing in South and Central America are known. Paprika cultivated for oleoresin production belongs to the C. annuum L. Typical cultivars in New Mexico are NuMex Sweet, NuMex Conquistador, B-18, Sonora, and NuMex R-Naky (Wall, 1998).

In August 1980 an expert group of genetic researchers met in Turrialba, Costa Rica, to discuss the status of Capsicum germplasm and develop a management system for the future collection of both domesticated and wild varieties. A significant accumulation of seeds resulted (Eshbaugh, 1993). These collections are an invaluable source to capsicum plant breeders worldwide. Academic research is conducted at New Mexico State University, Cornell, Texas A&M, and Oklahoma State University, among numerous other locations. Improving disease resistance, crop yield, pod color, and mechanical harvesting are being examined.

Agricultural practices vary depending on growing region. In the United States farmers in the southwest and California direct-seed the crop. The plant requires a long growing period of 8–9 months to reach fruit maturity. Fertilization and irrigation are timed to slow crop growth before the first frost to increase pod color development and to promote fruit dry-down

(Wall, 1998). Paprika will have several fruit sets; therefore, at harvest varying levels of pod maturity will exist on the plant. Ethephron, a growth regulator, is sometimes used to ripen the fruit (Matta et al., 1994). Prior to harvest, fruit drying can be aided by sodium chlorate treatment to the crop canopy. Harvesting has traditionally been a labor-intensive operation. The pods have been hand-picked for decades in Spain and the United States. More recently, farmers have begun to use mechanical harvesters. Harvested pods are sorted, washed, destoned, dehydrated in belt or tunnel driers, and ground. Once the pods have been dehydrated and ground, the stability of the carotenoid pigments decreases (Lee et al., 1992; Biacs et al., 1992; Isidoro et al., 1995). The dehydrated paprika must be handled carefully to ensure its quality.

Harvested paprika pods are destined for one of three markets: fresh produce, dehydrated-ground, or oleoresin extraction. Color, measured in ASTA units, is the most important characteristic for the farmer growing for the extraction market. The value of the crop is directly related to the pigment content of the dehydrated pods; higher ASTA paprika will procure a premium price. Preferred raw material contains very low levels of capsaicin, the pungent component in chiles, and carotenoid pigment concentrations greater than 0.5% based on dry weight.

EXTRACTION

Industrial production of paprika oleoresin is principally accomplished by organic solvent extraction. In the process, dehydrated ground paprika is fed into a continuous or batch-type extractor where solvents dissolve the pigments, waxes, oils, and minor components. The pigment-laden solvent, called miscella, is desolventized in thin film evaporators to produce an oily liquid containing up to 20% pigment. This raw material oleoresin can be further processed to remove pungency and other flavors. The final step is standardization of the ASTA value with food-grade diluents such as soybean oil, cottonseed oil, canola oil, polysorbates, lecithin, and fatty acid esters. The process is straightforward, but little information has been published regarding the specific techniques and equipment of modern extraction facilities. Paprika oleoresin production is highly competitive, and proprietary processes are carefully guarded by producers. Profitability requires careful attention to the quality of dehydrated raw material, grind size, solvent composition, residence time in extraction equipment, and desolventization technique. Verghese (1995) described production in a Mexican plant as follows:

> The raw material pelletised to 3–5mm size and with humidity 10–11% is fed into the extractor by rolling band. The spice receives successively a series of washing with miscella which each time is less rich in oil; the final wash is with pure solvent. Care is

taken to carry out extraction at temperatures not >45°C. The miscella from the extraction cycle containing 10–12% oil next passes through a basket filter; the fines are separated which otherwise would cause incrustation in the distillation equipment. From there it goes to a reflux system maintained at not >65° C/ 280 Torr maximum and the oil concentration is enhanced to 88–90%. Further enrichment takes place in the film vapouriser.

The effect of solvents on oleoresin yield and product quality has been studied by numerous researchers (Houser et al., 1975; Verghese, 1995). Solvent composition influences the extraction yield as well as the viscosity of the resultant oleoresin. Verghese (1995) reported that for extraction of the coloring principle the effectiveness of solvents is as follows: acetone > hexane > ethyl acetate > chloroform > ethylene dichloride > isopropyl alcohol > methyl ethyl ketone > ethanol. While acetone is the optimum extraction solvent, there are drawbacks to its industrial use. Use of acetone results in an oleoresin with high viscosity due to the formation of sludges upon storage. Oleoresin processors use separation techniques such as liquid-liquid partitioning and centrifugation to compensate for this problem. Hexane is also cited as one of the best solvents for extraction and has the added advantage of producing an oleoresin with low undissolved solids content.

Solvent choice is limited by governmental regulations. In the United States the solvents hexane, acetone, isopropyl alcohol, methyl alcohol, methylene chloride, ethylene dichloride, and trichloroethylene are permitted (CFR, 1999). Chlorinated solvents are allowed by the CFR, but they are seldom if ever used in paprika production. European Union (EU) regulations allow ethyl acetate, methanol, ethanol, acetone, hexane, and dichloromethane (Official Journal of the European Communities, 1995).

Alternative extraction procedures have been investigated. Coenen et al. (1981) patented a process for the extraction of the pigments from bell peppers with propane or butane that is in a supercritical state. The extraction takes place at a pressure of $>P_k$ to 350 bar and a temperature of $>T_k$ to 70°C. The extracted aromatics and/or dyestuffs are separated by lowering the density of the gas phase. Illes et al. (1999) looked at both supercritical CO_2 and subcritical propane and found that propane extracted more pigment than the supercritical CO_2—970–1370, ug vs. 290–390, ug per gram of ground paprika. Two processes using edible oil as the extraction medium are described in U.S. patents (Bennett and Wagner, 1987; Todd, 1996).

Secondary processing of paprika oleoresin to manufacture paprika extracts with specialized solubility, stability, and efficacy is commonly practiced in the industry. Removal of flavor and pungency components by CO_2 is described by Akira Mohri et al. (1993): 250 kg of hot red pepper oleoresin containing 3.98% capsaicin was extracted to produce two fractions—a natural coloring matter and a pungent taste component. Powder products with excellent water solubility and tinctorial strength are produced by spray drying oleoresin with water soluble carriers (Hansen, 1996). Saponification of paprika oleoresin is utilized in the chicken feed industry to produce pigments that have greater ability to color poultry skin (Bioquimex). Further processing can also include stabilization of the paprika pigments (Todd, 1992).

PIGMENTS

The carotenoid pigments in paprika and its oleoresin have been studied repeatedly and are well characterized (Curl, 1962; Phillip and Francis, 1971; Biacs et al., 1989; Mínguez-Mosquera et al., 1992). Two carotenoids, capsanthin and capsorubin, are unique to the *Capsicum* species (Fig. 5.1). These pigments are responsible for paprika's distinctive color. Capsanthin is the major pigment found in paprika and may be as much as 50% of the total pigments. The long chain of double bonds ending in one or two polar ketones efficiently absorbs green light to give a reflected red-orange hue (Judd and Wyszeki, 1975). Other significant pigments are violoxanthin, capsanthin-5, 6-epoxide, zeaxanthin, lutein, β-cryptoxanthin, and β-carotene. These pigments result in a more yellow-orange hue.

Formation of carotenoid pigments in general follows the same biosynthetic pathway in plants (Fig. 5.2). Starting with colorless precursors, lycopene, then α- and β-carotene, and then various oxidized and isomerized carotenoids are produced. It is the branching from this basic pathway that gives the several hundred identified botanical carotenoids. In the fruit of *Capsicum* species, enzymes isomerize antheraxanthin and violaxanthin to capsanthin and capsorubin, respectively. This enzyme has been isolated (Bouvier et al., 1994). The location of the gene producing this enzyme, capsanthin-capsorubin synthase, has also been identified (Camara and Kuntz, 1999). As knowledge of the genes of capsicums continues to accumulate, this brings the possibility for genetic modification of paprika for improved color, yield, stability, disease resistance, etc. However, consumer acceptance of such genetically modified organisms cannot be assumed.

Formation of high concentrations of carotenoids in capsicum fruit is concurrent with the ripening process. It is a visual indicator that a pod is ripening, but that is only part of the complex set of changes occurring in a plant. Various researchers have shown that the entire set of carotenoid genes in paprika chloroplasts begins to function with the onset of ripening (Deli et al., 1996; Almela et al., 1996; Gómez-Ladrón de Guevara and Pardo-González, 1996); that is, there is not a build-up of pigments early in the pathway followed by their later conversion to other pigments. Capsanthin is already found when paprika pods are only slightly orange, and the relative amounts of the different carotenoid pigments change little as a pod's color changes from orange to red. The total amount of pigment

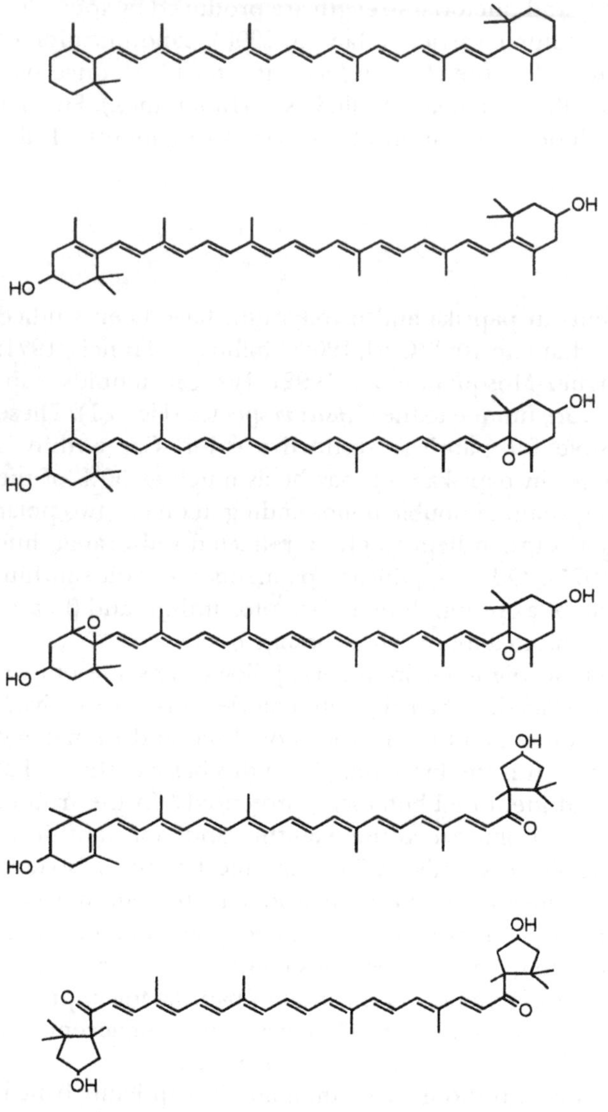

FIG. 5.1   Structures of carotenoid pigments in paprika.

obviously increases as ripening continues. The survey of changes in carotenoid content and identity for different cultivated varieties, species, and
growing conditions remains a very active area of research (Almela et al.,
1991, 1996; Levy et al., 1995; Goda et al.,1995; Deli et al., 1996; Gómez-
Ladrón de Guevara et al., 1996; Márkus et al., 1999).

  Although carotenoids are described as the pigments that produce color
in paprika and its oleoresin, it is important to remember that few free
carotenoids are actually present. The predominant structures are esters and

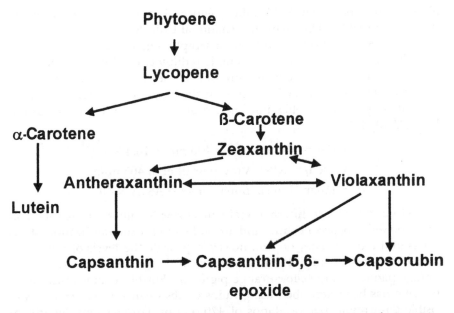

FIG. 5.2 Biosynthetic pathway for the formation of carotenoid pigments.

diesters of hydroxy carotenoids chemically bound to fatty acids. The fatty acids attached to capsanthin and capsorubin are predominantly saturated acids in contrast to the predominantly unsaturated fatty acids attached to other xanthophylls or in the vegetable oils extracted with the pigments and found in oleoresins (Biacs et al., 1989; Gross, 1991; Minguez-Mosquera and Hornero-Méndez, 1994).

Carotenoids and their esters accumulate in fibrils in chloroplasts (Steffen, 1955). These long, thin structures contain a core of pigments surrounded by galactolipids and phospholipids with an exterior coating of a protein called fibrillin (Deruère et al., 1994). These structures store large amounts of lipidic carotenoids and avoid destablilization of cell structure. During extraction, these structures are broken down and the pigments are released into the extraction solvent.

The pigment composition of an oleoresin is very similar to that of the paprika from which it is made. Extraction concentrates the carotenoids along with other lipidic materials but does not show selectivity for specific pigments.

ANALYTICAL METHODS

Analytical specifications for paprika oleoresin include color value, pigment concentration by high-performance liquid chromatography (HPLC), absorbance ratios, percent capsaicinoids, and sludge. While ground paprika is always traded based on ASTA value, oleoresins are sold based either on ASTA or the MSD-10 Standard International Unit (SIU).

In the ASTA method, 70–100 mg of sample is quantitatively transferred to a 100 mL volumetric flask. The sample is diluted to volume with acetone. A second dilution of 10.00 in 100 mL of acetone is made. The absorbance is read at 460 nm. A correction factor for the instrument is measured using a standard glass filter at 465 nm. The extractable color is calculated as follows (ASTA, 1999b):

$$ASTA = \text{Absorbance at 460 nm} \times 16.4 \times I_f$$

$$\text{where } I_f = \text{NIST A for glass filter at 465 nm.}$$

$$\text{Standard International Units} \cong ASTA \times 40$$

ASTA and Standard International Units provide values for the total coloring power of paprika extract and are sufficient for standardization of extracts in the trade. They do not, however, provide the levels of individual carotenoids or the hue. Hue is a function of the relative amounts of red-orange pigments and yellow-orange pigments. Further characterization of the hue has been described using ratios of absorbance values taken from visible absorption spectra. Ratios of 470/455 or 470/454 (wavelengths of absorbed light measured in namometers) are used to indicate the relative amounts of red to yellow pigments. The ratios for commercial varieties of paprika range from 0.960 (yellow-orange) to 1.01 (red-orange). Gomez et al. (1997) studied the relationship between the following measurements: total pigment content measured as ASTA units, level of red and yellow carotenoids, and the 470/455 ratio (Fig. 5.3). He found a strong correlation between red and yellow pigment content and ASTA value. However, no correlation was found between the hue ratio and ASTA value.

The 470/455 ratio is an important tool in describing the hue of paprika oleoresins, although it does not completely characterize its nature. If chlorophyll is present in the extract, it will contribute significantly to the hue. In fact, chlorophyll content, food-processing techniques, and variability in colors inherent in foods can have a greater impact on the hue of a food product colored with paprika oleoresin than the 470/455 ratio (KALSEC, 1999).

HPLC has been the most heavily used technique for separation and identification of carotenoid pigments in recent years (Ittah et al., 1993; Mínguez-Mosguera and Hornero-Méndez, 1993; Biacs and Daood, 1994; Cserháti et al., 1995; Schmitz et al., 1995; Weissenberg et al., 1997; Rodrigues et al., 1998). This instrumental technique has high sensitivity, allowing small sample sizes, and great flexibility, so that methods can be adapted to generate the specific data of interest. For example, techniques have been developed specifically to study cis-trans isomers (Saleh and Tan, 1991; Chen et al., 1994). Potential problems in HPLC method development for carotenoids have also been described (Scott, 1992).

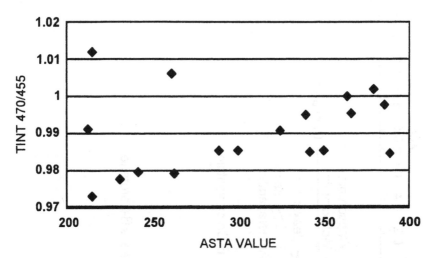

FIG. 5.3 Absorbance value ratios plotted versus ASTA values for different paprika varieties. (From Gomez et al., 1997.)

Since carotenoids are nonpolar, a majority of the published methods describe use of a reversed-phase column with gradients slowly decreasing the polarity of the eluent. This enables separations based upon small differences in structures. Saponification of paprika carotenoids is done prior to analysis when the carotenoid moieties and not their esters are of interest. Detection of carotenoids is usually done with a UV/Vis detector, although a photo-diode array detector to provide a spectrum of each chromatographic peak is of great benefit for compound identification. Fast atom bombardment (FAB) mass spectrometry has also been used for carotenoid identification (Breemen et al., 1995).

Figure 5.4 shows a HPLC chromatogram of a typical oleoresin paprika produced from capsicum grown in the southwest United States. The sample was saponified according to Baranyai et al. (1982) and Matus et al. (1981). The HPLC system consisted of the following Waters (Milford, MA) equipment: a 2690 solvent management system and a 996 photodiode-array detector. Detection with the 996 was a maxplot from 400 nm to 600 nm, which gives the maximum absorbance for the chosen range at a given point in time. Thus, all peaks in the chromatogram can be viewed.

A 5 mm Waters Symmetry C18 (250 × 4.6 mm) reverse-phase column was used combined with a Waters C18 guard column. Separations were achieved using gradient elution with HPLC grade methanol, water, and acetone. The initial solvent composition consisted of 25% water and 75% acetone. The solvent composition changed in a linear fashion to 20% MeOH, 5% water, and 75% acetone from 0–15 minutes. During the next 15 minutes the composition was held at 20% MeOH, 5% water, and 75% acetone. From 30 to 60 minutes the final composition was reached at 25%

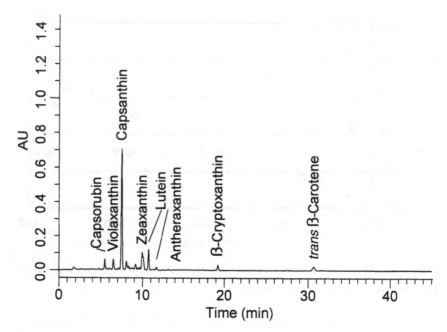

FIG. 5.4 HPLC chromatogram of oleoresin paprika grown in the U.S. southwest.

MeOH and 75% acetone. All solvent changes were made in a linear fashion. The relative amounts (%) in the extract are listed in Table 5.2 along with previously published data (Fisher and Kocis, 1987).

OXIDATIVE STABILITY

The oxidative stability of the carotenoid pigments in ground and oleoresin paprika has been the subject of extensive investigation (Ramakrishnan and Francis, 1973; Carnevale et al., 1980; Candela et al., 1984; Biacs et al., 1992; Minguez-Mosquera et al., 1994; Jarén-Galán et al., 1999). The stability of pigments in ground paprika is a function of cultivar, enzymatic activity, water activity, carotenoid level, and the endogenous ascorbic acid and tocopherol present at ripeness (Mínguez-Mosquera and Hornero-Méndez, 1994c; Wall, 1998). Color degradation occurs during grinding and storage due to pigment oxidation. Numerous antioxidants have been studied in an effort to increase the stability of the carotenoid pigments. Osuan-Garcia (1997) found ethoxyquin to be the most effective antioxidant in ground paprika.

Oleoresin paprika is also subject to degradation as a result of oxidation of the pigments. This color loss is a particular problem when the extract is

TABLE 5.2   Relative Amounts (%) of Carotenoid Pigment in Saponified
Oleoresin Paprika

	Kalsec (1999)	Almela (1991)	Camara (1978)	Curl (1962)
Capsanthin	42.65	47.81	33.33	34.7
β-Carotene	3.33	5.38	15.4	11.6
Violaxanthin	4.09	0.50	7.1	9.9
Cryptoxanthin	2.08	1.75	12.3	6.7
Capsorubin	3.54	6.63	10.3	6.4
Cryptocapsin	—	0.63	5.1	4.3
Zeaxanthin	7.20	4.26	3.1	2.3
Antheraxanthin	1.19	0.88	9.2	1.6

diluted for use in foods. One application of oleoresin paprika is the coloring of dry powders for use in breadings, seasonings, and snack coatings. A study comparing the stability of an oleoresin on five different substrates—flour, dextrose, maltodextrin, wheat starch, and salt—revealed that the pigments were most stable on flour and least stable on the salt. Patented systems without the use of ethoxyquin, an antioxidant that is not permitted in oleoresins, have been developed to stabilize the pigments against degradation (Todd, 1991, 1992). The efficacy of these systems is illustrated by the comparison of a dispersion of standard 2000 ASTA paprika oleoresin and a stabilized 2000 ASTA oleoresin on salt. After 28 days of light exposure the standard paprika retained only 10% of its original pigment concentration. The stabilized oleoresin retained 81% (KALSEC, 1999). Shackelford (1986) described a method for stabilizing oleoresin paprika by mixing it with an amount of dried yeast product and then mixing the stabilized additive into a dry food to form a dry food blend having a stabilized color.

COMMERCIAL PRODUCTS

The majority of paprika extract is sold in liquid form, either oil soluble or water dispersible. Typical strengths are 1000 ASTA (40,000 SIU), 2000 ASTA (80,000 SIU), and 2500 ASTA (100,000 SIU). Powders are manufactured by dry-blending or spray-drying with carriers and gums. Market forms of paprika extractives are listed in Table 5.3.

APPLICATIONS

Paprika is used in a wide variety of food products (Table 5.4). Selection of the most effective market form depends on food formulation, processing,

toluene as internal standard and extracted once with 10 ml of a mixture of methylene chloride-pentane (30:70). With this solvent system, less contaminants were extracted into the organic phase from the plasma. The column was a 25 m × 0.32 mm I.D. fused silica capillary column HP-1. Carrier gas was hydrogen at a flow rate of 15 ml/min. Column temperature was held at 96°C for 9 min and then increased at 4°C/min to 126°C. Two linear regions for the detectable concentrations of NG were found to be 0.025 to 0.3 ng/ml and 0.3 to 3 ng/ml, and for the DNGs they were 0.1 to 1 ng/ml and 1 to 10 ng/ml. The limits of detection for NG, 1,2-DNG and 1,3-DNG in plasma were 0.025, 0.1, and 0.1 ng/ml, respectively.

2. High Performance Liquid Chromatography (HPLC)

Spanggord and Keck[69] used HPLC and thermal energy analyzer (TEA®) detection for the analysis of NG and its metabolites in blood. N-nitrosodipropylamine was used as internal standard and was added to the blood sample before extraction. 100 μl of 1 M silver nitrate was added to 2 ml of blood to inhibit NG degradation. Extraction was achieved by rapid injection of 2 ml, then five times 1 ml of ethyl acetate into the blood solution. The organic layers were drawn off after each injection, combined, dried, and filtered through a short column of anhydrous sodium sulfate. The dried extract was concentrated to 500 μl using a gentle nitrogen stream. Then 500 μl of isooctane was added, and the solution was concentrated to 500 μl. The HPLC column was a 25 cm × 4.6 mm I.D. Lichrosorb Si-60 column. The mobile phase, isooctane-ethyl acetate-acetone (54:6:40), was run for 10 min isocratically and then programmed to (9:1:90) over 5 min, at a flow rate of 1.4 ml/min. Detection was done with a thermal energy analyzer using the following parameters: pyrolyzer temperature 450°C, argon flow rate 15 ml/min, and oxygen flow rate 5 ml/min. The detection limit per 100 μl injection was 0.5 ng for NG, 1.0 ng for 1,2- and 1,3-DNG, and 3.0 ng for 1- and 2-MNG. Linearity was obtained up to 1000 ng. Figure 9 shows a HPLC chromatogram of a dog's blood extract of NG and its metabolites, 10 min after the dog was dosed orally with NG.

HPLC with TEA® detection for the analysis of NG and its metabolites was also used by Yu and Goff.[70] Extraction was carried out with ethyl acetate. The column was a 25 cm × 4.6 mm I.D. Ultrasil NH$_2$ column. Mobile phase was isooctane-methylene chloride-methanol (80:13:7) at a flow rate of 2.0 ml/min. The TEA® detector was operated at a pyrolyzer temperature of 500°C and a nitrogen flow rate of 20 ml/min. The lowest detection limit for NG was 0.1 ng at a signal-to-noise ratio of 3.

Woodward et al.[71] also used HPLC with TEA® detection for the determination of NG and its dinitrate metabolites in human plasma. Internal standard was isosorbide dinitrate. Extraction was done with dichloromethane-ethyl acetate (1:1). The column was a 25 cm × 4.6 mm I.D. polar bonded-phase Zorbax CN column. Mobile phase was acetone-n-hexane (5:95) at a

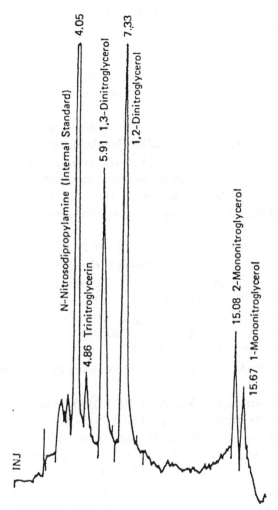

FIGURE 9. HPLC chromatogram of a dog's blood extract of NG and its metabolites. (From Spanggord, R. J. and Keck, R. G., *J. Pharm. Sci.*, 69, 444, 1980. With permission. Copyright, American Pharmaceutical Association).

flow rate of 2 ml/min. The TEA® detector was operated at a pyrolyzer temperature of 575°C, argon flow rate of 15 ml/min and oxygen flow rate of 25 ml/min. Retention times were 8.5 min for NG, 10.5 min for 1,3-DNG, and 11.5 min for 1,2-DNG. Lowest detection limits were 0.05 ng/ml for NG and 0.25 ng/ml for the DNGs. Linearity ranges were 0.1 to 2.0 ng/ml for NG and 0.5 to 10.0 ng/ml for the DNGs.

Noonan and Benet[49] used HPLC with UV detection at 214 nm for the analysis of the metabolites of NG in blood and plasma after incubation. Ether

Bouvier, F., Hugueney, P., d'Harlingue, A., Kuntz, M., and Camara, B. 1994. Xanthophyll biosynthesis in chromoplasts: Isolation and molecular cloning of an enzyme catalyzing the conversion of 5,6-epoxycarotenoid in ketocarotenoid. *Plant J.* 6(1): 45–54.

Breemen, R. B., Schmitz, H. H., and Schwartz, S. J. 1995. Fast atom bombardment tandem mass spectrometry of carotenoids. *J. Agric. Food Chem.* 43(2): 384–389.

Camara, B., and Moneger, R. Free and esterified carotenoids in green and red fruits of *Capsicum annuum.* 1978 *Phytochemistry* 17: 91–93.

Camara, B., and Kuntz, M. 1999. DNA constructs related to capsanthin capsorubin synthase, cells and plants derived therefrom. *U.S. Pat.* 5,880,332.

Candela, M. E., Lopez, M., and Sabater, F. 1984. Carotenoids from *Capsicum annuum* fruits: changes during ripening and storage. *Biol. Plant.* 26(6): 410–414.

Carnevale, J., Cole, E. R., and Crank, G. 1980. Photocatalyzed oxidation of paprika pigments. *J. Agric. Food Chem.* 28: 953–956.

Code of Federal Regulations (CFR). Paprika oleoresin. TITLE 21—Food and Drugs Part 73- Sec. 73.345.

Chen, B. H., Chen, T. M., and Chien, J. T. 1994. Kinetic model for studying the isomerizatin of α- and β-carotene during heating and illumination. *J. Agric. Food Chem.* 42(11): 2391–2397.

Chr Hansen. 1996. Micro-Cap™ Colours A Range to Top it All, The World of Ingredients.

Coenen, H., Hagen, R., and Knuth, M. 1981. Method for obtaining aromatics and dyestuffs from bell peppers. U.S. Pat. 4,400,398.

Cserháti, T., Forgács, E., and Kiss, V. 1995. High performance liquid chromatographic detection of a strongly retained pigment fraction in *Capsicum annuum. Die Nahrung* 39(4): 269–274.

Curl, A. C. 1962. The carotenoids of red bell peppers. *Agric. Food Chem.* 10(6): 504–509.

Deli, J., Matus, Z., and Tóth, G. 1996. Carotenoid composition in the fruits of *Capsicum annuum* Cv. Szentese Kosszarvú during ripening. *J. Agric. Food Chem.* 44: 711–716.

Deruère, J., Römer, S. d'Harlingue, A., Backhaus, R. A., Kuntz, M., and Camara, B. 1994. Fibril assembly and carotenoid overaccumulation in chromoplasts: a modes for supramolecular lipoprotein structures. *Plant Cell* 6: 119–133.

Eshbaugh, W. H. 1993. History and exploitation of a serendipitous new crop discovery. In *New Crops*, J. Janick and J. E. Simon (Eds.). Wiley, New York.

Fisher C., and Kocis, J. A. 1987. Separation of paprika pigments by HPLC. *J. Agr. Food Chem.* 35: 55–57.

Goda, Y., Sakamato, S., Nakanishi, T., Maitani, T., and Yamada, T. 1995.

Indentification of monoesterified capsanthin in paprika (*Capsicum annuum*): the nature of esterification of capsanthin. *Chem. Pharm. Bull,* 43(7): 1248–1250.

Gómez-Ladrón de Guevara, R., and Pardo-González, J. E. 1996. Evolution of color during the ripening of selected varieties of paprika pepper (*Capsicum annuum* L.). *J. Agric. Food Chem.* 44: 2049–2052.

Gómez-Ladrón de Guevara, R., Pardo-González, J., Varón-Castellanos, R., and Navarro-Albaladejo, F. 1996. Study of the color evaluation during maturation in selected varieties of paprika pepper. INF/COL II. Hamden, CT.

Gómez, R., Pardo, J., Navarro, F., and Varón, R. 1997. Methods for estimating color in paprika pepper varieties (*Capsicum annuum* L.). *Rivi. Sci. Aliment.* 26: 3–4.

Govindarajan, V. S. and Sathyanarajana, M. N. 1991. Capsicum—production, technology, chemistry, and quality. *Crit. Rev. Food. Sci. Nutr.* 29(6): 435–474.

Gross, J. 1991. *Pigments in Vegetables: Chlorophylls and Carotenoids.* Van Nostrand Reinhold, New York.

Heiser, C. B. 1976. Peppers, Capsicum (Solanaceae). In *The Evolution of Crop Plants*, N. W. Simmonds (Ed.). Longman Press, London.

Hornero-Mendez, D., Jaren-Galan, M., Garrido-Fernandez, J., and Minguez-Mosquera, M. I. 1993. *Color modification during paprika processing*. Presentation at INF/COL-I. Hamden, CT.

Houser, T. J., Biftu, T., and Hsieh, P. 1975. Extraction rate equations for paprika and turmeric with certain organic solvents. *Agric. Food Chemi.* 23(2): 353.

Illés, V., Daood, H. G., Biacs, P. A., Gnayfee, M. H., and Mészarós, B. 1999. Supercritical CO_2 and subcritical propane extraction of spice red pepper oil with special regard to carotenoid and tocopherol content, *J. Chromatogr. Sci.* 37: 345–352

Isidoro, E., Cotter, D. J., Fernandez, C. J., and Southward G. M. 1995. Color retention in red chile powder as related to delayed harvest. *J. Food Sci.* 60(5): 1075–1077.

Ittah, Y., Kanner, J., and Granit, R. 1993. Hydrolysis study of carotenoid pigments of paprika (*Capsicum annuum* L. variety Leheva) by HPLC/photodiode array detection. *J. Agric. Food Chem.* 42: 899–901.

Jarén-Galán, M., and Mínguez-Mosquera. 1997. β-Carotene and capsanthin co-oxidation by lipoixygenase. Kinetic and thermodynamic aspects of the reaction. *J. Agric. Food Chem.* 45: 4814–4820.

Jarén-Galán, M., Pérez-Gávez, A., and Mínguez-Mosquera. 1999. Prediction of decoloration in paprika oleoresins. Application to studies of stability in thermodynamically compensated systems. *J. Agric. Food Chem.* 47: 945–951.

Judd , D. B., and Wyszeki, G. 1975. *Color in Business Science and Industry.* John Wiley and Sons, New York.

KALSEC. 1999. Durabrite Paprika. Technical Bulletin. Kalamazoo, MI.

Lee, D. S., Chung, S. K., and Yam, K. L. 1992. Carotenoid loss in dried red pepper products. *Int. J. Food Sci. Technol.* 27: 179–185.

Levy, A., Harel, S., Palevitch, D., Akira, B., Menagem, E., and Kanner, J. 1995. Carotenoid pigments and β-Carotene in paprika fruits (*Capsicum* spp.) with different genotypes. *J. Agric. Food Chem.* 43: 362–366.

Márkus, F., Daood, H. G., Kapitány, J., and Biacs, P. A. 1999. Change in the carotenoid and antioxidant content of spice red pepper (paprika) as a function of ripening and some technological factors. *J. Agric. Food Chem.* 47: 100–107.

Matta, F. B., and Cotter, D. J. 1998. Chile production in north-central New Mexico. *http://www.cahe.nmsu.edu/pubs/_h/h/h-225.html.*

Matus, Z., Baranyai, M., Toth, G., and Szabolocs, J. 1981. *J. Chromatogr.* 14(6): 337–340.

Mínguez-Mosquera, M. I., and Hornero-Méndez, D. 1993. Separation and quantification of the carotenoid pigments in red peppers (*Capsicum annuum* L.), paprika, and oleoresin by reversed phase HPLC. *J. Agric. Food Chem.* 41: 1616–1620.

Mínguez-Mosquera, M.I. and Hornero-Méndez, D. 1994a. Formation and transformation of pigments during the fruit ripening of *Capsicum annuum* Cv. *Bola* and *Agridulce. J. Agric. Food Chem.* 42: 38–44.

Mínguez-Mosquera, M. I., and Hornero-Méndez, D. 1994b. Changes in carotenoid esterification during the fruit ripening of *Capsicum annuum* Cv. *Bola. J. Agric. Food Chem.* 42: 640–644.

Mínguez-Mosquera, M. I., and Hornero-Méndez, D. 1994c. Comparative study of the effect of paprika processing on the carotenoids in peppers (*Capsicum annuum*) of the Bola and Agridulce varieties. *J. Agric. Food Chem.* 42(7): 1555–1560.

Mínguez-Mosquera, M. I., Jarén-Galán, M., and Garrido-Fernándes, J. 1992. Color quality in paprika. *J. Agric. Food Chem.* 40: 2384–2388.

Mínguez-Mosquera, M. I., Jarén-Galán, M., and Garrido-Fernandez, J. 1994. Competition between the processes of biosynthesis and degradation of carotenoids during the drying of peppers. *J. Agric. Food Chem.* 42: 645–648.

Official Journal of the European Communities. 1995. Paprika, Color Purity Standards.

Osuna-Garcia, J. A. and Wall, M. M. 1997. Prestorage moisture content affects color loss of ground paprika (*Capsicum annuum* L.) under storage. *J. Food Qual.* 21: 251–259.

Phillip, T., and Francis, F. J. 1971. Isolation and chemical properties of capsanthin and derivatives. *J. Food Sci.* 36: 823–827.

Ramakrishnan, R. V., and Francis, F. J. 1973. Color and carotenoid changes in heated paprika. *J. Food Sci.* 38: 25–28.

Rodrigues, P., Morais, H., Mota, T., Olivera, S., Forgács. E., and Cserháti, T. 1998. Use of HPLC and multivariate methods for the evaluation of the stability of colour pigments of paprika *(Capsicum annuum)* powder. *Anal. Chim. Acta.* 372: 411–416.

Saleh, M. H. and Tan, B. 1991. Separation and identification of cis/trans isomers. *J. Agric. Food Chem.* 39: 1438–1443.

Schmitz, H. H., Emenhiser, C., and Schwartz, S. J. 1995. HPLC separation of geometric carotene isomers using a calcium hydroxide stationary phase. *J. Agric. Food Chem.* 43: 1212–1218.

Scott, K. J. 1992. Observations on some of the problems associated with the analysis of carotenoids in foods by HPLC. *Food Chem.* 45: 357–364.

Shackelford, J. R. 1986. Method for retarding degradation of food colors and flavors. *U.S. Pat.* 4,574,086.

Steffen, K. 1955. Die submikroscopische Struktur der Chromoplasten. *Ber. Deutsch. Bot. Ges.* 68: 23.

Todd, G. 1998. High temperature countercurrent solvent extraction of capsicum solids. *U.S. Pat.* 5,773,075.

Todd, P. H. 1991. Color stabilized paprika pigment compositions and foods colored therewith having increased resistance to oxidative color fading. *U.S. Pat.* 5,059,437.

Todd, P. H. 1992. Activated ascorbic acid antioxidant compositions and carotenoids, fats, and foods stabilized therewith. *U.S. Pat.* 5,084,293.

USDA. 1998. New Mexico chile production. *http://www.nass.usda.gov/nm/chile98.htm.*

USDA. 1999. United States: Imports of specified condiments, seasonings, and flavoring materials. *http://www.fas.usda.gov/htp/tropical/1999/99%2D03/troptoc.htm.*

Verghese, J. 1995. Capsicum oleoresin—II. Selection of solvent medium. *Indian Spices* 32 (3 & 4)

Wall, M. 1998. Postharvest handling of dehydrated chiles. New Mexico State University, Las Cruces, NM. *http://www.cahe.nmsu.edu/pubs/_h/h-236.html.*

Weissenberg, M., Schaeffler, I., Menagem, E., Barzilai, M., and Levy, A. 1997. Isocratic non-aqueous reversed-phase high-performance liquid chromatographic separation of capsanthin and capsorubin in red peppers *(Capsicum annuum* L.), paprika and oleoresin. *J. Chromatogr.* A 757: 89–95.

6
Annatto

Luis W. Levy and Diana M. Rivadeneira

Inexa, Industria Extractora C.A.
Quito, Ecuador

INTRODUCTION

The food color annatto is the extract of the outer layer of the seeds of *Bixa orellana*, the tropical tree named after the Spanish conquistador Francisco de Orellana, who discovered the Amazon River in 1541 starting his expedition from Quito, Ecuador. Patiño (1967) provides a complete description of the aboriginal uses of annatto seed. For centuries it was a traditional ingredient for food preparation and for cosmetic purposes in Central and South America. The natives prepared it by rubbing the seed with oils pressed from other plants. The Aztecs, in what is now Mexico, mixed it with cocoa to give a special taste and a more pleasing color to chocolate. A ceremonial drink with annatto resembled human blood, and its consumption became an important ritual for the Indians. This was probably also the basis for the cosmetic uses of annatto in pre-Spanish Latin America. Painting the human body with annatto was usually done by the women on the bodies of male warriors as a symbolic act representing the color of blood. During the Spanish colonization of America this ceremony was banned at times as a pagan ritual. However, the Spaniards also became intensely interested in this color.

The Dutch and British traders who arrived at ports of South America in the seventeenth century also became enchanted with annatto. They traded European products for annatto seed, which they took back to Europe. Annatto was the first vegetable color brought to Europe in large quantities after the discovery of America. Just from the Mexican port of Veracruz more than 35,000 pounds of annatto seed were shipped to Europe in the year 1644 (Patiño, 1967).

Annatto seeds and extracts have been used for over 100 years in Europe and North America to impart a yellow to red color to foods, especially dairy products such as cheddar and colby cheeses. According to Parish (1994), "The English started their own gold rush in the 18th century with the discovery that annatto was a useful coloring for cheese. The warm-toned cheeses of the Cheddar variety, Cheshire, Double Gloucester and Leicester, all get their blush from annatto. Before that, the English dairy maids wrung out (the juice of) scrapings of carrots, marigold leaves or saffron to get the light orange color." In the United States, the first modern annatto food color was exhibited at the Philadelphia World's Fair in 1876 by the Christian Hansen Laboratory.

In modern times, annatto ranks second in economic importance worldwide among all natural colors (Lauro, 1991) and it is the most frequently used natural colorant of the food industry in the United Kingdom (Scotter et al., 1998).

9'-cis-Bixin (Fig. 6.1) is the major color component of annatto seed, accounting for over 80% of the annatto pigment (Preston and Rickard, 1980; Lauro, 1991). Water-soluble annatto extracts are prepared by alkaline hydrolysis whereby bixin is converted to norbixin (Fig. 6.1). Reviews on the extraction, chemistry, and applications of annatto have been published by Dinesen (1974), Preston and Rickard (1980), Collins (1992), and Green (1995).

ANNATTO SEEDS

Commercial annatto extracts are obtained from the red oily layer covering the seeds of the tree *Bixa orellana* (Annatto), native of northern South America and introduced into Central America, India, and Africa (Heiser, 1965). It is one of the four species of the family Bixaceae, order Violales, subclass Archichlamydeae, division Angiosperms/Dicotyledons (Evans, 1996). The plant thrives from sea level to about 4000 feet altitude in a moist climate and prefers a deep, loamy, well-drained soil (Dendy, 1966). When large clusters of capsular fruits have formed at the end of the branches they are harvested. The capsules, about one inch in diameter, are cut off the stems. They contain 50–60 small seeds with a bright crimson oily covering, weighing 10–20 mg each. The capsules are allowed to dry in the sun, are

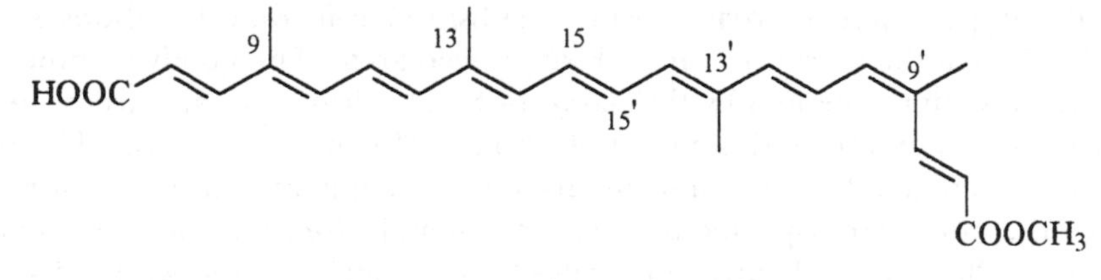

9' - *cis* – BIXIN

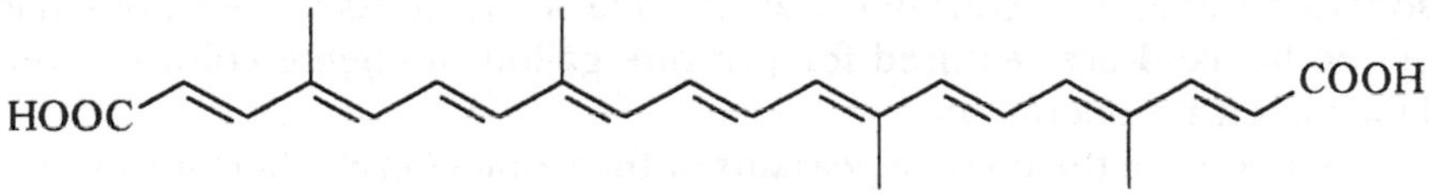

trans-NORBIXIN

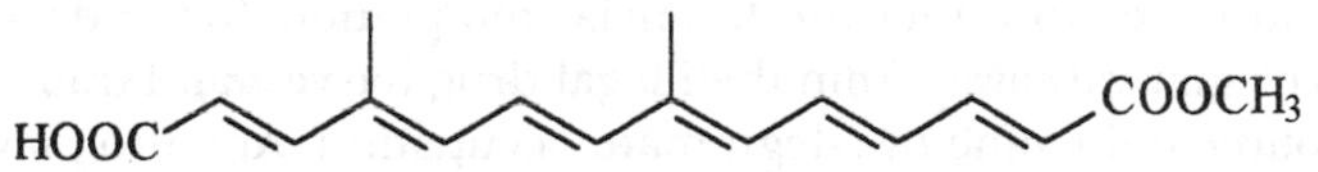

C-17 COMPOUND

FIG. 6.1 Molecular formulas of bixin, norbixin, and C-17 compound.

cracked open, and the small seeds are separated by hand from the husk. Dried annatto seeds and their extract are traded worldwide. The cultivation of annatto has been reviewed by Ingram and Francis (1969), Arce (1983), and Aparnathi et al. (1990).

Wild Annatto and Home-Grown Annatto— Their Social Significance

The major proportion of the world production of annatto seed comes from the collection of seed from wild trees or trees planted on family farms, especially in northern South America. Each mature tree may yield between 0.5 and 4 kg of seed. The collection and sale of seeds from wild annatto trees of the tropical jungle areas provide a subsistence income for thousands of families in the most economically depressed areas. The family income is frequently supplemented by the harvest of a few dozen annatto trees they have planted on the land around their home. The dried seeds are taken to an intermediary. After the intermediary has accumulated a few hundred pounds, they are transported to the next major city for sale to a wholesaler, who supplies the local extraction industry, or to an exporter, who waits to complete the 10–20 metric tons of seed necessary to fill a container for ocean shipment.

More than 2000 individual seeds have to be harvested to get just one gram of bixin. For the food technologist using annatto color, it may be a sobering thought to consider that 240,000 seeds in 4000 seed pods harvested by hand are required for just one gallon of cheese color—a very labor-intensive operation.

Thousands of the poorest peasants of the tropics derive their subsistence income from the collection and sale of annatto seed, which gives it enormous social and economic importance in the tropical areas of many countries. In the case of Peru, the collection of annatto seed is one of the few practical alternatives available to the production of illegal drugs, which are grown in the same areas. Major efforts of the so-called "war on drugs" are directed towards providing such alternative sources of income to the population in order to direct the financial temptation (or, indeed, their vital needs of survival) away from the illegal drug harvest and trade and replace that source of income by a legitimate occupation. Annatto plays an important role in this effort.

Plantation-Grown Annatto

A significant proportion of annatto seed is now produced in Ecuador, Central America, and West Africa on plantations or semi-plantations, with yields of 900–1500 kg of seed per hectare per year (Green, 1995). Possibly the first commercial annatto plantation based on modern agricultural techniques was established in Ecuador in 1986 by the firm INEXA, based on selected annatto varieties of high color content (Evans, 1996). The seeds are extracted in the INEXA industrial plant in Quito within 72 hours of harvest for maximum extraction yield and highest purity of the extract.

Stability of Annatto Seed

Prompt extraction after harvest is only possible when extraction facilities are located near the cultivation areas. If this is not the case, the seeds must be shipped overseas. Several months elapse before extraction. Cold storage is advisable to prevent the degradation of bixin through air oxidation. Annatto seeds that have been stored for more than 6 months in the tropics frequently contain peroxides, which can cause autocombustion. At times the peroxides can even carry over to the extracted bixin. A case of spontaneous combustion has been observed with bixin powder of low purity that had been extracted from one-year old annatto seed.

World Production and Commerce

The quantity of annatto seed harvested in the producing countries has been estimated at 6,500 metric tons (dry weight) per year (Wood et al., 1991) and 10,000 metric tons (Green, 1995). A more realistic estimate may be 14,500 metric tons because most statistics refer only to exports from the producing countries and do not include their local consumption. Such regional trade is considerable in some producer countries but is difficult to assess because these quantities do not go into the official statistics and because thousands of small producers are involved. An example is Brazil, where informal estimates put the harvest at the impressive figure of 5000 metric tons/year that escape the official statistics. They are for domestic use by the local population, mainly as the condiment called "Coloral," which is ground annatto seed and is as popular in Brazil for cooking as is pepper in the rest of the world. In Peru and Ecuador about 1000 metric tons of annatto seed may be consumed by the local population. Thus, the annatto export statistics, which are the only easily obtainable figures, seriously underestimate the total annatto production of the world. Table 6.1 shows estimated production figures of the main producing countries matched by internal consumption

TABLE 6.1   Estimated World Production and Commerce

Producing Countries	Metric tons dried seed	Importer	Metric tons dried seed[a]
Brazil	5,000	North America	3,000
Peru, Ecuador, Colombia, Bolivia	3,000	Europe	2,500
Kenya, Tanzania	2,500	Japan	1,500
Guatemala, Mexico, Caribbean	2,000	Other	500
Ivory Coast, Ghana	1,500		
		Total	7,500
India, Asia	500		
Total	14,500		
Of which:			
Domestic consumption	7,000		
Available for export	7,500		

[a]Or its equivalent in extract.
Source: Adapted from UNCTAD/GATT, 1990; Wood, 1991; Green, 1995; N. Dinesen, private communication.

and export. For the equivalence in terms of bixin, a good approximation is 20 kg of pure bixin per ton of seed.

The price history of annatto seed shows the typical roller coaster fluctuations for most agricultural commodities of the third world. The historical average price over the past 30 years is around $1,100 per metric ton, but with fluctuations from highs of $1,500–$2,500 per metric ton (1973, 1978, 1987) to lows of $500 per metric ton (1969, 1974, 1982, and 1999).

PRODUCTION OF ANNATTO EXTRACT

Extraction methods may either aim at the production of the native bixin from the annatto seed or may involve aqueous hydrolysis and simultaneous extraction of norbixin.

By Mechanical Means

The seventeenth-century Spanish conquerors of Ecuador and Peru discovered for Europe not only the annatto plant but also the first rudimentary extraction method, which to this day is practiced by the Colorado Indians of western Ecuador. For over 600 years the natives have been rubbing the annatto seed by hand to dislodge the red seed coating and obtain a powder or paste, which they apply to their skin and hair to dye them red (*colorado* in old Spanish) and as a dye or spice for cooking.

The same dry-extraction method in modern times uses equipment in which the annatto seeds are forced by an air current through a draft tube into a spouted bed extractor (Guimaraes et al., 1989; Massarani et al., 1992). The particle impact loosens the surface layer of the seed, yielding an average of 114 g of powder containing about 15% bixin (Passos et al., 1998).

The earliest "wet" industrial extraction method involved the same principle, but mechanical abrasion was provided by stirring the annatto seed in water to dislodge and separate the pigment layer, which was then filtered off and dried to give a paste with a bixin content of about 20%. For use as a food color the paste was then dissolved in vegetable oil to prepare dilute bixin solutions or mixed with hot aqueous alkali to hydrolyze the bixin and obtain dilute solutions of norbixin.

Directly Extracted Food Colors

In the original and traditional food recipes of Latin America, annatto seeds are heated with cooking oil, separated, and the colored oil used for preparing rice, soups, and tortillas. A similar direct-extraction system is applied industrially by immersion of the seed in vegetable oil to produce a bixin slurry, which is heated and filtered. The filtrate is marketed as a color for high-fat foods (Barnett and Espoy, 1957; Murthi et al., 1989). The use of alkaline propylene glycol instead of vegetable oil extended its application to both high-fat and low-fat foods and gave solutions of greater purity (Kocher, 1958). In all these cases, the extraction liquids, which contain up to 1.5% bixin, or in some cases oil suspensions with up to 8% bixin are marketed directly as oil-soluble annatto food color after standardization of color content.

For aqueous applications (e.g., cheese color) the extraction of the annatto seed is done with aqueous potassium hydroxide or sodium hydroxide, which hydrolyzes the bixin on the seed to the water-soluble norbixin salt that goes into aqueous solution. These solutions of 1–3% norbixin content are marketed directly as finished food colors after standardization of color content. Thus they are called directly extracted food colors.

Acid-Precipitated Norbixin Concentrates

Norbixin concentrates of between 25 and 50% purity can be prepared by acid precipitation of norbixin from the aqueous alkaline extraction liquid (Marcus, 1963; Dendy, 1966; Tadamasa, 1974) or by direct spray-drying (Park et al., 1990).

Solvent Extraction

Microcrystalline bixin products of 80–97% purity have been developed as a response to the modern need for more concentrated annatto extracts by extraction of annatto seed with volatile organic solvents and subsequent production of a solvent-free product, which is then processed by the manufacturers of food colors according to specific applications.

Numerous patents and research reports cover a variety of organic solvents for producing concentrates, such as chlorinated hydrocarbons (Marcus, 1956), mixtures of ethanol and chloroform (Alvarez-Smith, 1957), acetone (Todd, 1964), ethanol (San-Ei, 1975; DaSilva et al., 1994), ethyl acetate (Bahl et al., 1971), alcoholic sodium hydroxide (Tadamasa and Yasuda, 1985), or ethyl acetate (Srinivasulu and Mahapatra, 1989). It should be noted that chlorinated solvents are no longer suitable for food products. Several authors reported between 1991 and 1997 on the successful use of supercritical carbon dioxide fluid extraction (Jay et al., 1991; Chao et al., 1991; Degnan et al., 1991; El-Sharkawi et al., 1995; Anderson et al., 1997), but it is not known if this has been applied commercially so far.

The market trend is towards annatto extracts of the highest concentration and purity. The production processes now involve additional purification steps with different solvents.

MOLECULAR STRUCTURE OF BIXIN AND NORBIXIN

The groundbreaking work by McKeown and associates at the Department of National Health and Welfare of Canada provided much of the current basic knowledge about the composition of oil-soluble annatto food colors (McKeown 1961, 1963, 1965; McKeown and Mark, 1962). More recently, Scotter (1995) gave valuable insight into the mechanism of formation of a 17-carbon colored thermal degradation product of bixin, which is a common component of oil-soluble annatto food colors first reported by McKeown.

Bixin was used as annatto color long before its molecular structure and the intricacies of its isomeric composition were known. First isolated by Boussingault in 1825, its molecular formula ($C_{25}H_{30}O_4$) was established in 1917 by Heiduschka and Panzer. Much of the basic chemistry of bixin was developed by Karrer and associates (1929), who used such words as "magnificent" and "one of the most beautiful compounds." And yet, Karrer debated the molecular formula proposed by Heiduschka, erroneously claiming that it should be $C_{26}H_{50}O_4$.

Bixin is a half-ester carotenoid and more precisely a diapo-carotenoid, which means that for the purpose of IUPAC nomenclature it is considered a central part of a carotene molecule, i.e., without the terminal rings. The central carbons of the bixin molecule are thus numbered 15 and 15' towards the free carboxylic acid side and the methyl ester side, respectively. The carboxylic carbon atoms at each end of the molecule are assigned the numbers 6 and 6' and not the numbered 1 and 20, as would otherwise be the case (see Appendix). The two end groups of the bixin molecule are both designated with the Greek letter Ψ. (For further explanation of the IUPAC nomenclature of carotenoids, see Britton, 1998.)

Historically bixin was the first carotenoid in which geometrical isomerism was encountered (Karrer et al., 1929). Bixin, and even more so norbixin, are special among the carotenoids because their molecule contains two strongly polar groups. As with all the carotenoid molecules, the numerous conjugated double bonds may give rise to several geometrical isomers. Most carotenoids in nature have the all-*trans* configuration. Bixin is the exception; it is a *cis*-carotenoid in nature. But just exactly which of the nine double bonds is in the *cis* configuration was the subject of long and sometimes bitter controversy for almost 25 years (Zechmeister and Escue, 1944; Barber et al., 1961). The unambiguous and stereo-chemically controlled total synthesis of methylbixin and the comparison of its nuclear magnetic resonance spectrum with methylbixin derived from natural bixin (Pattenden et al., 1970) finally confirmed that natural bixin has the 9'-*cis* structure (at the time still called the *cis*-4 configuration because the carbon atom of the free carboxylic acid group was still assigned the number 1). Valuable insight into the electronic and molecular structure of bixin was obtained by NMR and X-ray crystallography (Kelly et al., 1996) and by resonance Raman spectroscopy (Oliveira et al., 1997). The first total synthesis of bixin has recently been accomplished (Häberli and Pfander, 1999)

Bixin is unique among the naturally occurring carotenoids not only because of its *cis* configuration, but also because the molecule has two car-boxylic groups, one of which is a methyl ester. Bixin is thus a half-ester. This gives it some liposolubility. By alkaline hydrolysis of this methyl ester group the water-soluble salt of the dicarboxylic acid norbixin ($C_{24}H_{28}O_4$) can be prepared (see Appendix). It is an important cheese color.

COLOR CHANGES CAUSED BY HEAT

The oil-solubility of the natural *cis*-bixin is improved by heating the sus-pension in vegetable oil to around 100°C, which causes an isomeric change to *trans*-bixin, which is more lipid-soluble. The commercial use of this iso-merization was described by Preston and Rickard (1980). However, it had long been observed that the heating of bixin in oil also changes the color to a more yellow tone, which is desirable for many applications. Iversen and Lamm (1953) suggested that this change involves a partial decomposition of the bixin molecule into a yellow decomposition product of (at the time) unknown molecular structure. McKeown (1963, 1965) later identified this product as the monomethyl ester of 4,8-dimethyltetradeca-hexanedioic acid. It is commonly called the C-17 compound (Fig. 6.1). It is formed by the loss of a xylene molecule from the bixin molecule and the molecular mechanism leading to its formation has recently been elucidated by Scotter (1995, 2000).

The formation of the yellow product upon heating bixin in vegetable oil has commercial importance because it allows the production of different color shades from orange-red to yellow to suit product requirements. The desired final color can be adjusted by controlling the degree of degradation (Barnett, 1957; Perret, 1958). Heating time is a critical factor in this thermal degradation (Prentice-Hernandez, 1993).

The change of color tone upon heating of solutions of bixin allows the food technologist great versatility to adjust the "hue" of the food product from reddish to orange to yellow as desired. For a more objective measure-ment of the red-to-yellow balance of annatto extracts, we have developed a hue index defined as the ratio of optical absorbances at 404 nm and 470 nm when the extract is dissolved in chloroform. Pure bixin has a hue index of 0.22. Depending on heating temperature and time, this value can go up to 1.0, as in oil-soluble annatto food colors used for processed cheese.

The proportion of yellow to red components in annatto food color can also be established by comparing reverse-phase HPLC chromatograms ob-tained at two wavelength settings of the detector. By this method we have found two groups of yellow components in annatto food colors. An early-eluting group of peaks are the C_{17}-type yellow compounds formed by the ac-tion of heat, whereas a group of late-eluting peaks, appearing about 9 min-utes after the *cis*-bixin peak, are natural, less polar yellow components from the annatto seed. The latter were first found in a sample of annatto food color by Scotter (1998) and marked as unknowns.

BIXIN DIESTERS

Esterification of the free carboxylic acid group of bixin yields the diester methylbixin. It is prepared from bixin by reaction with dimethyl sulfate (Zechmeister and Escue, 1944; Buchta and Andree, 1959; Jondiko and Pattenden, 1989). The use of ethylbixin has also been described (Geminder and McDonough, 1957). The improved liposolubility shown by the diesters has long been thought to increase their usefulness as food colors (Bahl et al., 1971), but the fact that in the strict sense they can no longer be considered natural compounds has limited their application.

It should be noted that traces of methylbixin were found in annatto extract for the first time by Mercadante et al. (1997b). These authors were in fact the first ones to find methylbixin in annatto, although they acknowledged the erroneous priority of Jondiko and Pattenden (1989). The latter, however, had prepared methylbixin from bixin. A useful distinction between methylbixin artificially prepared from natural bixin and natural methylbixin was given by Barber et al. (1961), who proposed the name "methyl-natural bixin" for the former and "natural methylbixin" for the latter, although at that time no natural methylbixin had yet been found.

ANALYSIS

Annatto Seeds

The usual solvents for analytical extraction of annatto seeds are chloroform (McKeown and Mark, 1962), chloroform/acetone (Ramamurty and Bhalerao, 1964), or chloroform/ethanol (Bahakar and Dubash, 1973). Supercritical carbon dioxide with soybean oil as entrainer has also been suggested (Degnan et al., 1991). Extraction at ambient temperature is preferred to prevent isomerization and other molecular changes of the cis-bixin.

The main problem of the analysis of annatto seed is the long time it takes to achieve complete extraction at ambient temperature, even with chloroform, which is the best solvent for bixin. Thus, for routine analytical work on numerous samples, many laboratories have resorted to hot extraction as a compromise between speed and stability.

Prior grinding of the seed sample has not been considered necessary because the pigment is located on the surface of the seed. However, the speed and completeness of extraction at ambient temperature is greatly improved by grinding the seed in a tissue homogenizer equipped with a porous glass filter (Avila et al., 1982). In our laboratory we use a high-frequency dispersing Polytron (model PT 10/35, Brinkman Instruments, Westbury, NY) positioned with the seed sample in an extraction thimble for easy filtration of the extraction liquid. We have found that the best extracting solvent is the 79:21 azeotropic mixture of chloroform and methanol (E. Regalado, unpublished), which has the added advantage of being recyclable by distillation without change in solvent composition.

Although chloroform or chloroform/methanol still appear to be the most efficient solvents for quantitative extraction of bixin from annatto seed, there have been repeated warnings against the use of chloroform as a solvent for carotenoids, due to the difficulty of ensuring the removal of all traces of HCl, which may cause molecular changes in carotenoids (Schiedt and Liaaen-Jensen, 1995). Severe degradation of carotenoids in chloroform have been reported (Scott, 1996), and dichloromethane has been suggested as an alternative (Furr, 1997). For the analytical extraction of bixin from annatto seed, however, no entirely satisfactory substitute for chloroform has been found.

An alternative procedure for routine color measurement of annatto seed is the determination of bixin as its norbixin derivative after hydrolytic extraction. The seed sample is boiled in aqueous potassium hydroxide (N. Dinesen and P. Collins, private communications). The spectrophotometric result must be multiplied by 1.037 to correct for the ratio of molecular weights of bixin and norbixin.

Optical absorbance for the determination of carotenoids is generally measured at one of the wavelengths of maximum absorption (see below). A suggestion has been made to replace optical spectrophotometry for the analysis of annatto seed by photoacoustic spectrophotometry (Haas and Vinha, 1995), but this system does not appear to be widely used.

Annatto Extracts and Formulations

For all practical purposes, the measurement of the color intensity of annatto food colors is the most important and at times the only criterion applied. Color intensity is the spectrophotometric absorbance of a 1 g/L solution of food color measured at 470 nm in chloroform/1% glacial acetic acid for oil-soluble annatto extracts and in 0.1 M sodium hydroxide at 453 nm for water-soluble extracts (Food Chemicals Codex, 1986) or at 454 nm in acetone/1% glacial acetic acid for oil-soluble extracts (Food Chemicals Codex, 1992). The absorbance is used as the final result. No extinction coefficient (specific absorbance) is used in the calculation, and thus the result does not reflect the actual bixin or norbixin content.

However, more precise methods are necessary for stating the bixin and norbixin content of annatto extracts, of food colors, and of foods containing them as more stringent specifications for these products are imposed and a better understanding for their chemistry and biochemistry is warranted (Mercadante et al., 1996).

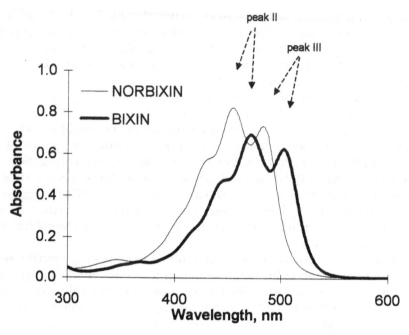

FIG. 6.2 Spectrometric curve of 2.5 mg 87.5% bixin/L, before and after hydrolysis.

Spectrophotometry. A precise examination of the entire visible spectrum provides important information on the qualitative composition of annatto extracts and food colors. Carotenoids usually show three peaks of absorption maxima. These peaks are numbered I, II, and III, going from the lower wavelength to the higher.

Chloroform has been used historically as the solvent for spectra of bixin despite its problems (see above). In the case of bixin in chloroform, peak I is actually only an inflection point at about 445 nm of the spectral curve. The main peak of *cis*-bixin is at 471 nm. It would normally be considered the first peak, but it is, by convention, designated as peak II. A lower but well-defined second peak is at about 503 nm and is designated by convention as peak III. For norbixin in 0.1 N NaOH, peak II is at 453 nm and peak III at 482 nm (Fig. 6.2).

Absorbance measurements at either peak II or peak III are used for quantitive determination of bixin and norbixin, although peak III is generally preferred because at its wavelength there is less interference from a possible presence of yellow decomposition products with overlapping spectral curves, which may require the application of a correction through an additional absorbance measurement at 404 nm (McGeown and Mark, 1962).

As in most carotenoids (Englert, 1995), the position of the wavelengths

TABLE 6.2   1% Extinction Coefficients[a]

Norbixin[b]		
Peak II (453nm)	Peak III (482nm)	Ref.
2850	2550	Rieth and Gielen (1971), FAO/WHO (1976), Smith (1983)
2818	2503	Scotter (1994)
3473	—	FAO/WHO (1981), Hirata (1989)
2620[c]	2290[c]	Smith (1983), Scotter (1998)
3208	—	Collins (private communication,1990)
—	2870	EEC (1995) for E160b, FAO/WHO (1996)

Bixin in chloroform[b]		
Peak II (470nm)	Peak III (501nm)	Ref.
—	2826	McKeown (1962), Dendy (1966)
3230	2880	Reith and Gielen (1971), FAO/WHO (1976), Smith (1983)
3130[d]	2790[d]	Reith and Gielen (1971), Smith (1983)
2826	—	FAO/WHO (1981), Hirata (1989)
3092	2773	Scotter (1994)
—	2870	EEC (1995) for E160b, FAO/WHO (1996)

[a]In chronological order of original research report.
[b]In 0.1 N KOH unless otherwise indicated.
[c]In chloroform + 0.5% acetic acid, peak II at 473 nm and peak III at 503 nm.
[d]In chloroform + 3.7% acetic acid, peak II at 474 nm and peak III at 505 nm.

of maximum absorption of the all-*trans* isomer shows a bathochromic shift of about 4 nm in respect to the *cis* isomer.

Conflicting Extinction Coefficients: Published 1% extinction coefficients (also called specific absorbance) or molar extinction coefficients are used for the calculation of results (McKeown and Mark, 1962; Dendy, 1966; Reith and Gielen, 1971; Avila et al., 1982; Smith et al., 1983; Hirata et al., 1989; Scotter 1994, 1998). Surprisingly we find a wide and disturbing variation among the published 1% extinction coefficients of bixin and norbixin (Table 6.2).

Depending on which extinction coefficient is chosen, the results of the determination of bixin or norbixin are obviously quite different. At least one of the erroneous values can be traced back to the extinction coefficient of norbixin published by Reith and Gielen (1971), which has been quoted and used extensively in most later publications. Their 1% extinction coefficients for norbixin of 2850 (at 453 nm) and 2550 (at 482 nm) in aqueous alkaline solution are in serious doubt as is the value of 3473 at 453 nm used in the earlier FAO/WHO specification (1981).

In 1990 P. Collins (private communication) was the first to question the extinction coefficient of 2850 for norbixin at peak II (453 nm). He purified a sample of norbixin by repeated recrystallization to a purity of 99.6% (HPLC, photodiode array detection) and found a 1% extinction coefficient of 3208 for peak II in aqueous 0.1 M potassium hydroxide, quite different from the published value of 2850. We have used the absorbance ratio of 0.894 between the norbixin peaks II (453 nm) and III (482 nm) to calculate from Collins's findings that the extinction coefficient of norbixin at peak III must be 2867 and not 2550 as published by Reith and Gielen. As was pointed out in 1990 by N. Dinesen (private communication), it must be a curious coincidence that Reith and Gielen's figure for peak II is actually rather close to the extinction for peak III as calculated on the basis of Collins's data.

The following practical conversion factors may be useful to correlate optical absorbances at peaks II and III. To convert absorbance at peak II in chloroform (472 nm) to absorbance at peak III in the same solvent (504 nm), multiply by 0.900. To convert norbixin absorbance at peak II in aqueous alkali to absorbance at peak III in the same medium, multiply by 0.894.

Correlation of Optical Absorbances of Norbixin and Bixin: When a sample of bixin is hydrolyzed, the optical absorbance increases. Figure 6.2 shows the spectrum of a 2.5 mg/L solution of bixin in chloroform before hydrolysis and after hydrolysis of the same sample of bixin (norbixin in aqueous alkali) without change in concentration. The norbixin curve is higher than was the curve of the same sample of bixin before hydrolysis, which confirms that the 1% extinction coefficient of norbixin is higher than that of bixin.

In an attempt to calculate the 1% extinction coefficient of bixin in chloroform on the basis of Collins's value for norbixin, we have made a statistical study with optical absorbance data recorded during 9 years for more than 1000 different samples of bixin before and after hydrolysis. This study shows that the difference between the two extinction coefficients is of the order of 6%. If the value of 3208 for norbixin at peak II (P. Collins, private communication) is adopted, then the 1% extinction coefficient of bixin in chloroform must be 3016 at peak II and 2714 at peak III. This is in acceptable agreement with the findings by Scotter (1994) of 3029 and 2773 at peaks II and III, respectively, for bixin in chloroform.

However, these values are in conflict with the extinction coefficient of 2870 at peak III for both bixin and norbixin as established in Europe (European Economic Communities, 1995; FAO/WHO, 1996). Their assumption of equal extinction coefficient for both bixin and norbixin may have been reasonable in theory because both molecules have the same chromophore. However, from our results it is clear that the ionization of the carboxylic groups of norbixin in aqueous alkali exerts a hyperchromic effect (as defined by Britton, 1995) and thus a revision of the European specification E160b may be in order.

Our findings also provide an explanation for the mention by Smith et al. (1983) that "unexplainably" a sample of oil-soluble annatto food color gave them a higher bixin assay result after saponification than before. These authors thought that perhaps some other component of the food color was responsible for this increase, but they did not at the time consider the possibility of a hyperchromic effect due to ionization.

It is obvious that further research is needed to finally establish the correct extinction coefficient of bixin. This requires a systematic purification of bixin by the methods of classical organic chemistry until achieving 100% purity as proven by a molecular weight determination and compliance of C, H, and O analyses with the molecular formula.

Never Use a Quartz Cuvette: An aspect often overlooked in the spectrophotometric analysis of bixin in chloroform solution is the type of cuvette used in the spectrophotometer. We have found that when using a quartz cuvette, photodegradation occurs at a surprisingly fast rate after placing the sample in the spectrophotometer solely by the action of the light energy of the instrument on the sample. The addition of antioxidants does not prevent this degradation. In as little as 10 seconds the absorbance of the sample is reduced by almost 5%, and the decrease continues thereafter. Figure 6.3 shows the spectral curves of a solution of 0.25 mg bixin in 100 mL of chloroform measured at 40-second intervals while in a quartz cuvette in the spectrophotometer.

As stated, this degradation occurs only when the sample is in a quartz cuvette, which transmits ultraviolet radiation below 300 nm. It does not occur when using glass cuvettes that do not transmit light in the UV area of the spectrum. It is therefore important to use only glass cuvettes for bixin analysis, a recommendation already made by Dendy (1966), which has been largely ignored. It is possible that some of the discrepancies in the extinction coefficients published by different authors may thus be explained.

Reverse-Phase HPLC.   Methods are available for the determination of the colored components of annatto extracts including the geometrical isomers of bixin and norbixin (Smith et al., 1983; Nishizawa et al., 1983; Amakawa et al., 1984; Rouseff, 1988; Luf and Brandl, 1988; Chatani and Adachi, 1988; Wood et al., 1991). The method of Scotter et al. (1994) is practical and very reliable. A reverse-phase HiRPB column is used with an isocratic (65%:35%) mobile phase of acetonitrile:2% aqueous acetic acid. We have found that a mobile phase with a 70%:30% ratio of the two components improves peak separations in some cases, especially when using a Spherisorb S50DS1 column.

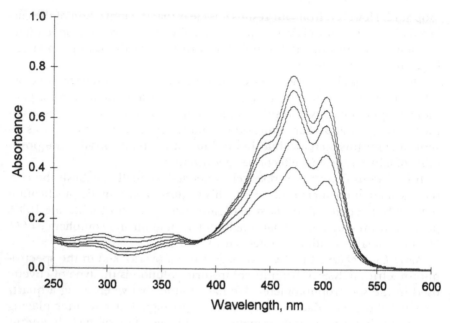

FIG. 6.3 Reduction of optical absorbance of bixin in chloroform in quartz cuvettes placed in spectrophotometer (40 sec interval between measurements).

An interesting feature of HPLC of annatto colors is the possibility to distinguish which of the peaks refer to the red and which to the yellow components if two detector settings are used. An aliquot of the sample is first injected with the detector set at 460 nm. Then an aliquot of the same sample if again injected but with the detector set at 404 nm. The two chromatograms are compared. Peaks that decrease in area between the first and the second chromatogram are the red components; peaks that increase in the second are the yellow components.

A method for the determination of bixin and norbixin in human plasma was recently published (Levy et al., 1997).

Methods for Specific Products

For mixtures of bixin/norbixin with other colors such as β-carotene or turmeric, open column chromatography or thin-layer separation must be used prior to spectrophotometric measurement because of overlapping spectral curves. An interesting method for direct measurement of annatto colors in the presence of β-carotene involves the use of derivative spectrophotometry (Luf, and Brandl, 1988). Photo-acoustic spectrometry has been proposed (Haas and Vinha, 1995).

Of special interest for the food technologist are specific methods for the determination of annatto in margarine, cheddar cheese, and boiled sweets (Smith et al., 1983) and in high-fat dairy products, margarine, and hard candy (Lancaster and Lawrence, 1995), the detection of bixin as adulterant in products derived from red pepper (Minguez-Mosquera et al., 1995), the determination of bixin in the presence of carminic acid (Lancaster and Lawrence, 1996), the group analysis of natural coloring matters in food products (Yamada, 1997), and the analysis of food coloring formulations (Scotter, 1998).

CONTAMINANTS

Tolerance levels in the United States for annatto extracts (U.S. Code of Federal Regulations, 21CFR73.30) are 3 mg/kg for arsenic and 10 mg/kg for lead. Maximum residual solvent levels shown in the Food Chemicals Codex (1996) are 0.003% for acetone, 0.0025% for hexane, 0.005% for methanol or isopropanol, and 0.003% for trichloroethylene or dichloromethane individually or in combination. The same limits are shown in the European specifications (1995) for "annatto, bixin, norbixin" (food color E160b) with additional limits for heavy metals (40 mg/kg expressed as lead), mercury (1 mg/kg) and cadmium (1 mg/kg) (European Economic Community, 1995).

There is a certain incongruence in the tolerance levels of contaminants in annatto food colors. In the European specification the same maximum levels of contaminants are shown for annatto extracts of the whole range of color concentration, i.e., for oil-extracted annatto of 0.1% bixin, for alkali extracted annatto of 0.1% norbixin, as well as for solvent-extracted annatto of 75% bixin, even though in practical use the dilutions at which these are incorporated into the food product vary from 1:70 (for the 0.1% bixin extract) to 1:40,000 (for the 75% bixin extract). The aim of establishing tolerances for contaminants is obviously to safeguard the health of the consumer of the ready-to-eat food product containing the annatto color. Such contaminants will appear in the final food product at different levels with each of the various annatto extracts, which nevertheless are used at widely differing dosages because of the range of bixin/norbixin concentrations available. The U.S. limits do not refer to any specific concentration of annatto color either. They refer simply to "annatto extract, including pigments precipitated therefrom" (21CFR73.30) and thus show the same incongruence.

In practice, a disturbing aspect of the analytical data of trace contaminants is the highly variable performance of different laboratories as shown in an assessment of 136 laboratories in 21 countries carried out by the World Health Organization. As many as 40% of the laboratories assessed showed unsatisfactory accuracy in the analytical trace element results they reported (Weigert et al., 1997). In the specific case of the determination of mercury at the 1 ppm level, we have found wide variations of results received from various internationally recognized independent laboratories to whom we sent aliquots of the same bixin sample. They reported results varying from 0.6 ("sample complies with international specifications") to 1.6 ppm ("sample does not comply"), all for the same sample.

STABILITY: LIGHT, OXYGEN, HEAT

Like all highly colored and highly unsaturated compounds, bixin and norbixin are unstable to light (Najar et al., 1988). This means, of course, that certain wavelengths of light are absorbed by the molecule, as we know from the spectrophotometric absorption curve, and it has led to the use of bixin or norbixin in sunscreens (Grollier et al., 1989). In practice, however, bixin and norbixin show better light stability than many other carotenoid colors.

Like all the other antioxidant carotenoids, bixin and norbixin are also unstable in the presence of atmospheric oxygen. Stabilization of bixin and norbixin can be achieved by the addition of antioxidants with higher affinity for oxygen (Hettiarachy et al., 1986; Ford and Mellor, 1987; Najar et al., 1988; Ono, 1988).

When bixin is formulated in a powder base, certain levels of moisture enhance color stability (Glória et al., 1995).

Normally no color fading is observed in practice in food products colored with annatto because bixin and norbixin are more stable than many other air-susceptible colors (Berset and Marty, 1986) and thus the products containing them have an excellent shelf life. In the few cases when fading has been observed, the cause was traced to quality defects of the food product such as rancidity of an ingredient, bacterial action, or defects in packaging materials (N. Dinesen, private communication). In fact, annatto colors tend to extend the shelf life of products because of their antioxidant action.

Bixin has even been used to stabilize vitamin D preparations (Yamada, 1988).

Relative to other carotenoids, bixin and norbixin have good heat stability during food processing. By heating bixin in oil, a partial thermal decomposition can be achieved intentionally to provide a purer yellow tone to the product.

USES AND APPLICATION RATES

The applications of oil-soluble and water-soluble annatto colors have been reviewed by Collins (1992). Between 0.1 and 50 parts per million of bixin (as pure color) provide pleasing "butter to egg-yolk shades" when applied to otherwise colorless food products. Bixin is completely soluble in fats and oils up to 0.75% by weight, whereas norbixin is soluble in water as the sodium or potassium salt up to 7% (Dinesen, 1974). Norbixin reacts with protein with a slight shift to a delicate peach-red color. This is seen in some cheeses colored with annatto. This binding property is useful in coloring products that must hold the color fast and not bleed into the surrounding medium (Lauro, 1991).

Commercial annatto food colors are available as water-soluble extracts with norbixin contents of 1.4% ("single-strength cheese color"), 2.8% ("double-strength"), or 3.8% ("triple strength") and as a 15% norbixin powder. Oil-soluble annatto colors contain between 0.2 and 5% bixin. Higher concentrations are available as emulsion or suspension colors. "Acid-proof" annatto colors are also available.

Annatto is especially suited for coloring cheese, meat, and fish because of its ability to combine with protein, thus creating a very stable color. The categories of foods in which annatto is used are (in alphabetical order): biscuit fillings, breakfast cereals, cakes, cheeses, dairy drinks, decorations and toppings, desserts, edible ices, fruit curds, imitation crab legs, jams, maraschino cherries, margarine, marmelade, marzipan, pickles, pie fillings, sauces and seasonings, sausage casing, smoked fish, snacks, soft drinks, and sugar confectionary (U.K. Food and Drink Federation, 1992). The use of annatto in surimi paste and in new surimi-based foods has been reviewed by Lauro (1999).

Numerous patents have been issued involving the preparation of annatto food colors and their applications (Kocher, 1958: annatto extract in propylene glycol; Todd, 1964: combination with turmeric; Marcus, 1963: margarine and cheese color; Sato, 1966: Vienna sausage; Unilever, 1981: margarine; San-Ei, 1983: acid-stable food color; Schmidt, 1985: acid-soluble annatto powder; Berset and Marty, 1986: extrusion cooking; Ford, 1987: stabilization with ascorbic acid for beverages; Ono, 1988: stabilization with quercitin; Maeda, 1988: bread manufacture, Winning and Isager, 1997: water-dispersible compositions). A drawback in the use of norbixin for coloring beverages is that it is water-soluble only at a high pH. To color products of lower pH, "acid-proof" forms been developed by the combination with gum arabic (Hettiarachy et al., 1986), modified food starch, milk proteins, or cyclodextrins and polysorbate-80, all of which produce stable emulsions in acidic media (Todd, 1991). A recent Japanese patent (Nakajima et al., 1999) describes the addition of unsaturated carboxylic acid esters to stabilize norbixin solutions at pH 3.

The ever-increasing reporting requirements to the governmental regulatory agencies have opened up the former industrial secrecy surrounding the application and usage level of many food ingredients. Thus the varied applications of bixin and norbixin and the allowed rates of application can now be known in greater detail than before. Table 6.3 has been compiled from such sources.

For cheese, the most important food product in which annatto is used,

TABLE 6.3 Maximum Application Levels for Annatto Colors in Various Food Products

Food product	mg/kg (as 100% bixin or norbixin)
Red Leicester cheese	50
Cake decorations, marzipan toppings	40
Mimolette cheese	35
Extruded or puffed breakfast cereals	25
Sauces, seasonings, pickles	20
Snacks, extruded snacks	20
Decorations and coatings, ice cream	20
Edible cheese rind	20
Process cheese	15
Margarine	10
Fine bakery wares, confectionary	10
Cakes, biscuit fillings, deserts, liqueurs	10
Nonextruded snacks	10
Sausage casings, smoked fish	10
Marmelade	3
Dairy drinks	2
Soft drinks	0.5

Source: Adapted from U.K. Food and Drink Federation, 1992, and European Economic Community, 1994.

the maximum application rate in Europe is 600 mg "annatto color" (probably based on 2–4% norbixin, thus equivalent to 12–24 mg pure pigment) per kg of cheese (Luf and Brandl, 1988), followed by baked goods, cake decorations, and breakfast cereals. Other applications for annatto are the coloring of sausage casings (Sato and Susuki, 1966), sausages and ham at up to 95 mg/kg (Chatani and Adachi, 1988), and the quality improvement of bread dough (Maeda, 1988). Lancaster and Lawrence (1995) reported on the norbixin and bixin content of commercial samples of various types of cheeses sold in Canada. The bixin content per kg was 5.9 mg in process cheese spread and 5.1 mg in Canadian wine cheese. The norbixin content in mg pure pigment per kg was 1.1 in process cheese spread, 2.5 in process cheese slices, 15.6 in skim milk cheese, 16.8 in Canadian wine cheese, 18.2 in medium Cheddar cheese, 21.2 in Canadian Colby, and 68.8 in sharp Cheddar cheese. Samples of commercial butter sold in Canada were found to contain 0.017–0.199 mg/kg of bixin and 0.033–0.914 mg/kg of norbixin.

Of interest to the food technologist is an abnormal color change that occurs occasionally in annatto-colored processed cheese commonly referred to as "pinking." Its relation to the processing techniques and to the types of annatto color used has recently been studied (Shumaker and Wendorff, 1998). It was found that the anomaly occurs more often with aqueous emulsions of annatto than with bixin-in-oil suspensions.

Annatto acts as an antioxidant in foods improving the stability of the product. Haila et al. (1996) have shown that the autoxidation of fats is significantly reduced with 30–60 mg bixin per kg. Unlike other carotenoids, such as β-carotene, lutein, or lycopene, no pro-oxidant action has been found for bixin.

The rather unusual application of cis-bixin as the rigid hydrophobic core of a bolaform amphiphile surface monolayer was reported by Fuhrhop et al. (1990), who also was able to integrate the gluconamide derivative of bixin into helical micellar fibers.

DAILY HUMAN INTAKE

This is an important consideration from the regulatory standpoint. Based on the application levels, many calculations have been attempted to establish the estimated daily human intake of bixin/norbixin through foods containing them. The estimates vary from 0.032 mg/kg body weight/day for bixin and 0.051 mg/kg body weight/day in the United States (N. Dinesen, private communication) to 0.065 mg in the United Kingdom (Ministry of Agriculture, Fisheries and Food, 1993). Calculated for a person of 60 kg average body weight, these figures represent an average daily bixin/norbixin intake of 1.9 mg for bixin and 3.1 mg for norbixin in the United States and of 3.9 mg in the United Kingdom. Another British study (UK Food and Drink Federation, 1992) suggests an 8 mg/day average intake, which appears to be greatly exaggerated. A recent French study (Verger et al., 1998) shows a theoretical maximum daily intake of 0.16 mg/day/kg body weight (2.5 mg/day for a 60 kg person) and a mean daily intake of 0.005 mg/day/ kg body weight (0.3 mg/day/person). The study is based on the additive list on the labels of foods, and thus the figures most probably are based on annatto color E160b as such and not on pure bixin or norbixin. If so, the results shown should at least be halved to reflect bixin or norbixin intakes.

The maximum acceptable daily intake (ADI) established by the Joint FAO/WHO Committee on Food Additives (JECFA, 1982) is 0.065 mg/ day/kg body weight expressed as bixin, which for the average human weight of 60 kg works out at 3.9 mg/day.

These calculations have achieved more significance since it has been shown that bixin is rapidly absorbed into the blood stream after ingestion (Levy et al., 1997). The experimental human ingestion of 16 mg bixin in a single dose gave rise to a maximum bixin/norbixin level in the plasma of 144 μg/L, which is comparable to the ranges of other carotenoids normally found in human plasma (e.g., β-carotene, 40–530; lutein, 90–140;

lycopene, 70–460 µg/L). Complete plasma clearance occurred for bixin by 8 hours after ingestion and for norbixin by 24 hours. This confirmed an anonymous report mentioned earlier by Preston and Rickard (1980).

SAFETY

Annatto is considered essentially nontoxic. In the United States it is a natural colorant "exempt from certification" (U.S. Code of Federal Regulations as of 1994). Its safety is based on its traditional use for many centuries as a food color by millions of people in South America and on the early work of Zbinden and Studer (1958) and van Esch et al. (1959), who conducted chronic feeding tests with aqueous alkaline annatto extracts on rats and found no adverse effects. A safety assessment of annatto extracts was made by the International Life Sciences Institute (1985). Based on prior work by Lück and Rickerl (Ghorpade et al., 1995) with rats and one-year chronic oral toxicity studies with dogs using solvent-extracted annatto and aqueous annatto extracts (J. H. Kay and J. C. Calandra, unpublished), as well as the genotoxicity tests of Haveland-Smith (1981), it was concluded that high levels of bixin or norbixin in the diet of experimental animals do not cause any toxicological or carcinogenic effects and that 0.065 mg/kg body weight (expressed as bixin) is an acceptable daily intake for humans.

Some care must be exercised when reviewing some of the toxicological information mentioned in the literature. Ghorpade et al. (1995), for example, erroneously quote Dunham and Allard (1960), saying that "annatto extracts" are antispasmodic and hypotensive, but they do not warn the reader that Dunham and Allard's work is irrelevant to annatto food color because it was done with extracts of the roots of the annatto plant, which contain no bixin.

More recently it was reported that petroleum ether extracts of annatto seed probably containing mainly nonbixin extractives have shown genotoxicity when applied to onion roots, as have total extracts with chloroform containing all the seed components soluble in this solvent (Aranez and Rubio, 1996). Also, a petroleum ether extract made from the chloroform extract given to male mice before mating appeared to produce more dead offspring in the females (Aranez and Bayot, 1997). A critical evaluation of this work is necessary. It would be important to find out if the observed result could be explained by the presence of residual chloroform in the sample.

Certainly the petroleum ether extract of annatto seeds contains only negligible amounts of bixin. In any event, these reports lack any practical importance because none of these solvents are used commercially for the extraction of annatto seed.

Occasional individual cases of allergic reactions to foods containing annatto color have been reported (Mikkelsen et al., 1977; Nish et al., 1991), but they are attributed to a trace of an annatto seed protein, which may have been in the annatto color as an impurity, rather than to bixin or norbixin.

The increased use of annatto colors means that the ADI needs to be raised. The international authorities have requested additional studies because all earlier studies had used addition rates for annatto now considered too low. The scientific and manufacturing groups involved in annatto are currently funding a major animal study at a well-known research laboratory in England.

NONBIXIN COMPOUNDS IN THE ANNATTO PLANT

Although bixin is the only component of *Bixa orellana* that currently has commercial importance, the plant is a treasure chest of exotic compounds, some found in no other plant, most of which were discovered during the past 10 years. Prominent among them are geranylgeraniol in the annatto seed coat (Jondiko and Pattenden, 1989) and ishwarane in annatto leaf oil (Lawrence and Hogg, 1973). The presence of these two compounds converts annatto seed into their richest known natural source. Recent work of the group of Prof. Pfander at the University of Bern (Switzerland) has enormously expanded our knowledge of the minor carotenoid components of annatto seed extract (Mercadante et al., 1996, 1997a, 1997b, 1999). Most of the compounds are found only in annatto and have not been reported previously. It has been postulated that the presence of geranylgeraniol is responsible for some of the "unusual properties" of annatto colors (Craveiro et al., 1989). Recently Mercadante et al. (1999) discovered geranylgeraniol esters of bixin in annatto seed. Many of these minor constituents of annatto may give new insight into the biosynthesis and metabolism of bixin in the annatto plant.

Compounds other than bixin that have been discovered in the annatto plant since 1965 are shown in chronological order of their discovery in Table 6.4.

PHARMACOLOGY

This is an entirely new field that is wide open for new research. Annatto seeds have long been used in traditional medicine of the South American Indians to promote the healing of wounds, against skin eruptions, and in the healing of burns "without a scar" and have been given internally to subdue diarrhea and asthma (Morton, 1989) and as an antipyretic (Terashima et al., 1991). It is not clear if these effects are attributable to bixin or to some other compounds in the annatto seed. Other botanical parts of the annatto

TABLE 6.4 Compounds Other than Bixin and Norbixin Identified in
Bixa orellana[a,b]

Compound	Ref.
Tomentosic acid (in roots)	Schneider, 1965
Ishwarane (in leaf oil)	Lawrence and Hogg, 1973
β,β-Carotene, cryptoxanthin, lutein zeaxanthin, methylbixin (all in traces)	Tirimanna, 1981
Geranylgeraniol	Jondiko and Pattenden, 1989 Craveiro et al., 1989
Geranylgeraniol formate, geranylgeraniol octadecanoate farnesylacetone, δ-tocotrienol	Jondiko and Pattenden, 1989
Methyl-9'-*cis*-apo-1-bixinal ester or Methyl-8-oxo-9'*cis*-8,6'-diapo-caroten-6'-oate[c]	Jondiko and Pattenden, 1989
α- and β-Pinene (in essential oil)	Rath, 1990
Iso-scutellarein, gallic acid, Pyrogallol (aqueous leaf extract)	Terashima et al., 1991
Methyl-9'-*cis*-apo-6'-lycopenoate[c]	Mercadante et al., 1996
Methyl-*trans*-8'-apo-á-caroten-8'-oate[d]	Mercadante et al., 1997a
Methyl-7,9,9'tri-*cis*-apo-6'-lycopenoate[c]	Mercadante et al., 1997a
Methyl-9-*cis*-apo-8'-lycopenoate[c]	Mercadante et al., 1997a
Methyl-*trans*—apo-8'-lycopenoate[c]	Mercadante et al., 1997a
Methyl-*trans*—apo-6'-lycopenoate[e]	Mercadante et al., 1997a
Methyl-9-*cis*-bixin	Mercadante et al., 1997b
Methyl-9,9'di-*cis* bixin[c]	Mercadante et al., 1997b
Methyl-9-*cis*-6'-oxo-6,5'-diapocaroten6-oate[c]	Mercadante et al., 1997b
Methyl-4-*cis*-4,8-dimethyl-12-oxo-dodecyl-2,4,6,8,10-penta-enoate[c,f]	Mercadante et al., 1997b
6-Geranylgeranyl-8'-methyl-6,8'-diapo6,8'dioate	Mercadante et al., 1999
Geranylgeraniol-*cis*-bixinate	Mercadante et al., 1999
Geranylgeraniol-*trans*-bixinate	Mercadante et al., 1999

[a]In chronological order of discovery.
[b]In seed extracts unless otherwise indicated.
[c]New carotenoid, found only in *B. orellana*.
[d]Previously found only in *Staphylococcus aureus* bacteria.
[e]Previously found only in *Sepherdia canadensis*.
[f]Two central methyl groups missing, thus no longer a carotenoid, possibly a metabolic product.

plant also contain physiologically active compounds. Annatto root extracts, for example, have been shown to be antisecretory, antispasmodic, and hypotensive (Dunham and Allard, 1960), and the aldose reductase inhibitor iso-scutellarein (a flavone) has been found in annatto leaf extract (Terashima et al., 1991). Annatto seed extract given to dogs showed the presence of a hyperglycaemic principle (Morrison et al., 1991).

Bixin has strong physical quenching activity of singlet molecular oxygen and thus may excert a protective action against some types of cancer (DiMascio et al., 1990). However, bixin did not prevent the formation of cancer cells in experimental carcinogenesis with methylcholanthrene (Bertram et al., 1991). As a protectant against biological membrane oxidation, bixin is a potent inhibitor of lipid peroxidation at the same level of lutein and canthaxanthin and is only surpassed by α-tocopherol (Zhang et al., 1991). Bixin acts as a lipoxydase inhibitor and modulates lipid hydroperoxide formation (Canfield and Valenzuela, 1993). Oral administration of bixin significantly reduced the otherwise increased level of lipid peroxides in serum and liver of rats caused by gamma radiation and can thus be considered a candidate drug for protection against the side effects in cancer patients undergoing therapeutic irradiation (Thresiamma et al., 1995, 1998). Bixin does not upregulate Connexin-43 gene expression as some other carotenoids do, but it is active in membrane protection (Zhang et al., 1992). Bixin increases the activity of xenobiotic metabolizing enzymes (Jewell and O'Brien, 1999). Food colors in general enhance immunoglobulin production by rat spleen lymphocytes, although bixin was not specifically included in the study (Kuramoto et al., 1996). Some interesting spectral changes have been observed during the interaction of bixin with respiring rat liver mitocondria (Inada et al., 1971; Hirose et al., 1972).

Compared to the extensive current research effort on the health-promoting effects of the other natural carotenoids (Krinsky, 1994), bixin has not yet received much attention. It may well be that many beneficial actions of bixin are yet to be discovered (Levy et al., 1997).

SPECULATIONS: ARE NATURAL COLORS GOOD FOR YOUR HEALTH?

The purpose for the use of colors in the food industry has been to make the product more pleasant to look at and, since "we eat through our eyes," to make it more palatable. Because most natural colors are also potent antioxidants, they also have the function of preservation of many nutritionally valuable food ingredients, increasing the shelf life of the product. Over the last few years more and more evidence has been accumulated that many natural antioxidants that are useful for the preservation of vitamins and lipids in foods also produce antioxidant activities in the human body after consumption of the foods (Lölinger, 1997). There is mounting evidence of the importance of the antioxidant carotenoids in human health. It is interesting to speculate that these colors not only make the food products look better and help to preserve it, but they may actually improve the health of the consumer.

Action of Natural Colors Against Free Radicals

There is increasing awareness of the importance of controlling the metabolic production of free radicals in our bodies to maintain good health. Ever since animals and humans appeared on this planet, this has been accomplished by their eating foods from the plant kingdom that are rich in antioxidants, such as fruits and vegetables. The statistical increase of human diseases such as cancer and atherosclerosis in this century has been related to the increase of stress, atmospheric pollution, pesticide contamination of foods, and ultraviolet radiation caused by the thinning of the protecting ozone layer in the stratosphere. All of these are facts of life for modern humans. All of these promote the formation of more free radicals in our bodies than may have been the case in former periods of history. Thus the current strong medical recommendation for all of us to increase our consumption of fruits and green vegetables containing carotenoids, flavonoids, and other natural antioxidants. It is believed that within the biological defense system against free radicals, unpaired electrons are taken up by vitamin E (tocopherol), carotenoids, and vitamin C (ascorbic acid) in a cascade-like fashion, where all of these act in combination (Truscott et al., 1996). Vitamin E, after taking up the electrons, becomes a transient tocopheryl free radical. Free electrons are then transferred to one after the other of the carotenoids through their conjugated double bonds. The first carotenoid of this cycle regenerates the tocopherol to its original non–free radical state, the next carotenoid regenerates the former carotenoid, and so on. In this process the excess energy is slowly dissipated during the trip of the electron through the various conjugated double-bond systems as thermal energy, and the electrons are finally transferred at the lipid/water interphase into the aqueous phase, where ascorbic acid takes final care of them.

Molecular Basis for Color and Free-Radical Scavenging

The color sensation of the human eye originates in the conjugated double-bond system of the carotenoids. A minimum of seven conjugated double bonds is necessary for a carotenoid molecule to absorb light of the visible spectrum (Britton, 1998). All carotenoid molecules have more than this number of double bonds. Free radical–scavenging activity is equally related to the number of conjugated double bonds in the carotenoid molecule.

Thus it can be postulated that both the color and the defense against free radicals originate on the same molecular basis.

Bioavailability of Carotenoid Colors

The bioavailability for humans of the carotenoids of foods is rather low. This means that only a small proportion of the carotenoids of fruits, vegetables, and spices reach the blood stream where they are most needed. Spinach, for example, is a rich source of the carotenoids β-carotene and lutein, but normally only a fraction of these is actually absorbed into the blood stream (Castenmiller et al., 1999). This is due to the "matrix" in which the carotenoid is locked in the spinach (i.e., the cell walls surrounding the carotenoid, which are broken down only partially by the act of chewing). The same is the case for lycopene in tomatoes (Gaertner et al., 1997) and other vegetables and fruits. Recommendations have been made to prepare the vegetables in a food blender in order to break down the cells mechanically prior to eating. Also, against all common wisdom, it was found that cooked foods actually provide more antioxidants than fresh, uncooked foods. The reason is the same: by softening the cell walls the antioxidants become more "available" for absorption into the blood stream during digestion. All this means that nutritional supplementation of carotenoids is an ever-increasing necessity for modern humanity. Such supplementation may come, in part, from carotenoid extracts added to food products, which then become fortified foods.

Artificial Dyes Versus Natural Colors in Relation to Cancer

The color effect of coal tar–derived artificial dyes is generally due to the electronic structure of the carbon-nitrogen or carbon-sulfur bond in their molecules, not by a system of conjugated double bonds as is the case with the carotenoids, which contain only carbon, hydrogen, and sometimes oxygen atoms. This fundamental molecular difference explains the apparent paradox that two substances having the same color—one an artificial dye and the other a carotenoid—have such radically different biological actions, such as one being a carcinogen and the other an anticarcinogen. Artificial dyes tend to produce free radicals, whereas the natural colors eliminate free radicals. For this reason artificial colors have been progressively banned for use in foods and are being replaced by natural colors.

How Does Bixin Fit in?

Bixin is one of the most powerful carotenoid antioxidants. Its conjugated double-bond system is ideally suited for scavenging free radicals. After consumption it is well absorbed into the blood stream. Its possible benefit for human health awaits further research. In summary, for all the natural colors, including bixin, it is entirely possible that the food technologist may soon watch in fascination as new scientific findings demonstrate that the natural food colors of today will be the vitamins of tomorrow.

ACKNOWLEDGMENT

The authors wish to express their gratitude to Mr. Niel Dinesen for his critical revision of the manuscript and for many helpful discussions and suggestions on several aspects of annatto chemistry and uses.

APPENDIX

Annatto

Synonyms: Achiote (South America), Onoto (Venezuela, Mexico, West Indies), Urucum (Brazil), Atzuete (Philippines), Rocou, Terre orellana, Orlean, Orange-3, CI Natural Orange 4

Chemical Abstract number: 1393-63-1

EINECS number: 215-735-4 (annatto), 289-561-2 (annatto extract)

Color Index (1982) number: 75120

European Community Identification number: Food color E-160b

Identity (USA): 21 CFR 73.30. Food Chemicals Codex, 4th ed., 1996

Bixin

$C_{25}H_{30}O_4$, 6,6'-diapo-Ψ, Ψ-carotenedioic acid monomethyl ester, or 6'-methylhydrogen 9'-*cis*-6,6' diapocarotene 6,6'-dioate (or *trans*), class: carotenoid, mol. wt. 394.51

New (2000) Chemical Abstracts nomenclature: 2,4,6,8,10,12,14,16,18 eicosanonaene dioic acid; 4,8,13,17 tetramethyl-monomethyl ester.

Chemical Abstract number 6983-79-5 and 39937-23-0 (the trans isomer

EINECS number: 230-248-7

Formerly called α-bixin or "unstable bixin" (the *cis* isomer) and β-bixin or "stable bixin" (the *trans* isomer)

Norbixin

$C_{24}H_{28}O_4$, 6,6'-diapo-Ψ, Ψ-carotenedioic acid, or 9'-*cis*-6,6'-diapocarotene-6,6'-dioic acid (or *trans*-), mol. wt. 380.48, class: carotenoid

New (2000) Chemical Abstracts nomenclature: 2,4,6,8,10,12,14,16,18 eicosanonaene dioic acid; 4,8,13,17 tetramethyl.

Chemical Abstract number 542-40-5 and 626-76-6 (the trans isomer)

Formerly called α-norbixin and β-norbixin (*cis*- and *trans*-isomers, respectively)

REFERENCES

Alvarez-Smith, M. A. 1957. Carotenoid recovery from B. orellana. Brit. Pat. 781,809.

Amakawa, A., Hirata, K., Ogiwara, T., and Ohnishi, K. 1984. Determination of oil-soluble natural dyes in food by HPLC. *Bunseki Kenkyu* 33: 586–590 (*Chem. Abstr.* 102: 22932)

Anderson, S. G., Nair, M. G., Chandra, A., and Morrison, E. 1997. Supercritical fluid carbon dioxide extraction of annatto seeds and quantification of bixin by HPLC. *Phytochem. Anal.* 8: 247–249. (*Chem. Abstr.* 127: 316457)

Aparnathi, K. D., Lata, R., and Sharma, R. S. 1990. Annatto, its cultivation, preparation and usage. *Int. J. Trop. Sci.* 8: 80–86.

Aranez, A. T., and Rubio, R. O. 1996. Genotoxicity of pigments from seeds of *B. orellana* determined by *Allium* test. *Philipp. J. Sci.* 125: 259–269. (*Chem. Abstr.* 126: 313518)

Aranez, A. T., and Bayot, E. 1997. Genotoxicity of pigments from seeds of *B. orellana* determined by lethal test. *Philipp. J. Sci.* 126: 163–173 (*Chem. Abstr.* 128: 305050)

Arce, J. 1983. El achiote, generalidades sobre el cultivo. *Rev. Cent. Agron. Trop. Invest. Enseñanza* (July/Sept): 8–9.

Avila, A. M., Barquero, L., and Calzada J. 1982. An improved method for determination of bixin in annatto seeds. *Ing. Cienc. Quim.* 6: 209–210. (*Chem. Abstr.* 99: 136324)

Bahakar, S. V., and Dubash, P. J. 1973. Methods of extraction of annatto from seeds of *B. orellana*. *Indian J. Dairy Sci.* 36: 157.

Bahl, C. P., Seshadri, T. R., and Vedantham, T. N. C. 1971. Preparation of bixin and methylbixin from Indian seeds of *B. orellana*. *Curr. Sci.* 2: 27–28.

Barber, M. S., Hardisson, A., Jackman L. M., and Weedon, B. C. L. 1961. Studies in nuclear magnetic resonance: Stereochemistry of the bixins. *J. Chem. Soc.* 1961: 1625–1630.

Barnett, H. M., and Espoy H. M., 1957. Extracting coloring matter from annatto seed. U.S. Pat. 2,815,287.

Berset, C., and Marty C. 1986. Potential use of annatto in extrusion cooking. *Lebensm. Wiss. Technol.* 19: 126–131. (*Chem. Abstr.* 105: 132370)

Bertram, J. S., Pung, A., Churley, M., Kappock, T. J., Wilkins, L. R., and

Cooney, R. V. 1991. Diverse carotenoids protect against chemically induced neoplastic transformation. *Carcinogenesis* 12: 671–678.

Boussingault, J. B. 1825. *Justus Liebig Ann. Chem.* 28: 440.

Britton, G. 1995. In *Carotenoids*, Vol.1B. *Spectroscopy*, G. Britton, S. Liaanen-Jensen, and H. Pfander (Eds.). Birkhaeuser, Basel.

Britton, G. 1998. General carotenoid methods. *Methods Enzymol.* 3: 113–149.

Buchta, E., and Andree, F. 1959. Eine Partialsynthese des all-*trans* Methylbixins und des all-*trans* 4,4′ des-dimethyl-methyl-Bixins. *Chem. Ber.* 92: 3111–3116.

Canfield, L. M., and Valenzuela J. G. 1993. Co-oxidations: Significance to carotenoid action in vivo. *Ann. NY Acad. Sci.* 691: 192–199.

Castenmiller, J. J., West, C. E., Linssen, J. P., van het Hof, K. H., and Voragen, A. G. 1999. The food matrix of spinach is a limiting factor in determining the bioavailability of β-carotene and to a lesser extent of lutein in humans. *J. Nutr.* 129: 349–355.

Chao, R. R., Mulvaney, S. J., Sanson, D. R., Hsieh F., and Tempesta, M. S. 1991. Supercritical carbon dioxide extraction of annatto pigments and some characteristics of the color extracts. *J. Food Sci.* 56: 80–83.

Chatani, Y., and Adachi, T. 1988. Determination of norbixin in foods by HPLC after protease digestion. *Kyoto-fu Eisei Kenkyusho Nenpo* 33: 36–40. (*Chem. Abstr.* 110: 191378)

Collins, P. 1992. The role of annatto in food colouring. *Food Ingredients Proc. Int.* (February): 23–27.

Craveiro, A. A., Oliveira, C. L. A., and Araujo F. W. L. 1989. The presence of geranyl-geraniols in *B. orellana*. *Quim. Nova* 12: 297–298. (*Chem. Abstr.* 112: 155274)

DaSilva, G. F., Cavalcanti, S. A., and Sobral, M. C. 1994. Extraction of annatto pigments. *Ann. Asoc. Bras. Quim.* 43: 58–64. (*Chem. Abstr.* 122: 185813)

Degnan, A. J., von Elbe, J. H., and Hartel, R. W. 1991. Extraction of annatto seed by supercritical carbon dioxide. *J. Food Sci.* 56: 1655–1659.

Dendy, D. A. V. 1966. Annatto, the pigment of *B. orellana*. *East Afr. Agric. For. J.* 32: 126–132.

DiMascio, P., Devasagayam, T. P. A., Kaiser, S., and Sies, H. 1990. Carotenoids, tocopherols and thiols as biological singlet molecular oxygen quenchers. *Biochem. Soc. Transact.* 18: 1054–1056.

Dinesen, N. 1974. Annatto. In *Encyclopedia of Food Technology*, A. H. Johnson and M. S. Peterson (Eds.). Avi Publishing Co., Westport, CT.

Dunham, N. W., and Allard, K. R. 1960. A preliminary pharmacologic investigation of the roots of *B. orellana*. *J. Am. Pharmac. Assoc.* 49: 218–219.

El-Sharkawi, S. H., Manaf-Ali, A., and Nashriyah, N. 1995. Supercritical fluid extraction of color from *B. orellana* seeds. *Alexandria J. Pharm. Sci.* 9: 155–158. (*Chem. Abstr.* 123: 337932)

Englert, G. 1995. In *Carotenoids*, Vol. 1B: *Spectroscopy*, G. Britton, S. Liaanen-Jensen, and H. Pfander (Eds.). Birkhaeuser, Basel.

European Economic Communities. 1994. Foodstuffs to which certain permitted colours may be added (Annex III), and Colours permitted for certain uses only: E160b, annatto, bixin, norbixin (Annex IV). *Off. J. Eur. Communities* L237/25.

European Economic Communities. 1995. *Off. J. Euro. Communities* No. L.226/33.

Evans, W. C. 1996. In *Trease and Evans' Pharmacognosy*, 14th ed. Saunders, London.

FAO/WHO. 1976. Specifications for the identity and purity of some food colors. World Health Organization Food Additive Series No. 7, Geneva.

FAO/WHO. 1981. Specifications for identity and purity of food colors. Joint FAO/WHO Expert Committee on Food Additives, Rome.

FAO/WHO. 1996. Specifications for the identity and purity of food colors. Codex Alimentarius Commission, Rome.

Food Chemicals Codex. 1996 (4th ed.), 1992 (3rd suppl. to 3rd ed.). 1986 (2nd suppl. to 3rd ed.). National Academy Press, Washington, DC.

Ford, M. A., and Mellor, C. 1987. Beverage containing ascorbate and carotenoids with good color stability. Brit. UK Pat. Appl. GB 2,190,822. (*Chem. Abstr.* 108: 93381)

Fuhrhop, J. H., Krull, M., Schulz, A., and Moebus, D. 1990. Bolaform amphiphiles with a rigid hydrophobic core in surface monolayers and lipid membranes. *Langmuir* 6: 497–505.

Furr, H. 1997. Chloroform as a solvent for carotenoids and retinoids. *Carotenoid News* 7(1): 7.

Gaertner, C., Stahl, W., and Sies, H. 1997. Lycopene is more bioavailable from tomato paste than from fresh tomatoes. *Am. J. Nutr.* 66: 116–122.

Geminder, J. J., and McDonough, E. E. 1957. Use of coloring ingredients in fatty food products, their physiology, chemistry and stability. *J. Am. Oil Chem. Soc.* 34: 314–318.

Ghorpade, V. M., Deshpande, S. S., and Salunkhe, D. K. 1995. Food colors. In *Food Additive Toxicology*, J. A. Maga and A. T. Tu (Eds.). Marcel Dekker Inc., New York.

Glória, M. B. A., Vale, S. R., and Bobbio, P. A. 1995, Effect of water activity on the stability of bixin in annatto extract-microcrystalline cellulose model system. *Food Chem.* 52: 389–391.

Green, C. L. 1995. Natural colorants and dyestuffs. Food and Agricultural Organization of the United Nations, Rome.

Grollier, J. F., Cotteret, J., and Rosenbaum, G. 1989. Sunscreens containing bixin, benzylidene camphor derivatives and benzophenones. German Offenbahrung DE 3,831,920. (*Chem. Abstr.* 112: 164736)

Guimaraes, I. S., Barbosa, A. L. S., and Massarani, G. 1989. Manufacture of a bixin concentrate in a spouted bed. *Rev. Bras. Eng. Quim.* 12: 22–23. (*Chem. Abstr.* 111: 56034)

Haas, U., and Vinha C. A. 1995. Qualitative and semiquantitative analysis of annatto and its content in food additives by photoacoustic spectrometry. *Analyst* 120: 351–354.

Häberli, A., and Pfander H. 1999. Synthesis of bixin and three minor carotenoids in annatto, *Helv. Chim. Acta* 82: 696–706.

Haila, K. M., Lievonen, S. M., and Heinonen, M. I. 1996. Effects of lutein, lycopene, annatto ad γ-tocopherol on autoxidation of triglycerides, *J. Agric. Food Chem.* 44: 2096–2100.

Haveland–Smith, R. B. 1981. Evaluation of the genotoxicity of natural food colors using bacterial assays. *Mutat. Res.* 91: 285–290 (*Chem. Abstr.* 95: 113545)

Heiser, C. B. 1965. Cultivated plants and cultural diffusion in nuclear America. *Am. Anthropol.* 67: 930–949.

Hettiarachy, N., Moffett, D. J., and Wedral, E. R. 1986. Stabilized natural pigment complexes useful in food and beverage manufacture. *Eur. Pat. Appl.* EP 200,043. (*Chem. Abstr.* 106: 83277)

Hirata, K., Hirokado, M., Uematsu, Y., Nakagima, K., Matsui, K., and Kazama, M. 1989. Analysis of color components in natural color preparations: Determination of bixin and norbixin in annatto extracts. *Kenkyu Nenpo-Tokyo-toritsu Eisei Kenkyusho* 40: 178–182. (*Chem. Abstr.* 112: 213894)

Hirose, S., Yaginuma, N., and Inada, Y. 1972. Energized state of mitochondria as revealed by the spectral change of bound bixin. *Arch. Biochem. Biophys.* 152: 36–43.

Inada, Y., Hirose, S., Yaginuma, N., and Yamashita, K. 1971. Spectral changes of bixin upon interaction with respiring rat liver mitochondria. *Arch. Biochem. Biophys.* 146: 366–367.

Ingram, J. F., and Francis, B. J. 1969. The annatto tree: Ocurrence, cultivation, preparation and uses, *Trop. Sci.* 11: 97–104.

International Life Sciences Institute. 1985. Annatto Literature Search, October 4, 1985.

Iversen, S., and Lamm J. 1953. Ueber den Farbstoff in Annatto Butter Farben. *Z. Lebensmittelunters. Forsch.* 97: 1–7.

Jay, A. J., Steytler, D. C., and Knights, M. 1991. Spectrophotometric studies of food colors in supercritical carbon dioxide. *J. Supercrit. Fluids* 4: 131–141. (*Chem. Abstr.* 115: 206397)

JECFA. 1982. Joint FAO/WHO Committee on Food Additives. Toxicological evaluation of certain food additives. WHO Food Add. Series No. 17.

Jewell, C., and O'Brien, N. M. 1999. *Br. J. Nutr.* 81: 235–242.

Jondiko, I. J. O., and Pattenden, G. 1989. Terpenoids and an apocarotenoid from seeds of *B. orellana*. *Phytochemistry* 28: 3159–3162.

Karrer, P., Helfenstein, A., Widmer, R., and van Itallie, T. B. 1929. Ueber Bixin. *Helv. Chim. Acta* 12: 741–756.

Kelly, D. R., Edwards, A. A., Parkinson, J. A., et al. *J. Chem. Res.* (S): 446–447.

Kocher, R. B. 1958. Edible annatto compositions. U.S. Pat. 2,831,775.

Krinsky, N. J. 1994. Biological properties of carotenoids. *Pure Appl. Chem.* 66: 1003–1010.

Kuramoto, Y., Yamada, K., Tsuruta, O., and Sugano, M. 1996. Effect of natural food colorings on immunoglobulin production in vitro by rat spleen lymphocytes. *Biosci. Biotechnol. Biochem.* 60: 1712–1713.

Lancaster, F. E., and Lawrence, J. F. 1995. Determination of annatto in high-fat dairy products, margarine and hard candy by solvent extraction followed by HPLC. *Food Addit. Contam.* 12: 9–19.

Lancaster, F. E., and Lawrence, J. F. 1996. HPLC separation of carminic acid, α- and β-bixin, α- and β-norbixin, and determination of carminic acid in foods. *J. Chromatogr. A.* 732: 394–398.

Lauro, G. J. 1991. A primer on natural colors. *Cereal Foods World* 36: 949–953.

Lauro, G. J. 1999. Natural colorants for surimi food. In *Surimi and Surimi Seafoods*, Jae Park (Ed.). Marcel Dekker, New York.

Lawrence, B. M., and Hogg, J. W. 1973. Ishwarane in *B. orellana* leaf oil. *Phytochemistry* 12: 2995.

Levy, L. W., Regalado, E., Navarrete, S., and Watkins, R. H. 1997. Bixin and norbixin in human plasma: Determination and study of absorption of a single dose of annatto food color. *Analyst* 122: 977–980.

Lölinger, J. 1997. Natural antioxidants for food and health, presentation at Workshop on Natural Antioxidants in Foods, project FAIR-CT 95–0158, Wageningen, Holland.

Luf, W., and Brandl, E. 1988. Zum Nachweis des Annattofarbstoffes Norbixin/Bixin in Käse unter Anwendung der Derivatspektroskopie sowie der HPLC. *Z. Lebensmittelunters. Forsch.* 186: 327–332.

Maeda, H., Kasuga, Y., and Maeda, Y. 1988. Bixin or norbixin as quality improvers for bread dough containing them, and manufacture of bread. Japan Kokai Tokkyo Koho JP-6359831 (88–59831). (*Chem. Abstr.* 110: 211328)

Marcus, F. K. 1956. Bixin recovery from annatto seed. German Pat. 950,165.

Marcus, F. K. 1963. Fabrication of oil and water-soluble coloring from annatto seeds used for coloring of margarine and cheese. *Germ. Pat.* 1,156,529. (*Chem. Abstr.* 60: 2260)

Massarani, G., Passos, M. L., and Barreto, D. W. 1992. Production of annatto concentrates in spouted beds. *Can. J. Chem. Eng.* 70: 954–959.

McKeown, G. G. 1961. Paper chromatography of bixin and related compounds. *J. Assoc. Offic. Analyt. Chem.* 44: 347–351.

McKeown, G. G., and Mark, E. 1962. The composition of oil-soluble annatto food colors. *J. Assoc. Offic. Analyt. Chem.* 45: 761–766.

McKeown, G. G. 1963. Composition of oil-soluble annatto food colors: Thermal degradation of bixin. *J. Assoc. Offic. Analyt. Chem.* 46: 790–796.

McKeown, G. G. 1965. Composition of oil-soluble annatto food colors: Structure of the yellow pigment formed by thermal degradation of bixin. *J. Assoc. Offic. Analyt. Chem.* 48: 835–837.

Mercadante, A. Z., Steck, A., Rodriguez-Amaya, D., Pfander, H., and Britton, G. 1996. Isolation of methyl 9'(Z)-apo-6'-lycopenoate from *B. orellana. Phytochemistry* 41: 1201–1203.

Mercadante, A. Z., Steck, A., and Pfander, H. 1997a. Isolation and identification of new apocarotenoids from annatto seeds. *J. Agric. Food Chem.* 45: 1050–1054.

Mercadante, A. Z., Steck, A., and Pfander, H. 1997b. Isolation and structure elucidation of minor carotenoids from annatto seeds. *Phytochemistry* 46: 1379–1383.

Mercadante, A. Z., Steck, A., and Pfander H. 1999. Three minor carotenoids from annatto (*Bixa orellana*). *Phytochemistry.* 52: 135–139.

Mikkelsen, H., Larsen, J. C., and Tarding, F. 1977. Hypersensitivity reactions to food colors with special reference to the natural color annatto extract (butter color). 19th Meeting of the European Society of Toxicology, Copenhagen, June 1977.

Minguez-Mosquera, M. I., Nornero-Mendez, D., and Garrido-Fernandez, J. 1995. Detection of bixin, lycopene, canthaxanthin and β-apo-8'-carotenal in products derived from red pepper. *J. Assoc. Offic. Analyt. Chem.* 78: 491.

Ministry of Agriculture, Fisheries and Food (UK). 1993. Dietary intake of food additives in the UK. Food Surveillance Paper 37 (*Chem. Abstr.* 120: 162012).

Morrison, E. Y., Thompson, H., Pascoe, K., West, M., and Fletcher, C. 1991. Extraction of an hyperglycaemic principle from the annatto, a medicinal plant in the West Indies. *Trop. Geogr. Med.* 43: 184–188.

Morton, J. 1989. Information within the book review on Potter's New Cyclopedia of Botanical Drugs and Preparations. *Econ. Bot.* 43: 280, 281.

Murthi, T. N., Devdhara, V. D., Punjrath, J. S., and Aneja, R. P. 1989. Extraction of annatto colors from seeds of *B. orellana* using edible oils. *Indian J. Dairy Sci.* 42: 750–756.

Najar, S. V., Bobbio, F. O., and Bobbio, P. A. 1988. Effects of light, air, antioxidants and pro-oxidants on annatto extracts. *Food Chem.* 29: 283–289.

Nakajima, K., and Moretome, N. 1999. Jpn. Pat. 11: 209,264.

Nish, W. A., Whisman, D. A., Goetz, D. W., and Ramirez, D. A. 1991. Anaphylaxis to annatto dye, a case report. *Ann. Allergy* 66: 121–131.

Nishizawa, M., Chonan, T., Sekijo, I., and Sugii, T. 1983. Studies on the analysis of natural dyes: Analysis of annatto extract and gardenia yellow dye in foods and natural dye preparations. *Hokkaidoritsu Eisei Kenkyushoho* 33: 32–34. (*Chem. Abstr.* 100: 207944)

Oliveira, L. F. C., Dantas, S. O., Velozo, E. S., Santos, P. S., and Ribeiro, M. C. C. 1997. Resonance Raman investigation and semi-empyrical calculation of the natural carotenoids bixin. *J. Mol. Struct.* 435: 101–107.

Ono, T. 1988. Prevention of carotenoid discoloration of food, pharmaceutical, cosmetic, and textile preparations by flavonoids. Japan Kokai Tokkyo Koho JP-62,243,655 (87,243,655). (*Chem. Abstr.* 108: 185517)

Parish, M. 1994. *The Milwaukee Journal* (October 23): 1–4.

Park, K. J., Prado-Cornejo, F. E., Nogeira, R. I., De Catro-Villaca, A., and Gama-Alves, I. T. 1990. Production of powdered norbixin by alkaline extraction and spray drying of *B. orellana* seeds. Braz. Pat. Appl. 89–5035. (*Chem. Abstr.* 115: 31115)

Passos, M. L., Oliveira, L. S., Franca, A. S., and Massarani, G.1998. Bixin powder production in conical spoouted bed units. *Drying Technol.* 16: 1855–1879.

Patiño, V. M. 1967. *Plantas Cultivadas en America Equinoccial*, Vol. 3. Imprenta Departamental Cali, Colombia.

Pattenden, G., Way, J. E., and Weedon, B. C. L. 1970. Carotenoids and related compounds: Synthesis of methyl natural bixin. *J. Chem. Soc.(C)* 1970: 235–241.

Perret, M. A. 1958. Food color and method of preparing same. *U.S. Pat.* 2,830,908.

Prentice-Hernandez, C., Rusig, O., and Nogueira-Carvalho, P. R. 1993. Influence of heating time on the thermal degradation of bixin in alkaline extracts of annatto. *Arq. Biol. Tecnol.* 36: 819–820. (*Chem. Abstr.* 121: 81328)

Preston, H. D., and Rickard, M. D. 1980. Extraction and chemistry of annatto. *Food Chem.* 5: 47–56.

Ramamurty, M. K., and Bhalerao, V. R. 1964. TLC method for identifying annatto and other food colors. *Analyst* 89: 740–744.

Rath, S. P., Sriniwasulu, C., and Mahapatra, S. N. 1990. *J. Indian Chem. Soc.* 67: 86.

Reith J. F., and Gielen, J. W. 1971. Properties of bixin and norbixin and the composition of annatto extracts. *J. Food Sci.* 36: 861–864.

Rouseff, R. L. 1988. HPLC separation and spectral characterization of the pigments in turmeric and annatto. *J. Food Sci.* 53: 1823–1826.

San-Ei Chemical Industries. 1975. Bixin suspensions as food colorants. *Chem. Abstr.* 83: 176993, 176994.

San-Ei Chemical Industries. 1983. Acid-stable annatto food coloring materials. Japan Kokai Tokkyo Koho JP-58-91768 (83–91768). (*Chem. Abstr.* 100: 50247)

Sato, T., and Suzuki, H. 1966. Coloring of Vienna sausage with water-soluble annatto. *Nippon Shokuhin Kogyo Gakkaishi* 13: 488–491. (*Chem. Abstr.* 66: 104045)

Schiedt, K., and Liaaen-Jensen, S. 1995. In *Carotenoids: Isolation and Analysis*, G. Britton, S. Liaaen-Jensen, and H. Pfander (Eds.), p. 83. Birkhaeuser, Basel.

Schmidt, T. R. 1985. Acid-soluble annatto colorant in powdered form. U.S. Pat. 4,548,822.

Schneider, W. P., Caron, E. L., and Hinman, J. W. 1965. *J. Org. Chem.* 30(8), 2856–2857.

Scott, K. J. 1996. Lycopene degradation in chloroform. *Carotenoid News* 6(2): 6.

Scotter, M. J. 1994. Characterization of the principal coloring components of annatto using HPLC with photodiode-array detection. *Food Addit. Contam.* 11: 301–305.

Scotter, M. J. 1995. Characterization of the colored thermal degradation products of bixin from annatto and a revised mechanism for their formation. *Food Chem.* 53: 177–185.

Scotter, M. J., Wilson, L. A., Appleton, G. P., and Castle, L. 1998. Analysis of annatto food coloring formulations: Determination of coloring components and colored thermal degradation products by HPLC with photodiode array detection. *J. Agric. Food Chem.* 46: 1031–1038.

Scotter, M. J., Wilson, L. A., Appleton, G. P., and Castle L. 2000. *J. Agric. Food Chem.* 48: 484–488.

Shumaker, E. K., and Wendorff, W. L. 1998. Factors affecting pink discoloration in annatto-colored pasteurized process cheese. *J. Food Sci.* 63: 828–831.

Smith P. R., Blake, C. J., and Porter, D. C. 1983. Determination of added natural colors in foods: Annatto. Leatherhead Food R. A., Research Report 431.

Srinivasulu, C., and Mahapatra, S. N. 1989. A process for isolation of bixin. *Res. India* 34: 137–138. (*Chem. Abstr.* 112: 6277)

Tadamasa, H. 1974. Bixin. Jpn. Pat. 7,489,732. (*Chem. Abstr.* 82: 110525)

Tadamasa, H., and Yasuda, A. 1985. Bixin coloring matter. Jpn. Pat. 60184566. (*Chem. Abstr.* 104: 70280)

Terashima, S., Shimizu, M., Horie, S., and Morita, N. 1991. Studies on aldose reductase inhibitors from natural products: Constituents and aldose reductase-inhibitory effect of *C. morifolium, B. orellana* and *I. batatas. Chem. Pharm. Bull.* 39: 3346–3347.

Thresiamma, K. C., Mathews, J. P., and Kuttan, R. 1995. Protective effect of curcumin, ellagic acid and bixin on radiation induced lipid peroxidation. *J. Exp. Clin. Cancer Res.* 14: 427–430.

Thresiamma, K. C., Josely G., and Kuttan, R. 1998. Protective effect of curcumin, ellagic acid and bixin on radiation induced toxicity. *J. Exp. Clin. Cancer Res.* 17: 431–434.

Tirimana, A. S. L. 1981. *Mikrochim. Acta* 2(1/2), 11–16.

Todd, P. H. 1964. Vegetable base food coloring for oleomargarine and the like. *U.S. Pat.* 3,162,538.

Todd, P. H. 1991. Norbixin adducts with water-soluble or water-dispersable proteins or branched-chain or cyclic polysaccharides. U.S. Pat. Appl. 07,426,578.

Truscott, T. G., Edge, R., and McGarvey, D. J. 1996, Carotene: Pro- and antioxidant reaction mechanisms and interactions with vitamins E and C, 11th International Symposium on Carotenoids, Leiden, Holland, August 1996.

U.K. Food and Drink Federation. 1992. Internal Memorandum on Annatto, London.

Unilever Ltd. 1981. Natural food color for margarine. Belgian Pat. (*Chem. Abstr.* 95: 148927)

U.S. Code of Federal Regulations. 1994. 21CFR73.30. Annatto extract.

Van-Esch, G. J., Van-Genderen, H., and Vink, H. H. 1959. Ueber die chronische Verträglichkeit von Annatto Farbstoff. *Z. Lebensmittelunters. Forsch.* 111: 93–108.

Verger, P., Chambolle, M., Babayou, S., and Volatier, J. L. 1998. Estimation of the distribution of the maximum theoretical intake for ten additives in France. *Food Addit. Contam.* 15: 759–766.

Weigert, P., Gilbert, J., Patey A. L., Key, P. E., Wood, R., and Barylko-Pikielna,

N. 1997. Analytical quality assurance for the WHO GEMS/Food-EURO programme: Results of 1993/1994 laboratory proficiency testing. *Food Addit. Contam.* 14: 399.

Winning, M., and Isager, P. P. 1997. Water-dispersable compositions containing natural hydrophylic, water-insoluble pigments, methods of preparing same, and their use. PCT Int. Appl. WO-97-26803. (*Chem. Abstr.* 127: 148578)

Wood, A., Baker, D. M., Coppen, J., and Green, C. L. 1991. Bixinoid assay in annatto seed and its extracts, lecture at First International Conference on Annatto, Campinas SP, Brazil.

Yamada, H. 1988. Stabilized active vitamin D-containing pharmaceuticals. Japan Kokai Tokkyo Koho JP-63,165,322 (88,165,322). (*Chem. Abstr.* 109: 216050)

Yamada, S. 1997. Group analysis of natural coloring matters in food products. *Foods Food Ingred. J. Japan* 172: 37–42. (*Chem. Abstr.* 127: 4302)

Zbinden, G., and Studer, A. 1958. Tierexperimentelle Untersuchung über chronische Verträglichkeit von β-Carotin und Bixin. *Z. Lebensmittelunters. Forsch.* 108: 113–134.

Zechmeister, L., and Escue, R. B. 1944. A stereochemical study of methylbixin. *J. Am. Chem. Soc.* 66: 322–330.

Zhang, L. X., Cooney, R. V., and Bertram J. S. 1991. Carotenoids enhance gap junctional communication and inhibit lipid peroxidation in C3H-10-T1-2 cells: Relationship to their cancer chemopreventive action. *Carcinogenesis* 12: 2109–2114.

Zhang, L. X., Cooney, R. V., and Bertram, J. S. 1992. Carotenoids upregulate Connexin-43 gene expression independent of their provitamin A or antioxidant properties. *Cancer Res.* 52: 5707–5712.

7
Lycopene

Minhthy L. Nguyen and Steven J. Schwartz
The Ohio State University
Columbus, Ohio

INTRODUCTION

Lycopene, the carotenoid present at high concentrations in tomatoes and tomato products, has attracted considerable attention recently as epidemiological evidence continues to suggest that it may provide protection against cancer and other degenerative diseases influenced by free radical reactions (Bendich, 1989; Block et al., 1992; Gey, 1993; Giovannucci et al., 1995). For example, a recent comprehensive review of 72 independent epidemiological studies revealed that intake of tomatoes and blood lycopene level are inversely associated with the risk of developing cancers at several anatomical sites including the prostate gland, stomach, and lung. Data were also suggestive of the inverse relationship between lycopene intake and the risk of breast, cervical, colorectal, esophageal, oral cavity, and pancreatic cancers (Giovannucci, 1999). These observational findings have been substantiated by studies with laboratory data from human cell cultures and animals that show lycopene to be very effective in trapping free radicals and quenching singlet oxygen–protecting oxidizable substrates from harmful degradative reactions (Wang et al., 1989; Countryman et al., 1991;

Conn et al., 1991; Nagasawa et al., 1995; Matsushima-Nishiwaki et al., 1995; Kim, 1995; Stahl and Sies, 1996).

This review will address key structural and functional characteristics of lycopene along with factors affecting its stability, chemistry, and biological functionalities. Special emphasis will be given to key aspects of the natural occurrence, biosynthesis, bioavailability, and physiological distribution of lycopene to provide the reader with a better perspective on the increasing interest in this compound and its potential role in human health.

BIOSYNTHESIS OF LYCOPENE AND OTHER TOMATO CAROTENOIDS

Lycopene belongs to the carotenoid family of pigments, a group of over 600 lipid-soluble phytochemicals, commonly known for the brilliant yellow, orange, and red colors they impart to plants and animals (Straub, 1987; Kull and Pfander, 1995). Carotenoids are found in both photosynthetic cells, being complexed with the chlorophylls, as well as nonphotosynthetic tissues. It has been suggested that carotenoids protect plants and algae against photosensitization and act as accessory pigments in photosynthesis (Moore et al., 1989; Bendich and Olson, 1989; Krinsky, 1998).

The biosynthesis of carotenoids, including that of lycopene, is extensively discussed in a number of books, symposia proceedings, and review articles (Goodwin, 1952, 1980; Hulme, 1970/1971; Czygan, 1980; Britton, 1983; Gross, 1991). Carotene biosynthesis in higher plants was reviewed by Jones and Porter (1986), while in vitro carotenoid biosynthesis was discussed by Bramley (1985). The overall characteristics of carotenogenesis are similar in higher plants, algae, fungi, and photosynthetic bacteria. This section discusses lycopene synthesis in higher plants using the tomato fruit system as a model.

Carotenogenesis consists primarily of the following stages (Fig. 7.1):

1. Formation of mevalonic acid.
2. Formation of isoprene precursor, isopentenyl pyrophosphate
3. Formation of geranylgeranyl pyrophosphate
4. Formation of phytoene
5. Desaturation of phytoene
6. Formation of lycopene
7. Cyclization
8. Formation of xanthophylls

Lycopene synthesis involves the first six steps. As is the case with all carotenoids, biosynthesis of lycopene begins with the reduction of β-hydroxy-β-

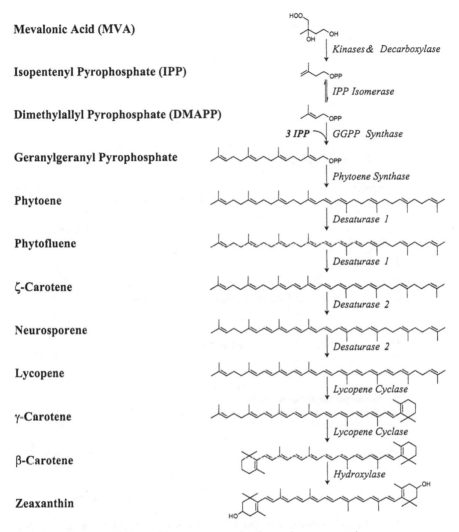

FIG. 7.1   Biosynthesis of lycopene and selected carotenoids.

methyl glutaryl-CoA derived from the condensation reaction between three molecules of acetyl-CoA. Isopentenyl pyrophosphate (IPP) serves as the biological isoprene precursor of carotenoids and terpenoids. It is derived from mevalonic acid (MVA). MVA is phosphorylated by kinases and ATP into mevalonic acid 5-pyrophospate and 5-pyrophosphate successively. MVA 5-pyrophosphate undergoes decarboxylation, yielding the isoprene unit IPP. The enzyme prenyl transferase then catalyzes the chain elongation from C_5 isopentenyl pyrophosphate (IPP) to C_{20} geranylgeranyl pyrophosphate (GGPP) (Gross, 1991).

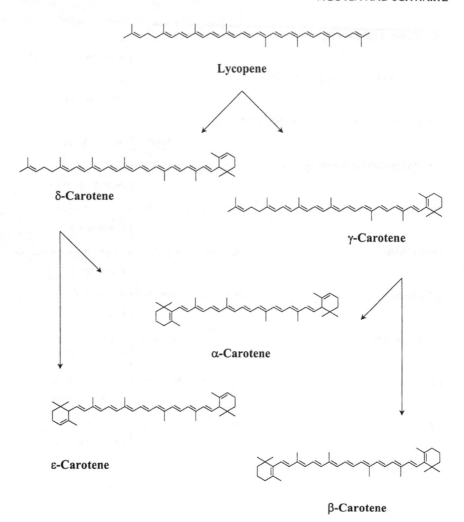

Lycopene

δ-Carotene

γ-Carotene

α-Carotene

ε-Carotene

β-Carotene

FIG. 7.2 General scheme for the biosynthesis of common β- and ε-ring cyclic carotenes. (Adapted from Britton, 1983; Gross, 1991.)

Two molecules of GGPP subsequently give rise to the first C_{40} compound in the biosynthetic pathway, phytoene, via a condensation reaction. Phytoene is eventually converted through a stepwise desaturation process in which the C_{40} chromophore is extended by two double bonds alternatively from both sides of the molecule in each step. This dehydrogenation process generates phytofluene, ζ-carotene, neurosporene, and finally lycopene, respectively. Ultimately, lycopene serves as the basic substrate for the formation of most cyclic carotenes via the formation of either β-ring or ε-ring at the terminal ends (Fig. 7.2). Following cyclization, the insertion of oxygen functional groups, e.g., hydroxy and epoxy, produces xanthophylls such as lutein and zeaxanthin. Hydroxylation occurs most often at the C-3 position, although other positions on the ring or within the aliphatic chain can also be hydroxylated and modified by other oxidation reactions (Gross, 1991).

Tomato Varieties with Unique Carotenoid Profile

While tomatoes represent the most common source of lycopene, some varieties of tomatoes have the capacity to produce other carotenoids that are either unique or in altered relative abandunce relative to the distributions in typical varieties. Mutant varieties such as Golden jubilee and Tangella biosynthesize several *cis* isomers of lycopene, which are not commonly found in other varieties (Zechmeister et al., 1941; Raymundo and Simpson, 1972; Glass and Simpson, 1976). In these varieties, the predominant carotenoid is prolycopene, a tetra-*cis* isomer of lycopene. Fig. 7.3 illustrates the biosynthetic pathway of prolycopene from 15-*cis*-phytoene.

Compared to all-*trans*-lycopene, prolycopene has two hindered *cis* double bonds (Clough and Pattenden, 1979; Englert et al., 1979) and a shortened chromophore, imparting the orange color to these tomato fruits that provides the basis for the common name, tangerine tomatoes. Zechmeister et al. (1943) reported spectroscopic characteristics of prolycopene to have an absorption maximum at 435 nm in hexane and a spectrum void of fine structure in contrast to those observed for lycopene and its mono-*cis* isomers (Fig. 7.4). In addition to having the absorption maximum shifted 35 nm upfield in wavelength compared to that for all-*trans*-lycopene at 470 nm, prolycopene also has a much lower molar absorptivity of 102,900 AU compared to 184,000 AU for its all-*trans* counterpart. Prolycopene also has been reported to be present in roots of *Brassica rutabaga* and *Brassica napus* (Joyce, 1954, 1959).

Several other minor carotenoids have also been isolated and identified in normal and mutant varieties of tomatoes. Zechmeister and Cholnoky (1936) reported the presence of the monohydroxy and dihydroxy derivatives of lycopene, including lycoxanthin (Ψ, Ψ-carotene-16-ol) and lycophyll (Ψ, Ψ-carotene-16,16'-diol). Both of these lycopene xanthophylls have been identified by other reseachers (Went et al., 1942; Curl, 1961, Ben-Aziz et al., 1973) in typical tomatoes, in Marzano tomatoes (Edwards and Reuter, 1967), and in cherry tomatoes (Laval-Martin et al., 1975). The chemical structures for these carotenoids are shown in Fig. 7.5 as reported by Markham and Liaaen-Jensen (1968) in addition to a series of epoxides of lycopene and its more saturated precursors, phytoene, phytofluene, and ζ-carotene (Britton and Goodwin, 1969; Ben-Aziz et al., 1973). The majority of the epoxides are 1,2-epoxy derivatives. However, the 5,6-epoxide of lycopene has also been detected along with apoderivatives of lycopene, namely apo-6'-lycopeneal and apo-8'-lycopenal (Khachik et al., 1998a,b, 1995).

All E-Phytoene

15Z-Phytoene

15Z, 9'Z-Phytofluene

9Z, 9'Z-ζ-Carotene

9Z, 7'Z, 9'Z-Neurosporene

7Z, 9Z, 7'Z, 9'Z-Lycopene

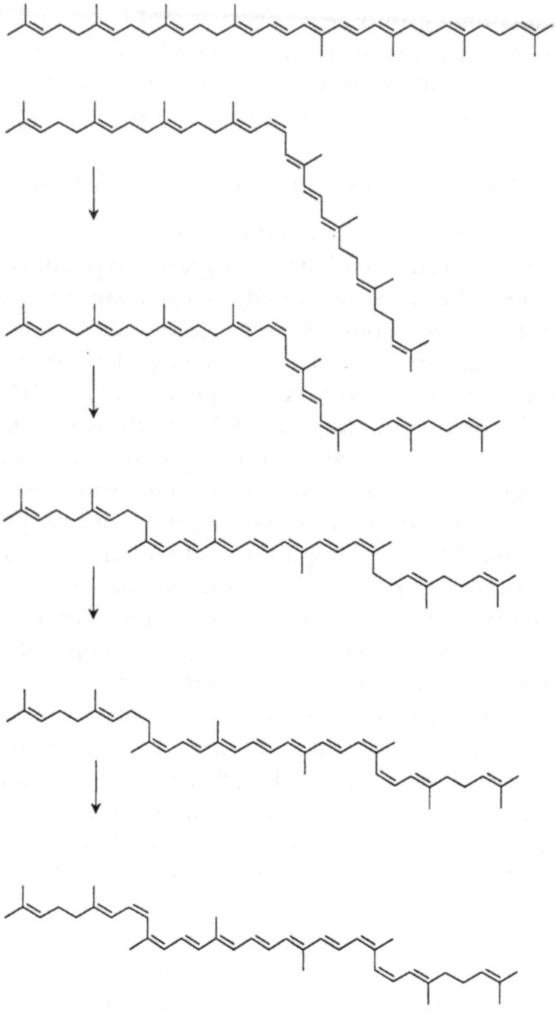

FIG. 7.3 Formation of prolycopene via the conversion of phytoene in the tangerine mutant of *Lycopersicum esculentum*. (Adapted from Britton, 1983; Goodwin, 1993.)

Genetics of Carotenoid Biosynthesis in Tomatoes

The color of the tomato fruit is dictated by Mendelian inheritance. For example, yellow tomato varieties differ from their red counterparts by a single recessive gene, r (Kirk and Tilney-Bassett, 1967; Giuliano et al., 1993). Red tomatoes have alleles of $r+/r+$ or $r+/r$ compared to the homozygous recessive alleles r/r in yellow tomatoes. As a result of this difference, yellow tomatoes lack phytoene and lycopene and are pale in color. Furthermore,

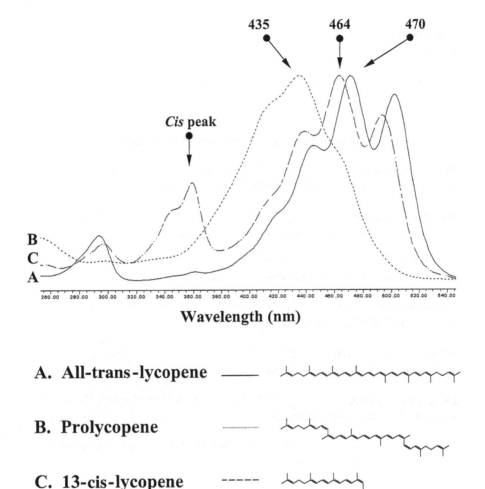

FIG. 7.4 Ultraviolet-visible spectrum of lycopene and selected geometrical isomers.

changes in a single gene in certain mutant varieties of tomatoes can dramatically alter the carotenoid composition and likewise the color of these fruit. For example, the expression of the dominant gene, *Del*, yields an increase in the activity of the enzyme lycopene epsilon-cyclase, which converts lycopene to δ-carotene. The accumulation of δ-carotene at the expense of

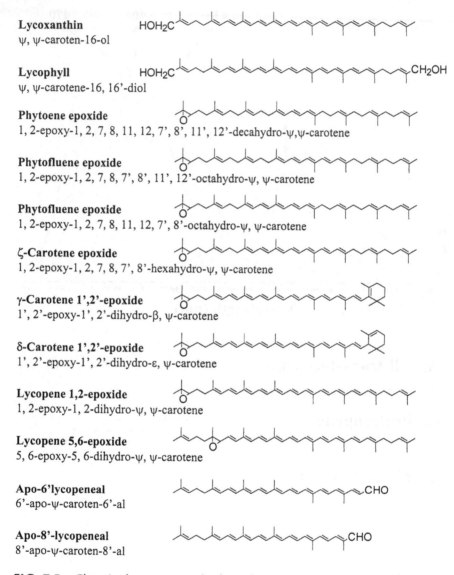

Lycoxanthin
ψ, ψ-caroten-16-ol

Lycophyll
ψ, ψ-carotene-16, 16'-diol

Phytoene epoxide
1, 2-epoxy-1, 2, 7, 8, 11, 12, 7', 8', 11', 12'-decahydro-ψ,ψ-carotene

Phytofluene epoxide
1, 2-epoxy-1, 2, 7, 8, 7', 8', 11', 12'-octahydro-ψ, ψ-carotene

Phytofluene epoxide
1, 2-epoxy-1, 2, 7, 8, 11, 12, 7', 8'-octahydro-ψ, ψ-carotene

ζ-Carotene epoxide
1, 2-epoxy-1, 2, 7, 8, 7', 8'-hexahydro-ψ, ψ-carotene

γ-Carotene 1',2'-epoxide
1', 2'-epoxy-1', 2'-dihydro-β, ψ-carotene

δ-Carotene 1',2'-epoxide
1', 2'-epoxy-1', 2'-dihydro-ε, ψ-carotene

Lycopene 1,2-epoxide
1, 2-epoxy-1, 2-dihydro-ψ, ψ-carotene

Lycopene 5,6-epoxide
5, 6-epoxy-5, 6-dihydro-ψ, ψ-carotene

Apo-6'lycopeneal
6'-apo-ψ-caroten-6'-al

Apo-8'-lycopeneal
8'-apo-ψ-caroten-8'-al

FIG. 7.5 Chemical structures of selected minor tomato carotenoids.

lycopene makes the former the predominant carotenoid and changes the fruit color from red to orange (Tomes, 1967, 1969; Ronen et al., 1999). Similarly, the *B* allele produces high-β-carotene tomato fruit that also is orange. The tangerine-type tomato, as discussed earlier, is rich in prolycopene, proneurosporene, and ζ-carotene, being homozygous fruits with *t/t* (Khudairi, 1972). Fig. 7.6 depicts a generalized scheme of the genetics of tomato carotenoid biosynthesis.

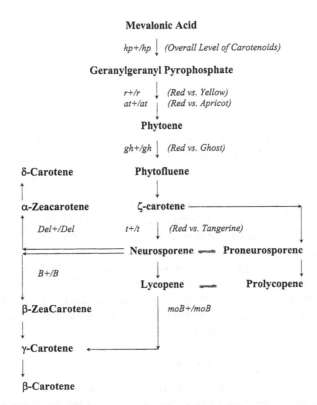

FIG. 7.6 Genetics of carotenoid biosynthesis in tomatoes. (Adapted from Khudairi, 1972.)

CHEMICAL PROPERTIES OF LYCOPENE

Natural Occurrence of Lycopene

Lycopene, found primarily in tomatoes and tomato products, is the most prevalent of the more than 50 dietary carotenoids in the western diet and the most abundant in human serum. It was first isolated by Hartsen (1873) from *Tamus communis* L. berries as a deep red crystalline pigment. Millardet (1875) obtained a crude mixture containing lycopene from tomatoes, referring to it as "solanorubin." Duggar (1913) referred to lycopene as "lycopersicon" in his work detailing the effects of growth conditions on its development. Schunck (1903) gave lycopene its name after showing that this pigment from tomato had a different absorption spectrum than carotenes from carrots.

In the common variety of tomatoes, *Lycopersicon esculentum*, lycopene is found predominantly in the all-*trans* configuration (Zechmeister et al., 1941) at concentrations of 3.1–7.7 mg/100 g of ripe fruit. Lycopene levels as high as 40 mg/100 g of tissue have been reported in certain varieties, such as *Lycopersicon pimpinellifolium,* accounting for 95–100% of the total carotenoid content of these tomatoes (Porter and Lincoln, 1950). Recent advances in isolation and chromatographic separation methodologies have shown that lycopene is much more widely distributed in nature than once thought (Nguyen and Schwartz, 1999a). Table 7.1 lists botanical and food sources of lycopene.

Lycopene is found predominantly in the chromoplast of plant tissues. In tomatoes, lycopene biosynthesis increases dramatically during the ripening process as chloroplasts undergo transformation to chromoplasts (Kirk and Tilney-Basset, 1978). Laval-Martin (1974) categorized tomato chromoplasts into two types. Globulous chromoplasts containing mainly β-carotene are found in the jelly part of the pericarp, while chromoplasts in the outer part of the pericarp contain voluminous sheets of lycopene. The development and ultrastructure of these sheets of lycopene were studied by Ben-Shaul and Naftali (1969) and named crystalloids. Mohr (1979) noted that the development of the pigment bodies are similar and follow the same sequence of granal membrane loss, increase in globule size and density, and deposition of crystal bodies along the extended thylakoid system in both normal red and high-lycopene varieties.

General Chemistry of Lycopene

Detailed information on the chemistry of lycopene is fairly complete, dating back to the beginning of this century. The molecular formula of lycopene, $C_{40}H_{56}$, was first determined when Willstatter and Escher (1910) presented their work showing that lycopene is an isomer of the carotenes. Karrer et al., (1930) published the chemical structure of lycopene, which was subsequently confirmed by Kuhn and Grundmann (1932) by identifying its degradation products after oxidation by chromic acid.

Zechmeister and coworkers made significant progress toward the isolation of lycopene, determination of spectrophotometric properties via iodine-catalyzed stereomutation, and establishment of the foundation for a better understanding of lycopene's chemical stability in terms of isomerization and oxidation (Zechmeister and Cholnoky, 1936; Zechmeister and Tuzson, 1938a,b; Zechmeister et al., 1941, 1943; Zechmeister and Polgar, 1944; Zechmeister, 1962). Many of these basic techniques and the fundamental considerations for the isolation, handling, and separation of lycopene are still applicable today.

The general structure of lycopene is an aliphatic hydrocarbon with 11 conjugated carbon–carbon double bonds, making it soluble in fats and lipids and red in color. Being acyclic, lycopene possesses symmetrical planarity and has no vitamin A activity. As a highly conjugated polyene, it is particularly susceptible to oxidative degradation. Physical and chemical factors

TABLE 7.1   Plant Sources of Lycopene

Common name	Type	Species
1. Aglaonema aroid	Fruit	*Aglaonema commutatum*
2. Apricot	Fruit	*Prunus armeniaca*
3. Arbutus	Leaf	*Arbutus unedo*
4. Berry	Fruit	*Tamus communis*
5. Bitter melon	Fruit	*Momordica charantia*
6. Bitter nightshade	Fruit	*Solanum dulcamara*
7. Calendula	Flower	*Calendula officinalis*
8. Carrot	Root	*Daucus carota*
9. Citrus	Fruit	*Citrus* spp.
10. Cuckoo pint	Tuber	*Arum maculatum*
11. Cowberry	Fruit	*Vaccinium vitis-iddaea* var. *minus*
12. Cloudberry	Fruit	*Rubus chamaemorus*
13. Cranberry	Fruit	*Vaccinium vitis*
14. Damask rose	Fruit	*Rosa damascena*
15. Date palm	Fruit	*Phoenix dactylifera*
16. Eggplant	Fruit	*Solanum melongena*
17. European nettle	Plant	*Urtica dioica*
18. Gazania	Plant	*Gazania rigens*
19. Grape	Fruit	*Vitis vinifera*
20. Grapefruit	Fruit	*Citrus paradisi*
21. Guava	Fruit	*Psidium guajava*
22. Mango	Fruit	*Mangifera indica*
23. Palm	Oil	*Elaeis* spp.
24. Papaya	Fruit	*Carica papaya*
25. Peach	Fruit	*Prunus persica*
26. Pepper	Fruit	*Capsicum annuum*
27. Pepper berry	Fruit	*Piper nigrum*
28. Persimmon	Fruit	*Diospyros kaki*
29. Plum	Fruit	*Prunus domestica*
30. Pumpkin	Fruit	*Cucurbita pepo*
31. Ramanas rose	Plant	*Rosa rugosa*
32. Red bryony	Plant	*Bryonia dioica*
33. Rosa mosqueta	Plant	*Rosa rubiginosa*
34. Rose	Fruit	*Rosa canica*
35. Rutabaga	Root	*Brassica napus* var. *napobrassica*
36. Saffron	Seed	*Crocus sativus*
37. Sallow thorn	Fruit	*Hippophae rhamnoides*
38. Tea	Leaf	*Camellia sinensis*
39. Tomato	Fruit	*Lycopersicon esculentum*
40. Turnip	Root	*Brassica rapa* var. *rapa*

TABLE 7.1 *(continued)*

Common name	Type	Species
41. Watermelon	Fruit	*Citrullus lanatus*
42. Yew	Fruit	*Taxus baccata*

Source: Joyce, 1954, 1959; List and Horhammer, 1979; Harborne and Baxter, 1983; Godoy and Rodriguez-Amaya, 1987; Variyar and Bandyopadhyay, 1990; Gross, 1991; Duke, 1992; Mangels et al., 1993; Wilberg and Rodriguez-Amaya, 1995; Choo et al., 1996; Cano et al., 1996; Muller, 1997; Duke and Beckstrom-Sternberg, 1998; Belakbir et al., 1998.

known to degrade other carotenoids, including elevated temperature, exposure to light, oxygen, extremes in pH, and active surfaces (Davies, 1976; Moss and Weedon, 1976; Scita, 1992; Crouzet and Kanasawud, 1992), apply to lycopene as well. Additionally, elevated temperature increases the rate of degradative reactions (Henry et al., 1998).

Another type of change that lycopene undergoes readily as a polyene is *cis-trans* isomerization. With very few exceptions, lycopene from natural plant sources exists predominantly in the all-*trans* configuration, the most thermodynamically stable form (Zechmeister et al., 1941; Wilberg and Rodriguez-Amaya, 1995; Emenhiser et al., 1995). As a result of the 11 conjugated carbon–carbon double bonds in its backbone, lycopene can theoretically assume 2^{11} or 2048 geometrical configurations. While a large number of geometrical isomers are theoretically possible for all-*trans* lycopene, according to Pauling (1939) and Zechmeister et al. (1941), only certain ethylenic groups of a lycopene molecule can participate in *cis-trans* isomerization because of steric hindrance. In fact, only about 72 lycopene *cis* isomers are structurally favorable (Zechmeister, 1962).

The interconversion of lycopene geometrical configuration can be induced in vitro using reagents such as iodine, which will release the π-bond of carbon atoms involved in double bonds. The free rotation about the axis of the remaining single σ-bond between these carbons, carrying substituted groups, and subsequent reformation of the double bond results in a change in configurational geometry (Zechmeister and Tuxson, 1939). Fig. 7.7 illustrates the structural distinctions of the predominant lycopene geometrical isomers. As has been discussed in earlier section, by applying rules of designation similar to those involving *R* and *S* chirality, the stereochemical prefixes *E* and *Z* are used whenever the corresponding *trans* and *cis* designations are not sufficient to denote the priority of the two substituents on the individual carbon atoms involved in the double bond (Weedon and Moss, 1995). All-*trans* lycopene and its 5-*cis* geometrical isomer, therefore, can also be referred to as the all-*E* and 5-*Z* configurations. Likewise, the all-*trans* configuration is implied in the absence of a geometrical designation (IUPAC, 1975).

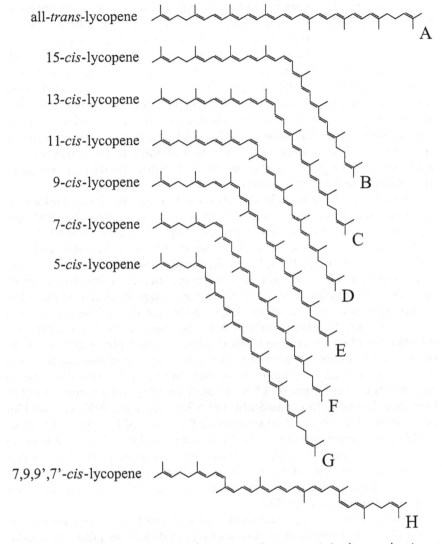

FIG. 7.7 Lycopene geometrical isomers. Structure A is the predominant species in tomatoes and tomato products (~95%). Structures B, C, E, and G are found in human serum, comprising more than half of the total lycopene content. Structure H (Zechmeister, 1941) is the naturally occurring form of lycopene in tangerine-type tomatoes.

Cis isomers of lycopene have physical and chemical properties distinctly different from their all-*trans* counterparts. Some of the differences observed as a result of a *trans*-to-*cis* isomerization reaction include decreased color intensity, lower melting points, smaller extinction coefficients, a shift in the λ_{max}, and the appearance of a new maximum in the ultraviolet spectrum (Zechmeister and Polgar, 1944). The decrease in color intensity is of paramount importance as far as the perception of quality is concerned. Smaller extinction coefficients for *cis* isomers of lycopene need to be taken into account during quantitative analysis of lycopene isomers to avoid underestimation. The appearance of the new maxima in the ultraviolet region, so-called "*cis*-peaks," and their relative intensity are useful in tentative identification of lycopene isomers.

Lycopene, as the predominant carotenoid in human plasma and a variety of anatomical sites, has been shown to exist in several geometrical configurations, where the *cis* isomer content ranges from 50 to 88% of the total lycopene level (Krinsky et al., 1990; Schmitz et al., 1991; Stahl and Sies, 1992; Stahl et al., 1992, 1993; Emenhiser et al., 1995; Clinton et al., 1996). The presence of elevated levels and larger number of *cis*-lycopene isomers in biological samples than in fresh tomato suggests that isomerization occurs either during food processing or after ingestion of the carotenoid-containing food. Another possibility is that lycopene in the *cis* configuration is better absorbed or preferentially deposited under physiological conditions than its all-*trans* counterpart. The prevailing hypothesis assumes that the high percentage of lycopene *cis* isomers in human biological samples is due in part to consumption of heat-treated tomato products containing *cis* isomers of lycopene (Sies and Stahl, 1982; Schierle et al., 1997; van het Hof et al., 1998). The accumulating evidence for a role of lycopene in human health implications and the markedly different profile of lycopene isomers in humans compared to fresh tomatoes has prompted considerable interest in elucidating the roles of dietary lycopene, its stability in foods and the significance of isomer formation on the biological activities of this carotenoid (Nguyen and Schwartz, 1999b).

The extraction, storage, handling, and analysis of lycopene must be carried out under controlled environmental conditions to minimize oxidative degradation and to avoid the introduction of artifactual levels of isomers. Exposure of extracted lycopene to light should be avoided, and only gold, yellow, or red lights should be used (Landers and Olson, 1986). Antioxidants such as butylated hydroxytoluene (BHT) should be employed in solvents used for the extraction and separation of lycopene to control oxidation and isomerization reactions (Nguyen and Schwartz, 1998a). In addition, nitrogen or argon headspace can be employed to keep exposure to atmospheric oxygen to a minimum. Saponification with methanolic potassium hydroxide can be performed to enhance the analysis of lycopene by eliminating chlorophyll and lipid materials, which can interfere with its chromatographic elution and detection (Kimura et al., 1990).

High-performance liquid chromatography (HPLC) is the conventional separation method of choice for lycopene. In general, lycopene is separated from other carotenoids using reversed-phase C_{18} columns. Variations in the properties of the silica packing material in terms of particle size, porosity, carbon load, end-capping technique, and polymerization can greatly influence the sensitivity and selectivity of lycopene analysis (Sander and Wise, 1987; Craft, 1992; Epler et al., 1992; Sander et al., 1994).

Reversed-phase C_{30} columns are often employed to achieve superior selectivity of lycopene isomers compared to conventional C_{18} reversed-phase and silica normal-phase columns (Sander et al., 1994; Emenhiser et al., 1996). The polymerically synthesized C_{30} columns not only provide excellent separation of the all-*trans* lycopene isomers from the *cis* counterpart, but they also exhibit remarkable selectivity among the individual *cis* isomers (Emenhiser et al., 1996; Rouseff et al., 1996). Recently, another HPLC method using multiple columns in series has also been shown to comparably resolve *cis* and *trans* lycopene isomers (Schierle et al., 1997).

Stability During Food Processing

The intensity of the red color is used as an index of quality for tomato products. Therefore, minimizing the loss of lycopene throughout the production process and during storage has always been an important consideration for food producers. Thermal treatments during food-processing operations induce well-documented changes in the physiochemical stability of carotenoids. Moreover, heat treatment enhances the absorption of carotenoids in humans. This has been demonstrated for both β-carotene from spinach and carrots as well as lycopene from tomatoes (Gardner et al., 1997; Rock et al., 1998). The increased absorption of these carotenoids is believed to be the result of heat-induced cellular disruption, which in turn facilitates the release of carotenoids from the plant tissue matrix (Erdman et al., 1993). Therefore, heat treatments represent a value-added and desirable process for increasing carotenoid bioavailability from foods.

Thermal processing has also been shown to induce the isomerization of certain carotenoids in several fruit and vegetables (Panalaks and Murray, 1970; Sweeney and Marsh, 1971; Tsukida et al., 1981; Bushway, 1985; Quackenbush, 1987; Chandler and Schwartz, 1988; Lessin et al., 1997). This conversion of the all-*trans* carotenoids to various *cis* configurations can result in products with different biological activities. For example, in the case of provitamin A carotenoids such as β-carotene and α-carotene, thermal treatment increases the *cis* isomer content of these carotenoids in carrots, spinach, sweet potatoes, and tomatoes. Since *cis* isomers of these carotenoids are converted to vitamin A less efficiently than their all-*trans* counterparts, thermal treatments may reduce the overall provitamin A activity of fruits and vegetables (Quackenbush, 1987; Chandler and Schwartz, 1988).

We have evaluated recently the thermal stability of lycopene during food processing and its bioavailability from a variety of sources. These studies have been enhanced by the combination of highly selective carotenoid separation technologies and sensitive detection methodologies. We have found that food processing does not have a significant impact on lycopene's stability independent of product type, container type, moisture content, tomato variety, and severity of heat treatments (Nguyen and Schwartz, 1998a,b,c, 1999a,b). Preliminary results from in vitro studies also have revealed that lycopene remains stable during gastric and small intestinal phases of digestion (Schwartz et al., 1999).

As discussed earlier, the loss of lycopene can be attributed to both oxidation and/or isomerization. Cole and Kapur (1957a,b) examined the kinetics of lycopene degradation by studying the effects of oxygen, temperature, and light intensity on the formation of its volatile oxidation products. Confirming Monselise and Berk's (1954) report of oxidative degradation of lycopene in heat-treated tomato puree, Cole and Kapur (1957b) showed significant losses of lycopene in serum-free tomato pulp samples following thermal treatment at 100°C in the presence of oxygen with or without light. The intensity of illumination and temperature were found to be in direct correlation with lycopene degradation in the presence of oxygen.

Noble (1975) found that concentration of tomato pulp by heating resulted in approximately 57% loss of lycopene. Sharma and Le Maguer (1996) reported the kinetics of lycopene degradation in tomato pulp solids to be a pseudo first-order reaction. Boskovic (1979) observed a reduction of all-*trans* lycopene content by up to 20% following processing and extended storage of dehydrated tomato products. Food-processing operations, such as freezing and canning, led to a significant decrease in lycopene and total carotenoid content of papaya slices (Cano et al., 1996).

In contrast, several studies have found that hydrocarbon carotenoids such as lycopene, α-carotene, and β-carotene in processed fruits and vegetables are relatively heat-resistant (Khachik et al., 1992a,b). According to Khachik et al., (1992a), most of these carotenoids remain stable following bench-top food preparation. Saini and Singh (1993) also reported that the lycopene content was not altered by thermal processing in juices made from several high-yield tomato hybrids. Zanoni et al. (1998) recently reported that lycopene exhibited high stability during drying of tomato halves even though the ascorbic acid and 5-hydroxymethyl-2-furfural (HMF) values reflected the thermal and oxidative severity of the drying process. It is difficult to reconcile the differences in these findings, as a number of factors may contribute to the problems in accounting for the changes in lycopene levels. The inability to distinguish between chemical degradation and geometrical isomerization of the parent molecule, for example, is a common limitation.

Isomerization of fruit and vegetable carotenoids resulting from thermal treatments during food processing and preparation are well known, especially for β-carotene. For example, Lessin et al., (1997) reported that canning fresh tomatoes increases the β-carotene *cis* isomer content from 12.9 to 31.2%. Likewise, the levels of *cis* isomers of lycopene were increased by heating tomato juice (Stahl and Sies, 1992) and bench-top preparation of a spaghetti sauce from canned tomatoes (Schierle et al., 1997). However, Khachik et al. (1992b) indicated that common heat treatments during food preparation such as microwaving, boiling, steaming, and stewing did not significantly alter the distribution of carotenoids in green vegetables and tomatoes. Other studies have also reported low levels of lycopene *cis* isomers in thermally processed tomato products (Clinton et al., 1996; Emenhiser et al., 1996; Nguyen and Schwartz, 1998a). Lycopene isomer profiles in dietary and biological samples (Fig. 7.8) show similar all-*trans* to *cis* isomer ratios in fresh tomato and processed tomato soup samples (A1 vs. A2), while human serum (B), chylomicron fraction of human plasma (C), and prostate tissue (D) distributions have a much higher *cis* isomer content. Experimental data from our laboratory have recently confirmed that thermal treatments during typical food preparation or commercial production processes do not result in significant losses of lycopene or a shift in the distribution of *cis*-lycopene isomers (Nguyen and Schwartz, 1998a). Thus, lycopene in the diet from both fresh and processed foods is consumed predominantly as the all-*trans* configuration. The elevated levels of *cis* isomers observed in human biological samples cannot be attributed to consumption of thermally processed food but rather to in vivo mechanisms (Schwartz et al., 1999), which are still unclear at this time.

Much information remains to be gathered on the thermal behavior of lycopene before definitive answers can be provided regarding its physical state and stability. Nevertheless, it is apparent that lycopene is more stable in native tomato fruit tissues and matrices than in isolated or purified form (Simpson et al., 1976) as a result of the protective effects of cellular constituents such as water. Therefore, care must be exercised to minimize the loss of lycopene through oxidation or isomerization during extraction, storage, handling, and analysis to accurately account for cause-and-effect changes.

LYCOPENE ABSORPTION AND DISTRIBUTION IN HUMANS

The relevance of carotenoids to human nutrition and health was historically confined to those possessing provitamin A activity such as α-carotene and β-carotene. However, other carotenoids have also emerged as important dietary phytochemicals. Among these carotenoids with potentially beneficial biological activities beyond their traditional role as vitamin A precursors, lycopene in particular arguably has the most promising implications for human nutrition and health. In humans, lycopene along with

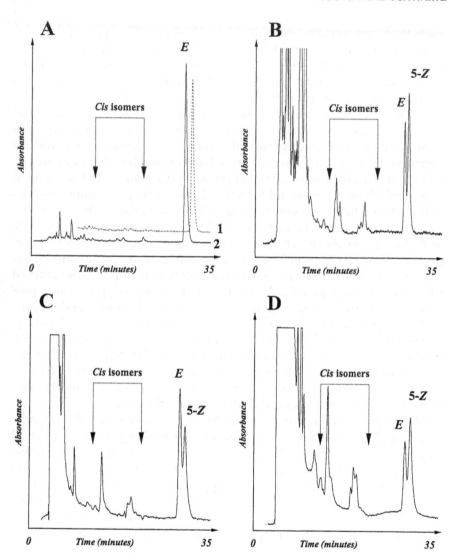

FIG. 7.8 Representative C30 RP-HPLC chromatograms illustrating *cis* and *trans* isomers of lycopene in fresh tomato, tomato soup, chylomicron fraction of human plasma, human serum, and human prostate. (A) 1. Fresh tomato. 2. Tomato sauce. (B) Chylomicron fraction. (C) Serum. (D) Prostate tissue. *E* = All-*trans*-lycopene; 5-*Z* = lycopene 5-*cis* isomer.

β-carotene, lutein, β-cryptoxanthin, and α-carotene account for about 90% of circulating carotenoids, with lycopene being the most predominant carotenoid in the blood stream and at most anatomical sites. The prevalence of lycopene in humans is due in part to consumption of a variety of tomato products such as tomatoes, spaghetti sauce, tomato soup, salsa, and

TABLE 7.2   Common Food Sources of Lycopene

Food	Type	Amount per serving (mg/100 g wet wt.)	(mg)	Serving size
Apricots	Fresh	0.005[a]	0.007	140 g
Apricots	Canned, drained	0.065[a]	0.091	140 g
Apricots	Dried	0.86[a]	0.34	40 g
Chili	Processed	1.08–2.62[a]	1.40–3.41	130 g
Grapefruit	Pink, fresh	3.36[a]	4.70	140 g
Guava	Pink, fresh	5.40[a]	7.56	140 g
Guava juice	Pink, processed	3.34[a]	8.35	240 mL (~250 g)
Ketchup	Processed	16.60[a]	3.32	1 tbsp. (~20 g)
Papaya	Red, fresh	2.00–5.30[b]	2.8–7.42	140 g
Pizza sauce	Canned	12.71[a]	15.89	125 g
Pizza sauce	From pizza	32.89[c]	9.867	slice (~30 g)
Rosehip puree	Canned	0.78[a]	0.47	60 g
Salsa	Processed	9.28[d]	3.71	2 tbsp. (~40 g)
Spaghetti sauce	Processed	17.50[d]	21.88	125 g
Tomatoes	Red, fresh	3.1–7.74[c]	4.03–10.06	130 g
Tomatoes	Whole, peeled, processed	11.21[c]	14.01	125 g
Tomato juice	Processed	7.83[c]	19.58	240 mL (~250 g)
Tomato soup	Canned, condensed	3.99[c]	9.77	245 g
Tomato paste	Canned	30.07[c]	9.02	30 g
Watermelon	Red, fresh	4.10[a]	11.48	280 g
Vegetable juice	Processed	7.28[c]	17.47	240 mL (~250 g)

[a]From USDA, 1998.
[b]From Mangels et al., 1993.
[c]From Nguyen and Schwartz, 1998a.
[d]From Nguyen and Schwartz, 1999a.

ketchup, all of which are good sources of lycopene. Table 7.2 lists some common food sources of lycopene, taking serving size into consideration.

Lycopene follows the same intestinal absorption path as dietary fat since it is very lipophilic. Since lycopene and β-carotene are both hydrocarbon carotenoids, it is expected that they share similar absorption patterns. Once absorbed, however, β-carotene can follow a different metabolic path as it possesses provitamin A activity. Comprehensive reviews by Wang (1994) and Parker (1996) on β-carotene metabolism are recommended for more details.

Lycopene absorption consists of four main phases: digestion of and release from the food matrix; incorporation into mixed micelles; uptake by small intestinal absorptive epithelial cells; and transport to the blood stream. Masticative and digestive actions facilitate the release of lycopene from the food matrix, while the presence of fat and conjugated bile acids facilitates its absorption. The efficiency of this process is influenced by several factors, including the physical state of lycopene in the food matrix, particle size before and after mastication, and various digestive processes (Johnson, 1998). In the small intestine, ingested lycopene is incoporated into micelles formed via the interaction of the products of lipid digestion (monoacylgrycerides, fatty acids, cholesterol, and phospholipids) and bile acids (Sies and Stahl, 1998). Micelle formation is necessary for lycopene absorption since the absorption of carotenoids is very low when ingested in the absence of adequate lipid (Prince and Frisoli, 1993; Olson, 1994). Dietary fiber, which has been shown to interfere with micelle formation (Rock and Swenseid, 1992), can also decrease the overall lycopene uptake. The absence of bile and pancreatic lipase and the loss of the integrity of the villi will also reduce the amount of lycopene being absorbed by the intestinal mucosal cells. Because lycopene is a fat-soluble compound, its absorption is affected by the presence and amount of dietary lipid as well as by factors influencing lipid absorption itself.

Uptake of lycopene has been suggested to occur via passive transport as mixed micelles come into contact with the brush border surface of intestinal mucosal cells (Erdman et al., 1993). Other mechanisms of cellular uptake have not been suggested, although interactions between lycopene and other carotenoids administered in combined oral doses have been noted (Johnson et al., 1997). Fig. 7.9 illustrates that following acquisition by the cell, lycopene can be metabolized or directly packaged into chylomicrons for transport to the blood stream via the lymphatics (Johnson, 1998). Similar to most other carotenoids, lycopene molecules, which are not packaged into chylomicrons for distribution, will be lost after about 3 days upon the sloughing off of mucosal cells into the lumen. In the plasma, lycopene is carried exclusively by lipoproteins, and its appearance in lipoprotein fractions follows the same time-course as triacylglycerols ingested in the meal. Specifically, the majority of the absorbed lycopene molecules are associated with low-density lipoproteins (LDL). Only a minor fraction is found in high-density (HDL) and very-low-density lipoproteins (VLDL) (Krinsky 1998). Likewise, lycopene distribution among lipoproteins is similar to β-carotene and exhibits the same relative profile in both sexes (Forman et al., 1998).

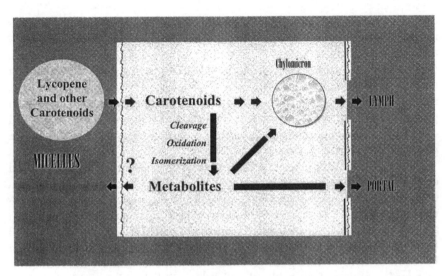

FIG. 7.9   Schematic of lycopene absorption in the human intestinal epithelial cell.

Carotenoid Bioavailability

Accumulating epidemiological and experimental findings have directed considerable interest towards the bioavailability of dietary carotenoids, especially in terms of enhancing their uptake from the diet. Ironically, food processing is in fact a value-added step since more lycopene becomes available for absorption following thermal treatment. Heating of tomato juice was shown to result in an improvement in uptake of lycopene in humans (Stahl and Sies, 1992). Gartner et al. (1997) showed that the bioavailability of lycopene from tomato paste, a processed product, was higher than from fresh tomatoes when both are consumed in conjunction with corn oil. This appears to be the result of thermal disruption and weakening of lycopene-protein complexes as well as the rupturing of cell walls. Likewise, various food-processing operations such as chopping and pureeing, which result in a reduction in physical size of food particle, will also enhance the bioavailability of carotenoids including lycopene (Erdman et al., 1988, 1993; Rock et al., 1998).

Carotenoid bioavailability, including that of lycopene, is generally assessed by monitoring the postprandial level in the plasma following the administration of either purified carotenoids, natural carotenoid-rich sources, or a high-carotenoid meal (Micozzi et al., 1992; de Pee et al., 1995; Oshima et al., 1997). Likewise, other investigators such as Gartner et al. (1997) monitored the response in the chylomicron fraction of lycopene after consumption of tomato products. Research studies on the bioavailability of carotenoids are often limited by several factors. The use of human subjects is costly as well as time and labor intensive. In addition, most laboratory animals, with the exception of primates, ferrets, and preruminant calves, convert provitamin A carotenoids quite efficiently into retinol, making it difficult to account for carotenoids in terms of mass balance. Furthermore, conventional detection methods are limited to measuring changes in plasma levels (circulating levels) instead of accounting for newly absorbed carotenoids, which are present at much lower relative concentration in the plasma chylomicron fraction.

The recent development of a C_{30} reversed-phase gradient HPLC method coupled with a coulometric electrochemical (EC) array detector offers a much lower dectection limit (Ferruzzi et al., 1998) and an unique opportunity to quantify low levels of carotenoids in the plasma chylomicron fraction and tissue samples. The enhanced sensitivy of this HPLC-EC method (1–10 fmol) may allow for the direct comparison of bioavailable carotenoids as a function of fat levels and postingestion time while facilitating the reduction in sample volume of each blood collection. This highly sensitive method will be useful in future studies on key factors influencing the bioavailability of major carotenoids and their absorption from the diet.

Distribution in Biological Fluids and Tissues

The carotenoid profile of human serum is quite extensive and complex. At least 18 different carotenoids have thus far been identified in human serum, with β-carotene and lycopene being the most prominent (Krinsky et al., 1990; Khachik et al., 1992a, 1995). Similarly, Peng and Peng (1992) found lycopene to be the predominant carotenoid present in buccal mucosal cells at 15.54 ng/10^6 cells. The advent of new and refined separation and detection technologies has greatly increased the available information regarding the distribution of lycopene in human serum and selected tissues

TABLE 7.3   Serum Concentrations (μmol/L) of Selected Carotenoid Species in Different Groups

Carotenoid	Young men	Young women	Older men	Older women
α-Carotene	0.09 ± 0.01	0.18 ± 0.04	0.27 ± 0.07	0.20 ± 0.04
All-*trans*-β-Carotene	0.44 ± 0.07	0.80 ± 0.17	1.51 ± 0.41	0.78 ± 0.10
13-*cis*-β-Carotene	0.03 ± 0.01	0.05 ± 0.01	0.09 ± 0.02	0.04 ± 0.01
All-*trans*-Lycopene	0.38 ± 0.05	0.37 ± 0.06	0.41 ± 0.06	0.28 ± 0.02
9-*cis*-Lycopene	0.07 ± 0.01	0.07 ± 0.01	0.08 ± 0.01	0.06 ± 0.00
13-*cis*-Lycopene	0.12 ± 0.02	0.10 ± 0.02	0.14 ± 0.02	0.09 ± 0.01
15-*cis*-Lycopene	0.01 ± 0.001	0.01 ± 0.001	0.01 ± 0.003	0.01 ± 0.00

Source: Yeum et al., 1996.

TABLE 7.4 Tissue Concentrations of Lycopene

Tissue	Concentration (nmol/g wet weight)				
	Kaplan et al., 1990	Schmitz et al., 1991	Nierenberg and Nann, 1992	Stahl et al., 1992	Khachik et al., 1998
Adipose	1.30	—	—	0.20	—
Adrenal	21.60	—	—	1.90	—
Brain	—	—	—	—	2.55[a]
Breast	—	—	0.78	—	0.43
Cervix	—	—	—	—	0.18
Colon	—	—	0.31	—	—
Kidney	0.39	0.62	—	0.15	—
Liver	2.45	5.72	—	1.28	0.65
Lung	—	0.57	0.22	—	0.56
Ovary	0.28	—	—	0.25	—
Prostate	—	—	—	—	0.63[b]
Skin	—	—	0.42	—	—
Stomach	—	—	—	—	0.20[c]
Testes	21.36	—	—	4.34	—

[a]From Craft et al., 1998.
[b]From Clinton et al., 1996.
[c]From Clinton, 1998.

(Bieri et al., 1985; Kaplan et al., 1990; Schmitz et al., 1991; Stahl et al., 1993; Yeum et al., 1996; Clinton et al., 1996). Yeum et al. (1996) reported the distribution of 13 carotenoids and carotenoid isomers in the plasma of both men and women (Table 7.3).

Lycopene has been shown to exist in several geometrical configurations in human plasma and in a variety of tissue samples, where the *cis* isomer content ranges from 50 to 88% of the total lycopene level (Krinsky et al., 1990; Schmitz et al., 1991; Stahl et al., 1992, 1993; Emenhiser et al., 1996; Clinton et al., 1996). The two most predominant lycopene isomers are the all-*trans* and its 5-*cis* counterpart (Schierle et al., 1997). Lycopene levels in various human organs and tissues have also been studied (Parker, 1988; Kaplan et al., 1990; Schmitz et al., 1991; Nierenberg and Nann, 1992; Stahl et al., 1992). The findings were summarized by Stahl and Sies (1996) and are adapted in Table 7.4. It is important to note the large differences between one type of tissue and another, as well as the relative amounts reported by independent investigators. The latter differences suggest that the absolute value being reported is only meaningful in the context of individual studies and methodologies.

The deposition of lycopene in a variety of tissues reflects the effective transfer from plasma lipoproteins. The mechanism responsible for the deposition and immobilization of carotenoids in adipose and other tissues as well as intracellular transport of carotenoids are poorly understood. Lycopene is present at relatively high concentration in testes and adrenals, where a high rate of LDL uptake has been noted (Spady et al., 1985). Furthermore, Clinton et al. (1996) compared major carotenoids levels in normal versus malignant human prostate tissue. Lycopene and other major carotenoids were reportedly present in higher concentration in the malignant prostate tissue. In general, while lycopene can be found in a large number of human tissues and organs, the relative amounts have not been systematically correlated to the types or functions of these biological tissues.

BIOLOGICAL AND CLINICAL IMPLICATIONS

The majority of studies involving lycopene prior to the past two decades primarily focused on its physical and chemical properties in the context of color stability in food processing. The shift in emphasis to the biological effects of lycopene is a fairly recent phenomenon, with the most dramatic rise in interest occurring after the publication of a number of recent key findings. Di Mascio et al. (1989) discovered that lycopene is the most efficient singlet oxygen quencher of the biological carotenoids, including β-carotene. Levy et al. (1995) showed lycopene to be a more potent inhibitor of human cancer cell proliferation than either α-carotene or β-carotene. Epidemiological support of these laboratory findings was provided by the Harvard Health Professionals Follow-Up Study (Giovannucci et al., 1995), in which the relationship between intake of various carotenoids, retinol, fruits, and vegetables and the reduced risk of prostate cancer was examined for a cohort of 47,894 male subjects. Giovannucci et al. (1995) concluded that consumption of fresh tomatoes, tomato sauce, and pizza, which account for the bulk of dietary lycopene intake, is significantly related to a lower incidence of prostate cancer.

Prior to the latter study, accumulated human epidemiological evidence indicated that diets high in tomatoes may reduce the risk of developing cervical, colon, esophageal, rectal, and stomach cancers (Bjelke, 1974; Cook-Mozaffari et al., 1979; Tajima and Tominaga, 1985; Batieha et al., 1993; Ramon et al., 1993; Potischman et al., 1994). In another case-control study, a high intake of fresh tomatoes was linked to a protective effect of the digestive tract against the risk of cancer (Franceschi et al., 1994). A high tomato intake in an elderly American population was likewise associated with a 50% reduction in mortality from cancers at all sites (Colditz et al., 1985).

Palan et al. (1996) reported a decrease in plasma lycopene level, along with β-carotene, canthaxanthin, retinol, and α-tocopherol, in women from New York City diagnosed with cervical intraepithelia neoplasia (CIN) or cervical cancer. The authors suggested that the lowered plasma antioxidant levels might play a role in the pathogenesis of CIN and carcinoma of the cervix. Similarly, Ha et al. (1996) observed lowered plasma lycopene in patients with chronic renal failure compared to the control group. Rao and Agarwal (1998) reported that while the serum lycopene in nonsmokers is comparable to that in habitual smokers, levels in smokers were lowered by 40% after smoking three regular cigarettes. This suggests a significant influence of oxidative stress, in the form of smoking, on circulating lycopene status. Likewise, supplementing the diet in a group of male nonsmokers with tomato juice (40 mg of lycopene) resulted in a significant decrease in endogenous levels of lymphocyte DNA breakage, as measured by the COMET assay (Pool-Zobel et al., 1997).

Brady et al. (1996) indicated that lifestyle and physiological factors such as gender, smoking, alcohol consumption, and body mass index were not related to serum lycopene levels in a population-based group of 400 individuals. Nonetheless, lower serum lycopene appears to be associated with older age and lower non-HDL cholesterol. Furthermore, Snowdon et al. (1996) also found a corresponding correlation between high blood lycopene level and a positive influence on the functional capacity of the elderly, such as the ability to perform self-care tasks. Higher lycopene concentration in body fat, on the other hand, was associated with a lower risk of heart attack in a group of 1379 European men; the findings from this EURAMIC study are among the first to link lycopene to protection against heart disease (Kohlmeier et al., 1997).

Lycopene has been shown to inhibit carcinogenesis in animal model systems (Wang et al., 1989; Nagasawa et al., 1995; Narisawa et al., 1996) as well as human cell lines (Countryman et al., 1991; Bertram, 1993; Kim, 1995; Matsushima-Nishiwaki et al., 1995). For example, the development of spontaneous mammary tumors in SHN virgin mice was greatly reduced in the group of mice receiving lycopene-enriched food compared to the control group (Nagasawa et al., 1995). According to Mitamura et al. (1996), inhibitory effect of lycopene on mammary tumorigenesis in mice may be its ability to decrease the gene expression of transforming growth factor alpha (TGF-α). Narisawa et al. (1996) reported the prevention of colon carcinogenesis in rats given small doses of lycopene and lutein. Recently, physiological concentration of lycopene was shown to act synergistically with α-tocopherol to inhibit the proliferation of human prostate carcinoma cells (Pastori et al., 1998). While all these studies continue to show that lycopene plays a role in preventing certain types of cancers, the mechanisms of action remain unclear.

The ability of lycopene to act as an antioxidant and scavenger of free radicals is potentially a key to the mechanism for its beneficial effects on human health (Khachik et al., 1995). As a result of having an extensive chromophore system of conjugated carbon-carbon double bonds, lycopene can accept energy from various electronically excited species (Fig. 7.10). This is

$$Sensitizer \ + \ Light \ \rightarrow \ {}^{1}Sensitizer$$

$$^{1}Sensitizer \ \ \ \ \ \ \ \ IC\rightarrow \ \ \ {}^{3}Sensitizer$$

$$^{3}Sensitizer \ + \ {}^{3}O_{2} \ \ \rightarrow \ \ \ Sensitizer \ \ + {}^{1}O_{2}$$

$$Lycopene \ + \ {}^{1}O_{2} \ \ \rightarrow \ \ \ {}^{3}Lycopene \ \ + {}^{3}O_{2}$$

$$^{3}Lycopene \ \ \ \ \ \ \ R,V\rightarrow \ \ \ \ Lycopene$$

FIG. 7.10 Singlet-oxygen quenching by lycopene. IC: Intersystem crossing; R and V: rotational and vibrational interactions with the solvent. (Adapted from Krinsly, 1998.)

attributable to its ability to quench singlet oxygen (Di Mascio et al., 1989), formed by energy transfer from a meta-stable excited photosensitizer (Krinsky, 1998). Lycopene may prevent carcinogenesis and atherogenesis by interfering passively with oxidative damage to DNA and lipoproteins (Gester, 1997; Clinton, 1998). It may also inhibit the formation of oxidized products of LDL cholesterol; these oxidation products have been suggested to participate in the early stages of coronary heart disease (Ojima et al., 1993; Diaz et al., 1997; Weisburger, 1998). Lycopene's protective effects against oxidative stress also were illustrated when lycopene was found to be preferentially destroyed relative to β-carotene as human skin is irradiated with UV light (Ribayo-Mercado et al., 1995).

Despite the overwhelming evidence linking lycopene to various beneficial bioactivities, a number of inconsistencies exist in the epidemiological data regarding the role of lycopene in the prevention of diseases. For example, Steinmetz et al. (1993) found no association between lung cancer risk and either tomatoes or three carotenoid-rich food groups. Likewise, Jarvinen et al. (1997) found that lycopene intake was not significantly related to the occurrence of breast cancer in a Finnish prospective cohort study. Differences in the oxidative environment of the lung compared to other cancer sites and the uniqueness of breast cancer carcinogenesis have been cited as key factors influencing lycopene's effectiveness in these cases. Obviously, further research on lycopene is needed to discover its mode of action and to better understand the scope of its effectiveness.

SUMMARY

Over the past 25 years, the diet–health paradigm of foods being the source of essential nutrients to sustain life and growth has gradually evolved into one in which foods also are recognized as sources for the management or prevention of diseases. The paradigm shift is due in part to the body of epidemiological evidence associating diets rich in fruits and vegetables with the reduced risk of developing certain types of cancer and other chronic diseases. Lycopene, as discussed in this review, is among a number of compounds in fruits and vegetables that are potentially responsible for such associations.

It has been suggested that as much as 20–42% of all cancer deaths in the United States are preventable by changes in dietary intake (Willett, 1995). In general, the median intake of fruits and vegetables in the United States falls far short of the recommendation by the National Cancer Institute (Subar et al., 1995). Fortunately, the consumption of tomatoes and tomato products is relatively high in the United States, where the average daily dietary intake of lycopene is 3.7 mg (Forman et al., 1993)—more than three times the average intake in England, for example (Scott et al., 1996). In fact, the mean lycopene intake between 1987 and 1992 has increased among adults aged 18–69 years by 5–6% (Nebeling et al., 1997).

The most recent comprehensive review of 72 independent epidemiological studies revealed that intake of tomatoes and tomato-based products is inversely associated with the risk of developing cancers at several anatomical sites including the prostate gland, stomach, and lung. Data were also suggestive for the reduced risk of breast, cervical, colorectal, esophageal, oral cavity, and pancreatic cancers in individuals ingesting lycopene rich foods (Giovannucci, 1999). Nonetheless, the observational nature of epidemiological associations precludes conclusions about the preventive role of lycopene in reducing cancer risks until more definitive results are obtained from long-term intervention trials. Until such direct correlations from double-blind, placebo-controlled studies are made, findings from human cell culture and animal studies will continue to be relied upon to substantiate the potential role of lycopene in the observed relationship between higher intake of tomato products and decreased risks of cancer.

At the very least, lycopene can be considered a nutritional indicator of good dietary habits and healthy lifestyles. Lycopene is present in a variety of fruit and vegetables, which upon consumption will not only enhance plasma level of lycopene but will also address the recommendations of the National Cancer Institute to increase dietary intake of fruit and vegetables to promote better health. Meanwhile, further research on lycopene in terms of its stability, bioavailability, and metabolism continue to be the next logical step to better understand its role in human health.

REFERENCES

Batieha, A. M., Armenian, H. K., Norkus, E. P., Morris, J. S., Spate, V. E., and Comstock, G. W. 1993. Serum micronutrients and subsequent risk of cervical cancer in a population-based nested case-control study. *Cancer Epidemiol. Biomarkers Prev.* 2: 335–339.

Belakbir, A., Ruiz, J. M. and Romero, L. 1998. Yield and fruit quality of pepper (*Capsicum annuum* L.) in response to bioregulators. *HortScience* 33(1): 85–87.

Ben-Aziz, A., Britton, G., and Goodwin, T. W. 1973. Carotene epoxides of *Lycopersicon esculentum. Phytochemistry* 12: 2759–2764.

Bendich, A. 1989. Carotenoids and the immune response. *J. Nutr.* 119: 112–115.

Bendich, A., and Olson, J. A. 1989. Biological actions of carotenoids. *FASEB J.* 3: 1927–1932.

Ben-Shaul, A., and Naftali, Y. 1969. The development and ultrastructure of lycopene bodies in chromoplasts of *Lycopersicum esculentum. Protoplasma* 67: 333–344.

Bertram, J. S. 1993. Cancer prevention by carotenoids: Mechanistic studies in cultured cells. *Ann. NY Acad. Sci.* 691: 177–191.

Bieri, J. G., Brown, E. D., and Smith, J. C. 1985. Determination of individual carotenoids in human plasma by high performance liquid chromatography. *J. Liq. Chromatogr.* 8: 473–484.

Bjelke, E. 1974. Case-control study in Norway. *Scand. J. Gastroenerol.* 9: 42–49.

Block, G. 1994. Nutrient sources of provitamin A carotenoids in the American diet. *Am. J. Epidemiol.* 139: 290–293.

Block, G., Patterson, B., and Subar, A. 1992. Fruit, vegetables, and cancer prevention: A review of the epidemiological evidence. *Nutr. Cancer* 18: 1–29.

Boskovic, M. A. 1979. Fate of lycopene in dehydrated tomato products: Carotenoid isomerization in food system. *J. Food Sci.* 44: 84–86.

Brady, W. E., Mares-Perlman, J. A., Bowen, P., and Stacewicz-Sapuntzakis, M. 1996. Human serum carotenoid concentrations are related to physiologic and lifestyle factors. *J. Nutr.* 126(1): 129–137.

Bramley, P. M. 1985. The in vitro biosynthesis of carotenoids. In *Advances in Lipid Research*, Vol. 21, R. Paoletti and D. Kritchevsky (Eds.), pp. 243–279. Academic Press, New York.

Britton. G. 1983. *The Biochemistry of Natural Pigments*. Cambridge University Press, Cambridge.

Britton. G. 1995. *Carotenoids: Spectroscopy. 1B*. Birkhauser Verlag, Boston.

Britton, G. and Goodwin, T. W. 1969. The occurrence of phytoene 1,2-oxide and related carotenoids in tomatoes. *Phytochemistry* 8: 2257–2258.

Bushway, R. J. 1985. Separation of carotenoids in fruits and vegetables by high performance liquid chromatography. *J. Liq. Chromatogr.* 8: 1527–1547.

Cano, M. P., Ancos, B., Lobo, G., Monreal, M., and De-Ancos, B. 1996. Effects of freezing and canning of papaya slices on their carotenoid composition. *Z. Lebensm. Unters. Forsch.* 202(4): 279–284.

Chandler, L. A., and Schwartz, S. J. 1988. Isomerization and losses of *trans* β-carotene in sweet potatoes as affected by processing treatments. *J. Agric. Food Chem.* 36: 129–133.

Choo, Y. M., Yap, S. C., Ooi, C. K., Ma, A. N., Goh, S. H., and Ong, S.H. 1996. Recovered oil from palm-pressed fiber: A good source of natural carotenoids, vitamin E, and sterols. *J. Am. Oil Chem. Soc.* 73: 599–602.

Clinton, S. K. 1998. Lycopene: Chemistry, biology, and implications for human health and disease. *Nutr. Rev.* 56(2): 35–51.

Clinton, S. K., Emenhiser, C., Schwartz, S. J., Bostwick, D. G., Williams, A. W., Moore, B. J., and Erdman, J. W. 1996. *Cis-trans* lycopene isomers, carotenoids, and retinol in the human prostate. *Cancer Epidemiol. Biomarkers Prev.* 5: 823–833.

Clough, J. M., and Pattenden, G. 1979. Naturally-occurring poly-*cis* carotenoids. Stereochemistry of poly-*cis* lycopene and its congeners in 'tangerine' tomato fruits. *J. Chem. Soc. Chem. Comm.* 14: 616–619.

Colditz, G. A., Branch, L. G., and Lipnick, R. J. 1985. Increased green and yellow vegetable intake and lowered cancer deaths in an elderly population. *Am. J. Clin. Nutr.* 41: 32–36.

Cole, E. R., and Kapur, N. S. 1957a. The stability of lycopene. I. Degradation by oxygen. *J. Sci. Food Agric.* 8: 360–365.

Cole, E. R., and Kapur, N. S. 1957b. The stability of lycopene. II. Oxidation during heating of tomato pulps. *J. Sci. Food Agric.* 8: 366–368.

Conn, P. F., Lambert, C., Land, E. J., Schalch, W., and Truscott, T. G. 1992. Carotene-oxygen radical interactions. *Free Rad. Res. Comms.* 16: 401–408.

Cook-Mozaffari, P. J., Azordegan, F., and Day, N. E. 1979. Oseophageal cancer studies in the Caspian Litoral of Iran: Results of a case-control study. *Br. J. Cancer* 39: 292–309.

Countryman, C., Bankson, D., Collins, S., Man, B., and Lin, W. 1991. Lycopene inhibits the growth of the HL-60 promyelocytic leukemia cell line. *Clin. Chem.* 37: 1056.

Craft, N. E. 1992. Carotenoid reversed–phase high-performance liquid chromatography methods: Reference compendium. *Meth. Enzymol.* 213: 185–205.

Craft, N., Garnett, K., Hedley-Whyte, E. T., Fitch, K., Haitema, T., and Dorey, C. K. 1998. Carotenoids, tocopherols, and vitamin A in human brain. *FASEB J.* 12: AS601, Part 2.

Crouzet, J., and Kanasawud, P. 1992. Formation of volatile compounds by thermal degradation of carotenoids. *Meth. Enzymol.* 213: 54–62.

Curl, A. L. 1961. The xanthophylls of tomatoes. *J. Food Sci.* 26: 106–111.

Czygan, F. C. 1980. *Pigments in Plants*, 2nd ed. Fischer, Stuttgart.

Davies, B. H. 1976. Carotenoids. In *Chemistry and Biochemistry of Plant Pigments*, Vol. 2, 2nd ed., Goodwin, T. W. (Ed.), pp. 38–165. Academic Press, New York.

de Pee, S., West, C. E., Karyadi, D., and Huatvast, J. 1995. Lack of improvement in vitamin A status with increased consumption of dark-green leafy vegetables. *Lancet* 346: 75–81.

Deuel, H. J., Johnston, C. H., Meserve, E. R., Polgár, A., and Zechmeister, L. 1945. Stereochemical configuration and provitamin A activity. IV. Neo-α-carotene and neo-β-carotene. *Arch. Biochem.* 7: 247–255.

Diaz, M. N., Frei, B., Vita, J. A., and Keaney, J. F. 1997. Antioxidants and atherosclerotic heart disease. *N. Engl. J. Med.* 337: 408–416.

Di Mascio, P., Kaiser, S., and Sies, H. 1989. Lycopene as the most efficient biological carotenoid singlet oxygen quencher. *Arch. Biochem. Biophys.* 274: 532–538.

Duggar, B. M. 1913. Lycopersicon: The red pigment of the tomato and the effects of conditions on its development. *Washington Univ. Stud.* 1: 22–45.

Duke, J. A. 1992. *Handbook of Phytochemical Constituents of GRAS Herbs and Other Economic Plants*. CRC Press, Boca Raton, Florida.

Duke, J. A., and Beckstrom-Sternberg, S. M. 1998. Plants containing lycopene. Phytochemical database. USDA-NCI Carotenoid Food Composition Database. Agric. Res. Service, U.S. Dept. of Agriculture, Beltsville, MD.

Edwards, C. G., and Reuter, F. H. 1967. Pigment changes during the maturation of tomato fruits. *Food Technol. Aust.* 19: 352–357.

Emenhiser, C., Sander, L. C., and Schwartz, S. J. 1995. Capability of a polymeric C_{30} stationary phase to resolve *cis-trans* carotenoid isomers in reversed-phase liquid chromatography. *J. Chromatogr. A* 707: 205–216.

Emenhiser, C., Simunovic, N., Sander, L. C., and Schwartz, S. J. 1996. Separation of geometric isomers in biological extracts using a polymeric C_{30} column in reversed-phase liquid chromatography. *J. Agric. Food Chem.* 44: 3887–3893.

Englert, G., Brown, B. O., Moss, G. P., Weedon, B. C. L., Briton, G., Goodwin, T. W., Simpson, K. L., and Williams, R. S. H. 1979. Prolycopene, a tetra-*cis*-carotene with two hindered *cis* double bonds. *J. Chem. Soc. Chem. Commun.* 7: 545–547.

Epler, K. S., Sander, L. C., Ziegler, R. G., Wise, S. A., and Craft, N. E. 1992. Evaluation of reversed-phase liquid chromatographic columns for recovery and selectivity of selected carotenoids. *J. Chromatogr.* 595: 89–101.

Erdman, J. W., Poor, C. L., and Dietz, J. M. 1988. Factors affecting the bioavailability of vitamin A, carotenoids, and vitamin E. *Food. Technol.* 42: 214–221.

Erdman, J. W., Bierer, T. L., and Guggar, E. T. 1993. Absorption and *trans-port* of carotenoids. *Ann. NY Acad. Sci.* 691: 76, 85.

Ferruzzi, M. G., Sander, L. C., Rock, C. L., and Schwartz, S. J. 1998. Carotenoid determination in biological microsamples using liquid chromatography with a coulometric electrochemical array detector. *Anal. Biochem.* 256: 74–81.

Forman, M. R., Lanza, E., Yong, L. C., Holden, J. M., Graubard, B. I., Beecher, G. R., Meltiz, M., Brown, E. D., and Smith, J. C. 1993. The correlation between two dietary assessments of carotenoid intake and plasma carotenoid concentrations: Application of a carotenoid food-composition database. *Am. J. Clin. Nutr.* 58: 519–524.

Forman, M. R., Johnson, E. J., Lanza, E., Graubard, B. I., Beecher, G. R., and Muesing, R. 1998. Distribution of individual carotenoids in lipoproteins of premenopausal women: A controlled dietary study. *Am. J. Clin. Nutr.* 67: 81–87.

Franceschi, S., Bidoli, E., La Vecchia, C., Talamini, R., D'Avanzo, B., and Negri, E. 1994. Tomatoes and risk of digestive-tract cancers. *Int. J. Cancer* 59(2): 181–184.

Gartner, C., Stahl, W., and Sies, H. 1997. Lycopene is more bioavailable from tomato paste than from fresh tomatoes. *Am. J. Clin. Nutr.* 66: 116–122.

Gester, H. 1997. The potential role of lycopene for human health. *J. Am. Coll. Nutr.* 16(2): 109–126.

Gey, K. F., Moser, U. K., Jordan, P., Stähelin, H. B., Eichholzer, M., and Lüdin, E. 1993. Increased risk of cardiovascular disease at suboptimal plasma concentrations of of essential antioxidants: An epidemiological update with special attention to b-carotene and vitamin C. *Am. J. Clin. Nutr.* 57: 787S–797S.

Giovannucci, E. L. 1999. Tomatoes, tomato-based products, lycopene, and cancer: Review of the epidemiologic literature. *J. Natl. Cancer Inst.* 91(4): 317–329.

Giovannucci, E. L., Ascherio, A., Rimm, E. B., Stampfer, M. J., Colditz, G. A., and Willett, W. C. 1995. Intake of carotenoids and retinol in relationship to risk of prostate cancer. *J. Natl. Cancer Inst.* 87: 1767–1776.

Giuliano, G., Bartley, G. E., and Scolnik, P. A. 1993. Regulation of carotenoid biosynthesis during tomato development. *Plant Cell* 5: 379–387.

Glass, R. W., and Simpson, K. L. 1976. The isolation of γ-carotene and poly-cis-γ-carotene from the Tangerine tomato. *Phytochem.* 15: 1077–1078.

Godoy, H. T., and Rodriguez-Amaya, D. B. 1987. Changes in individual carotenoids on processing and storage of mango (*Mangifera indica*) slices and puree. *Int. J. Food Sci. Technol.* 22: 451–460.

Goodwin, T. W. 1952. *The Comparative Biochemistry of the Carotenoids.* Chapman and Hall, London.

Goodwin, T. W. 1980. *The Biochemistry of the Carotenoids*, Vol. 1, 2nd ed. Chapman and Hall, London.

Goodwin, T. W. 1993. Biosynthesis of carotenoids: An overview. In *Methods in Enzymology*, Vol. 214, Lester Packer (Ed), pp. 330–340. Academic Press, Inc., San Diego.

Gross, J. 1991. *Pigments in Vegetables: Chlorophylls and Carotenoids*, pp. 148–249. Van Nostrand Reinhold, New York.

Ha, T. K. K., Sattar, N., Talwar, D., Cooney, J., Simpson, K., O'-Reilly, D., and Lean, M. E. J. 1996. Abnormal antioxidant vitamin and carotenoid status in chronic renal failure. *Q. J. M.* 89: 765–769.

Harborne, J. B., and Baxter, H. 1983. *Phytochemical Dictionary. A Handbook of Bioactive Compounds from Plants.* Taylor & Frost, London.

Hartsen. 1873. *Chem. Centr.* 204.

Henry, L. K., Catignani, G. L., and Schwartz, S. J. 1998. Oxidative degradation kinetics of lycopene, lutein, 9-*cis* and all-*trans* β-carotene. *J. Am. Oil Chem. Soc.* 75: 823–829.

Hulme, A. C. 1970/1971. *The Biochemistry of Fruits and Their Products*, Vol. 1, 2. Academic Press, London.

IUPAC. 1975. Nomenclature of carotenoids. IUPAC Commission on the Nomenclature of Organic Chemistry and IUPAC-IUB Commission on Biochemical Nomenclature. *Pure Appl. Chem.* 41: 407–419.

Jarvinen, R., Knekt, P., Seppanen, R., and Teppo, L. 1997. Diet and breast cancer risk in a cohort of Finnish women. *Cancer Lett.* 114: 251–253.

Johnson, E. J. 1998. Human studies on bioavailability and plasma response of lycopene. *Proc. Soc. Exp. Biol. Med.* 218: 115–120.

Johnson, E. J., Qin, J., Krinsky, N. I., and Russel, R. M. 1997. Ingestion by men of a combined dose of β-carotene and lycopene does not affect the absorption of β-carotene but improves that of lycopene. *J. Nutr.* 127: 1833–1837.

Jones, B. L., and Porter, J. W. 1986. Biosynthesis of carotenes in higher plants. *CRC Crit. Rev. Plant Sci.* 3: 295–324.

Joyce, E. 1954. Some polyenes of *Brassica rutabaga. Nature* 173: 311–312.

Joyce, E. 1959. Carotenoids of *Brassica napus. J. Sci. Food Agric.* 10: 342–348.

Kaplan, L. A., Lau, J. M., and Stein, E. A. 1990. Carotenoid composition, concentrations, and relationships in various human organs. *Clin. Physiol. Biochem.* 8: 1–10.

Karrer, P., and Jucker, E. 1950. *Carotenoids.* Elsevier Publishing Company, Inc. Amsterdam.

Karrer, P., Helfenstein, A., Wehrli, H., and Wettstein, A. 1930. Pflanzenfarbstoffe. XXV. Ueber die Konstitution des Lycopins und Carotins. *Acta* 14: 154–162.

Khachik, F., Beecher, G. R., Lusby, W. R., and Smith, J. C. 1992a. Separation and identification of carotenoids and their oxidation products in the extracts of human plasma. *Anal. Chem.* 64: 2111–2122.

Khachik, F., Goli, M. B., Beecher, G. R., Holden, J., Lusby, W. R., Tenorio, M. D., and Barrera, M. R. 1992b. Effect of food preparation on qualitative and quantitative distribution of major carotenoid constituents of tomatoes and several green vegetables. *J. Agric. Food Chem.* 40: 390–398.

Khachik, F., Beecher, G. R., and Smith, J. C. 1995. Lutein, lycopene, and their oxidative metabolites in chemoprevention of cancer. *J. Cell Biochem. Suppl.* 22: 236–246.

Khachik, F., Cohen, L. A., and Zhao, Z. 1998a. Metabolism of dietary carotenoids and their potential role in prevention of cancer and age-related macular degeneration. In *Functional Foods for Disease Prevention. I: Fruits, Vegetables, and Teas*, T. Shibamoto, J. Terao, and T. Osawa, (Eds.), pp. 71–85, Oxford University Press, Washington, DC.

Khachik, F., Steck, A., Niggli, U. A., and Pfander, H. 1998b. Partial synthesis and structural elucidation of the oxidative metabolites of lycopene identified in tomato paste, tomato juice, and human serum. *J. Agric. Food Chem.* 46(12): 4874–4884.

Khachik, F., Pfander, H., and Traber, B. 1998c. Proposed mechanisms for the formation of synthetic and naturally occurring metabolites of lycopene in tomato products and human serum. *J. Agric. Food Chem.* 46(12): 4885–4890.

Khudairi, A. K. 1972. The ripening of tomatoes: A molecular ecological approach to the physiology of fruit ripening. *Am. Sci.* 60: 696–707.

Kim, H. 1995. Carotenoids protect cultured rat hepatocytes from injury caused by carbon tetrachloride. *Int. J. Biochem. Cell Biol.* 27: 1303–1309.

Kimura, M., Rodriguez-Amaya, D. B., and Godoy, H. T. 1990. Assessment of the saponification step in the quantitative determination of carotenoids and provitamins A. *Food Chem.* 35: 187–195.

Kirk, J. T. O., and Tilney-Basset, R. A. E. 1978. *The Plastids. Their Chemistry, Structure, Growth, and Inheritance*, 2nd ed. Elsevier, Amsterdam.

Kohlmeier, L., Clark, J. D., Gomez-Gracia, E., Martin, B. C., Steck, S. E., Kardinaal, A. F., Ringstad, J., Thamm, M., Masaev, V., Riemersma, R., Martin-Moreno, J. M., Huttunen, J. K., and Kok, F. J. 1997. Lycopene and myocardial infarction risk in the EURAMIC study. *Am. J. Epidemiol.* 146: 618–626.

Krinsky, N. I. 1998. Overview of lycopene, carotenoids, and disease prevention. *Proc. Soc. Exp. Biol. Med.* 218: 95–97.

Krinsky, N. I., Russett, M. D., Handelman, G. J., and Snodderly, D. M. 1990.

Structural and geometrical isomers of carotenoids in human plasma. *J. Nutr.* 120: 1654–1662.

Krinsky, N. I., Wang, X. D., Tang, G., and Russell, R. M. 1993. Mechanism of carotenoid cleavage to retinoids. *Ann. NY Acad. Sci.* 691: 167–176.

Kuhn, R., and Grundmann, C. 1932. Die Konstitution des Lycopins. *Ber. Deutsch. Chem. Ges.* 65: 1880–1889.

Kull, D., and Pfander, H. 1995. Appendix: List of new carotenoids. In *Carotenoids: Isolation and Analysis*, Vol. 1A, pp. 295–317. Birkhäuser, Basel.

Landers, G. M., and Olson, J. A. 1986. Absence of isomerization of retinyl palmitate, retinol, and retinal in chlorinated and unchlorinated solvents under gold light. *J. Assoc. Offic. Anal. Chem.* 69: 50–55.

Laval-Martin, D. 1974. La maturation du fruit de tomate "cerise": Mise en evidence, par cryodecapage de l'evolution des chloroplastes en duex types de chromoplates. *Protoplasma* 82: 33–59.

Laval-Martin, D., Quennemet, J., and Moneger, R. 1975. Pigment evolution in *Lycopersicon esculentum* fruits during growth and ripening. *Phytochemistry* 14: 2357–2362.

Lessin, W. J., Catigani, G. L., and Schwartz, S. J. 1997. Quantification of *cis-trans* isomers of provitamin A carotenoids in fresh and processed fruits and vegetables. *J. Agric. Food Chem.* 45: 3728–3732.

Levy, J., Bosin, E., Feldman, B., Giat, Y., Munster, A., Danilenko, M., and Sharoni, Y. 1995. Lycopene is a more potent inhibitor of human cancer cell proliferation than either α-carotene or β-carotene. *Nutr. Cancer* 24: 257–266.

List, P. H., and Horhammer, L. 1979. *Hager's Handbuch der Pharmazeutischen Praxis*, Vols. 2–6. Springer-Verlag, Berlin.

Mangels, A. R., Holden, J. M., Beecher, G. R., Forman, M., and Lanza, E. 1993. Carotenoid content of fruits and vegetables: An evaluation of analytical data. *J. Am. Dietet. Assoc.* 93: 284–296.

Markham, M. C., and Liaaen-Jensen, S. 1986. Carotenoids of higher plants. I. The structures of lycoxanthen and lycophyll. *Phytochemistry* 7: 839–844.

Matsushima-Nishiwaki, R., Shidoji, Y., Nishiwaki, S., Yamada, T., Moriwaki, H., and Muto, Y. 1995. Suppression by carotenoids of microcystin-induced morphological changes in mouse hepatocytes. *Lipids* 30: 1029–1034.

Micozzi, M. S., Brown, E. D., Edwards, B. K., Bieri, J. G., Taylor, P. R., Khachik, F., Beecher, G. R., and Smith, J. C. 1992. Plasma carotenoid response to chronic intake of selected foods and β-carotene supplements in men. *Am. J. Clin. Nutr.* 55: 1120–1125.

Millardet. (1875). *Bull. Soc. Sci. Nancy* 2(1): 21.

Mitamura, T., Tsunoda, S., and Nagasawa, H. 1996. Lycopene, a carotenoid, suppresses TGFalpha mRNA expression in spontaneous mammary tumours of SHN mice. *Bull. Faculty Agric. Meiji Univ.* 108: 45–48.

Mohr, W. P. 1979. Pigment bodies in fruits of crimson and high pigment lines of tomatoes. *Ann. Bot.* 44: 427–434.

Monselise, J. J., and Berk, Z. 1954. Oxidative destruction of lycopene during the manufacturing of tomato puree. *Bull. Res. Council Isr.* 4: 188–191.

Moore, T. A., Gust, D., and Moore, A. L. 1989. The function of carotenoid pigments in photosynthesis and their possible involvement in the evolution of higher plants. In *Carotenoids: Chemistry and Biology*, N. I. Krinsky, M. M. Mathews-Roth, and R. F. Taylor (Eds.), pp. 223–228. Plenum Press, New York.

Moss, G. P., and Weedon, B. C. L. 1976. Chemistry of the carotenoids. In *Chemistry and Biochemistry of Plant Pigments*, Vol. 1, 2nd ed., T. W. Goodwin, (Ed.), pp. 149–224. Academic Press, New York.

Muller, H. 1997. Determination of the carotenoid content in selected vegetables and fruit by HPLC and photodiode array detection. *Z. Lebensm. Unters. Forsch.* 204(2): 88–94.

Nagasawa, H., Mitamura, T., Sakamoto, S., and Yamamoto, K. 1995. Effects of lycopene on spontaneous mammary tumour development in SHN virgin mice. *Anticancer Res.* 15: 1173–1178.

Narisawa, T., Fukaura, Y., Hasebe, M., Ito, M., Aizawa, R., Murakoshi, M., Uemura, S., Khachik, F., and Nishino, H. 1996. Inhibitory effects of natural carotenoids, α-carotene, β-carotene, lycopene and lutein, on colonic aberrant crypt foci formation in rats. *Cancer Lett.* 107: 137–142.

Nebeling, L. C., Forman, M. R., Graubard, B. I., and Snyder, R. A. 1997. Changes in carotenoid intake in the United States: The 1987 and 1992 National Health Interview Surveys. *J. Am. Diet. Assoc.* 9: 991–996.

Nguyen, M. L., and Schwartz, S. J. 1999a. Lycopene: Chemical and biological properties. *Food Technol.* 53(2): 38–45.

Nguyen, M. L., and Schwartz, S. J. 1999b. Stability of lycopene and other tomato carotenoids following in vitro digestion. Paper presented at Experimental Biology 1999, A888, 670.7. Washington, DC.

Nguyen, M. L., and Schwartz, S. J. 1998a. Lycopene stability during food processing. *Proc. Soc. Exp. Biol. Med.* 218: 101–105.

Nguyen, M. L., and Schwartz, S. J. 1998b. Effects of industrial thermal treatments on carotenoid geometrical isomers in fresh and processed fruits and vegetables. Paper presented at the Annual Meeting of the Institute of Food Technologists, Atlanta.

Nguyen, M. L., and Schwartz, S. J. 1998c. Thermal isomerization of carotenoids in different tomato varieties. Paper presented at The 3rd Worldwide Congress Processing Tomato, Pamplona, Spain.

Nierenberg, D. W., and Nann, S. L. 1992. A method for determining concentrations of retinol, tocopherol, and five carotenoids in human plasma and tissue samples. *Am. J. Clin. Nutr.* 56: 417–426.

Noble, A. C. 1975. Investigation of the color changes in heat concentrated tomato pulp. *J. Agric. Food Chem.* 23: 48–49.

Ojima, F., Sakamoto, H., Ishiguro, Y., and Terao, J. 1993. Consumption of carotenoids in photosensitized oxidation of human plasma and plasma low-density lipoprotein. *Free Radic. Biol. Med.* 15: 377–384.

Olson, J. A. 1994. Absorption, transport, and metabolism of carotenoids in humans. *Pure Appl. Chem.* 66: 1011–1015.

Oshima, S., Sakamoto, H., Ishiguro, Y., and Terao, J. 1997. Accumulation and clearance of capsanthin in blood plasma after the ingestion of paprika juice in men. *J. Nutr.* 127: 1475–1479.

Palan, P. R., Mikhail, M. S., Goldberg, G. L., Basu, J., Runowicz, C. D., and Romney, S. L. 1996. Plasma levels of β-carotene, lycopene, canthaxanthin, retinol, and α- and τ-tocopherol in cervical intraepithelial neoplasia and cancer. *Clin. Cancer Res.* 2(1): 181–185.

Panalaks, T., and Murray, T. K. 1970. The effect of processing on the content of carotene isomers in vegetables and peaches. *Can. Inst. Food Sci. Technol. J.* 3: 145–151.

Parker, R. S. 1988. Carotenoid and tocopherol composition in human adipose tissue. *Am. J. Clin. Nutr.* 47: 33–36.

Parker, R. S. 1996. Absorption, metabolism, and transport of carotenoids. *FASEB J.* 10: 542–551.

Pastori, M., Pfander, H., Boscoboinik, D., and Azzi, A. 1998. Lycopene in association with α-tocopherol inhibits at physiological concentrations proliferation of prostate carcinoma cells. *Biochem. Biophys. Res. Comm.* 250(3): 582–585.

Pauling, L. 1939. Recent work on the configuration and electronic structure of molecules with some applications to natural products: Isomerism and the structure of carotenoids. *Fortschr. Chem. Org. Naturstoffe* 3: 227–229.

Peng, Y. S., and Peng, Y. M. 1992. Simultaneous liquid chromatographic determination of carotenoids, retinoids, and tocopherols in human buccal mucosal cells. *Cancer Epidemiol. Biomarkers Prev.* 1: 375–382.

Pfander, H. 1987. *Key to Carotenoids*, 2nd ed. Birkhauser Verlag, Basel.

Pool-Zobel, B. L., Bub, A., Muller, H., Wollowski, I., and Rechkemmer, G. 1997. Consumption of vegetables reduces genetic damage in humans: First results of a human intervention trial with carotenoid-rich foods. *Carcinogenesis* 18: 1847–1850.

Porter, J. W., and Lincoln, R. E. 1950. *Arch. Biochem.* 27: 390.

Potischman, N., Hoover, R. N., Brinton, L. A., Swanson, C. A., Herrero, R., Tenorio, F., de Britton, R. C., Gaitan, E., and Reeves, W. C. 1994. The relations between cervical cancer and serological markers of nutritional status. *Nutr. Cancer* 21(3): 193–201.

Prince, M. R., and Frisoli, J. K. 1993. β-carotene accumulation in serum and skin. *Am. J. Clin. Nutr.* 57: 175–181.

Quackenbush, F. W. 1987. Reversed-phase HPLC separation of *cis* and *trans*-carotenoids and its application to food materials. *J. Liq. Chromatogr.* 10: 643–653.

Ramon, J. M., Serra, L., Cerdo, C., and Oromi, J. 1993. Dietary factors and gastric cancer risk. *Cancer* 71: 1731–1735.

Rao, A. V., and Agarwal, S. 1998. Effect of diet and smoking on serum lycopene and lipid peroxidation. *Nutr. Res.* 18: 713–721.

Raymundo, L. C., and Simpson, K. L. 1972. The isolation of a poly-*cis*-ζ-carotene from the tangerine tomato. *Phytochemistry* 11: 397–400.

Ribayo-Mercado, J. D., Garmyn, M., Gilchrest, B. A., and Russell, R. M. 1995. Skin lycopene is destroyed preferentially over β-carotene during ultraviolet irradiation in humans. *J. Nutr.* 125: 1854–1859.

Rock, C. L., and Swendseid, M. E. 1992. Plasma β-carotene response in humans after meals supplemented with dietary pectin. *Am. J. Clin. Nutr.* 55: 96–99.

Rock, C. L., Lovalvo, J. L., Emenhiser, C., Ruffin, M. T., Flatt, S. W., and Schwartz, S. J. 1998. Bioavailability of β-carotene is lower in raw than in processed carrots and spinach in women. *J. Nutr.* 128(5): 913–916.

Ronen, G., Cohen, M., Zamir, D., and Hirschberg, J. 1999. Regulation of carotenoid biosynthesis during tomato fruit development: expression of the gene for lycopene ε-cyclase is down-regulated during ripening and is elevated in the mutant Delta. *Plant J.* 17(4): 341–351.

Rouseff, R., Raley, L., and Hofsommer, H. J. 1996. Application of diode array detection with a C_{30} reversed phase column for the separation and identification of saponified orange juice carotenoids. *J. Agric. Food Chem.* 44: 2176–2181.

Saini, S. P. S., and Singh, S. 1993. Thermal processing of tomato juice from new hybrids. *Res. Industry* 38(3): 161–164.

Sander, L.C., and Wise, S. A. 1987. Effect of phase length on column selectivity for the separation of polycyclic aromatic hydrocarbons by reversed-phase liquid chromatography. *Anal. Chem.* 59: 2309–2313.

Sander, L. C., Sharpless, K. E., Craft, N. E., and Wise, S.A. 1994. Development of engineered stationary phases for the separation of carotenoid isomers. *Anal. Chem.* 66: 1667–1674.

Schierle, J., Bretzel, W., Buhler, I., Faccin, N., Hess, D., Steiner, K., and Schuep, W. 1997. Content and isomeric ratio of lycopene in food and human blood plasma. *Food Chem.* 96: 459–465.

Schmitz, H. H., Poor, C. L., Wellman, R. B., and Erdman, J. W. 1991. Concentrations of selected carotenoids and vitamin A in human liver, kidney and lung tissue. *J. Nutr.* 121: 1613–1621.

Schunck, C. A. 1903. *Proc. Royal Soc. London* 72: 165.

Schwartz, S. J., Nguyen, M. L., Ferruzzi, M. G., and Francis, D. M., 1999. Lack of isomerization of all-*E* lycopene in different tomato varieties during thermal food processing and digestion. Paper presented at the International Carotenoid Conference, Cairns, Australia.

Scita, G. 1992. Stability of β-carotene under different laboratory conditions. *Meth. Enzymol.* 213: 175–185.

Scott, K. J., Thurnham, D. I., Hart, D. J., Bingham, S. A., and Day, K. 1996. The correlation between the intake of lutein, lycopene and β-carotene from vegetables and fruits, and blood plasma concentrations in a group of women aged 50–65 years in the UK. *Br. J. Nutr.* 75: 409–418.

Sharma, S. K., and Le Maguer, M. 1996. Kinetics of lycopene degradation in tomato pulp solids under different processing and storage conditions. *Food Res. Int.* 29: 309–315.

Sies, H., and Stahl, W. 1998. Lycopene: Antioxidant and biological effects and its bioavailability in humans. *Proc. Soc. Exp. Biol. Med.* 218: 121–124.

Simpson, K. L., Lee, T. C., Rodriguez, D. B., and Chichester, C. O. 1976. Metabolism in senescent and stored tissues. In *Chemistry and Biochemistry of Plant Pigments*, Vol. 1, 2nd ed., T. W. Goodwin (Ed.), pp. 779–842, Academic Press, New York.

Snowdon, D. A., Gross, M. D., and Butler, S. M. 1996. Antioxidants and reduced functional capacity in the elderly: Findings from the Nun Study. *J. Gerontol. A. Biol. Sci. Med. Sci.* 51(1): M10–16.

Spady, D. K., Turley, S. D., and Dietschy, J. M. 1985. Receptor-independent low-density lipoprotein transport in the rat in vivo. *J. Clin. Invest.* 76: 1113–1122.

Stahl, W., and Sies, H. 1992. Uptake of lycopene and its geometrical isomers is greater from heat-processed than from unprocessed tomato juice in humans. *J. Nutr.* 122: 2161–2166.

Stahl, W., and Sies, H. 1996. Lycopene: A biologically important carotenoid for humans? *Arch. Biochem. Biophys.* 336: 1–9.

Stahl, W., Schwarz, W., Sundquist, A. R., and Sies, H. 1992. *Cis-trans* isomers of lycopene and β-carotene in human serum and tissues. *Arch. Biochem. Biophys.* 294(1): 173–177.

Stahl, W., Sundquist, A. R., Hanusch, M., Schwarz, W., and Sies, H. 1993. Separation of β-carotene and lycopene geometrical isomers in biological samples. *Clin. Chem.* 39: 810–814.

Steinmetz, K. A., and Potter, J. D. 1996. Vegetables, fruit, and cancer prevention: A review. *J. Am. Dietet. Assn.* 96: 1027–1039.

Steinmetz, K. A., Potter, J. D., and Folson, A. R. 1993. Vegetables, fruit, and lung cancer in the Iowa Women's Health Study. *Cancer Res.* 53: 536–543.

Straub, O. 1987. In *Key to Carotenoids*, 2nd ed, H. Pfander (Ed.). Birkhäuser, Basel.

Subar, A. F., Heimendinger, J., Patterson, B. H., Krebs-Smith, S. M., Pivonka, E., and Kessler, R. 1995. Fruit and vegetable intake in the United States: The baseline survey of the Five-A-Day-for-Better-Health Program. *Am. J. Health Promot.* 9: 352–360.

Sweeney, J. P., and Marsh, A. C. 1971. Effect of processing on provitamin A in vegetables. *J. Am. Diet. Assoc.* 59: 238–243.

Tajima, K., and Tominaga, S. 1985. Dietary habits and gastrointestinal cancers: A comparative case-control study of stomach and large intestinal cancer in Nagoya, Japan. *Jpn. J. Cancer Res.* 76: 705–716.

Tomes, M. L. 1967. The competitive effect of the β- and δ-carotene genes on α- or β-ionone ring formation in the tomato. *Genetics* 56: 227–331.

Tomes, M. L. 1969. δ-Carotene in tomato. *Genetics* 62: 769–780.

Tsukida, K., Saiki, K., Takii, T., and Koyama, Y. 1981. Structural elucidation of the main *cis*-β-carotenes. *J. Nutr. Sci. Vit.* 27: 551–561.

USDA. 1998. USDA-NCI Carotenoid Database for U.S. Foods. Nutrient Data Lab., Agric. Res. Service, U.S. Dept. of Agriculture, Beltsville Human Nutrition Research Center, Riverdale, MD.

van het Hof, K. H., Gartner, C., West, C. E., and Tijburg, L. B. M. 1998. Potential of vegetable processing to increase the delivery of carotenoids to man. *Int. J. Vit. Nutr. Res.* 68: 366–370.

van Poppel, G. 1993. Carotenoids and cancer: An update with emphasis on human intervention studies. *Eur. J. Cancer* 29A: 1335–1344.

Variyar, P. S., and Bandyopadhyay, C. 1990. On the carotenoids of ripened pepper berries (*Piper nigrum* L.) *J. Food Sci. Technol.* 27(5): 294–295.

Wang, C. J., Chou, M. Y., and Lin, J. K. 1989. Inhibition of growth and development of the transplantable C-6 glioma cells inoculated in rats by retinoids and carotenoids. *Cancer Lett.* 48(2): 135–142.

Wang, X. D. 1994. Review: Absorption and metabolism of β-carotene. *J. Am. Coll. Nutr.* 13: 314–325.

Weedon, B. C. L. 1971. Carotenoids: Occurrence. In *Carotenoids*, O. Isler (Ed.), pp. 267–324. Birkhäuser, Basel.

Weedon, B. C. L., and Moss, G. P. 1995. Structure and nomenclature. In *Carotenoids: Spectroscopy, 1B*, G. Britton (Ed.), pp. 27–70, Birkhauser Verlag, Boston.

Weisburger, J. H. 1998. Evaluation of the evidence on the role of tomato products in disease prevention. *Proc. Soc. Exp. Biol. Med.* 218: 140–143.

Went, F. W., Le Rosen, A. L., and Zechmeister, L. 1942. Effects of external factors on tomato pigments as studied by chromatographic methods. *Plant Physiol.* 17: 91–100.

Wilberg, V. C., and Rodriguez-Amaya, D. B. 1995. HPLC quantitation of major carotenoids of fresh and processed guava, mango, and papaya. *Lebensm. Wiss. Technol.* 28: 474–480.

Willett, W. C. 1995. Diet, nutrition and avoidable cancer. *Environ. Health Perspect.* 103: 165–170.

Willstatter, R., and Escher, H. H. Z. 1910. *Physiol. Chem.* 64: 47.

Yeum, K. J., Liu, C., Tang, G. W., Booth, S. L., Sadowski, J. A., Krinsky, N. I., and Russell, R. M. 1996. Human plasma carotenoid response to the ingestion of controlled diets high in fruits and vegetables. *Am. J. Clin. Nutr.* 64: 594–602

Zanori, B., Peri, C., Nani, R., and Lavelli, V. 1998. Oxidative heat damage of tomato halves as affected by drying. *Food Res. Int.* 31(5): 395–401.

Zechmeister, L. 1962. *Cis-Trans Isomeric Carotenoids, Vitamins A and Arylpolyenes.* Academic Press, New York.

Zechmeister, L., and Cholnoky, L. 1936. Lycoxanthin und Lycophyll, zwei natürliche Derivate des Lycopins. *Ber. Deutsch. Chem. Ges.* 69: 422–429.

Zechmeister, L., and Polgar, A. 1944. *Cis-trans* isomerization and *cis*-peak effect in the α-carotene set and in some other stereoisomeric sets. *J. Am. Chem. Soc.* 66: 137–144.

Zechmeister, L. and Tuzson, P. 1938a. Spontaneous isomerization of lycopene. *Nature* 141: 249–250.

Zechmeister, L., and Tuzson, P. 1938b. Isomerization of carotenoids. *Biochem. J.* 32: 1305–1311.

Zechmeister, L., and Tuzson, P. 1939. Umkerbare Isomerrisierung von Carotinoid durch Jod-katalyse. *Ber. Deutsch. Chem. Ges.* 72: 1340.

Zechmeister, L., Le Rosen, A. L., Went, F. W., and Pauling, L. 1941. Prolycopene, a naturally-occurring stereoisomer of lycopene. *Proc. Natl. Acad. Sci. USA* 27: 468–474.

Zechmeister, L., Le Rosen, A. L., Schroeder, W.A., Polgar, A., and Pauling, L. 1943. Spectral characteristics and configuration of some stereoisomeric carotenoids including prolycopene and pro-γ-carotene. *J. Am. Chem. Soc.* 65: 1940–1951.

8
Phycobilins

Jennifer D. Houghton

Ard Sgoil Port Righ Biology
Portree, Isle of Skye, Scotland

PHYCOBILINS IN NATURE

Structure

Think of the intense red of blood, and you are thinking of one of the most visually striking groups of pigments in nature. Heme and its related chrompohores provide some of the most intense natural pigments known and are distributed far more widely than we might suppose. Probably every living organism has the ability to synthesize heme, and many utilize the light-absorbing properties of its derivatives to their own ends. Perhaps the greatest quantities of heme-related pigments on the planet are in our lakes and oceans, where they are used by algae as light-harvesting and reaction center pigments in photosynthesis.

To start at the beginning we need to consider the family group of pigments known as the porphyrins. These large heterocyclic ring structures have a central theme in biology as energy-transducing compounds; they are involved in the processes of both respiration and photosynthesis and as such all life is dependent on them—indeed the porphyrins have been called the "pigments of life." Although porphyrins carry out many roles in biology and exist in many different forms, some common features are

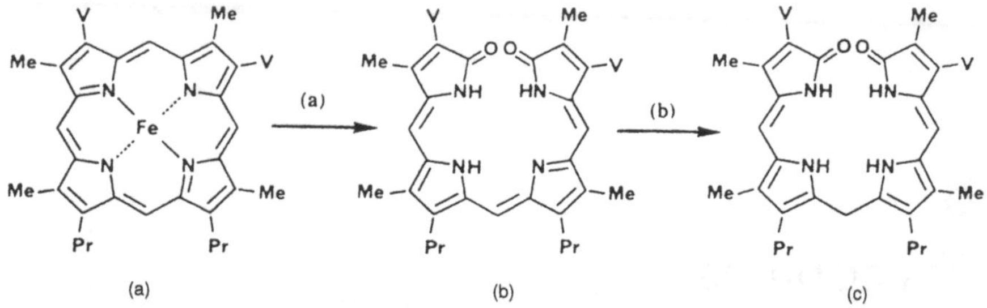

FIG. 8.1   The pathway of degradation of (a) heme to form (b) biliverdin and (c) bilirubin. $V = CHCH_2$, $Me = CH_3$, $Pr = CH_2CH_2CH_2OH$.

characteristic of them all. All porphyrins consist of four nitrogen-containing pyrrole rings joined together to produce a tetrapyrrole. This structure is reduced to form a series of delocalized double and single bonds not unlike those of a benzene ring. This sharing of electrons throughout the tetrapyrrole ring produces not only a remarkably stable structure but also gives the molecules their light-absorbing properties. The exact wavelengths of light absorbed depend on the chemical substituents surrounding the ring and the environment of the molecule itself. The structure of heme, for example (Fig. 8.1), has two propionic acid side chains, two vinyl groups, and four methyl groups, which act together on the tetrapyrrole ring to produce absorption in the blue and yellow parts of the spectrum, leaving red light alone to pass through and give heme its characteristic hue.

In our own bodies heme is broken down constantly as part of the turnover of old red blood cells in our spleens. However, we can also see the process of heme breakdown all too clearly in the development of a subcutaneous bruise. The natural progression of colors in a maturing bruise exactly reflects the pathway by which heme is broken down not only in our bodies but also in every living organism so far studied. The familiar changes from purple to green and finally yellow show the transformation of heme to biliverdin and bilirubin under our very eyes. The pigments produced are still tetrapyrroles but now have open ring structures, which change their light-absorbing properties and therefore their colors. These open ring structures are called bilins or bile pigments (Fig. 8.1) The green pigment biliverdin is produced by oxidation of the tetrapyrrole ring and is relatively unstable, breaking down to form the water-soluble yellow bilirubin, which is ultimately removed and excreted from our bodies.

In the plant kingdom, however, this breakdown process is more than just an excretory pathway; the breakdown of heme provides the biosynthetic route for synthesis of some of the most ubiquitous pigments on earth—the plant bilins. These pigments—phycocanin and phycoerythrin (Fig. 8.2)—are derived from biliverdin; the nature of their substituents renders phycocyanin blue and phycoerythrin red.

(a)

(b)

(c)

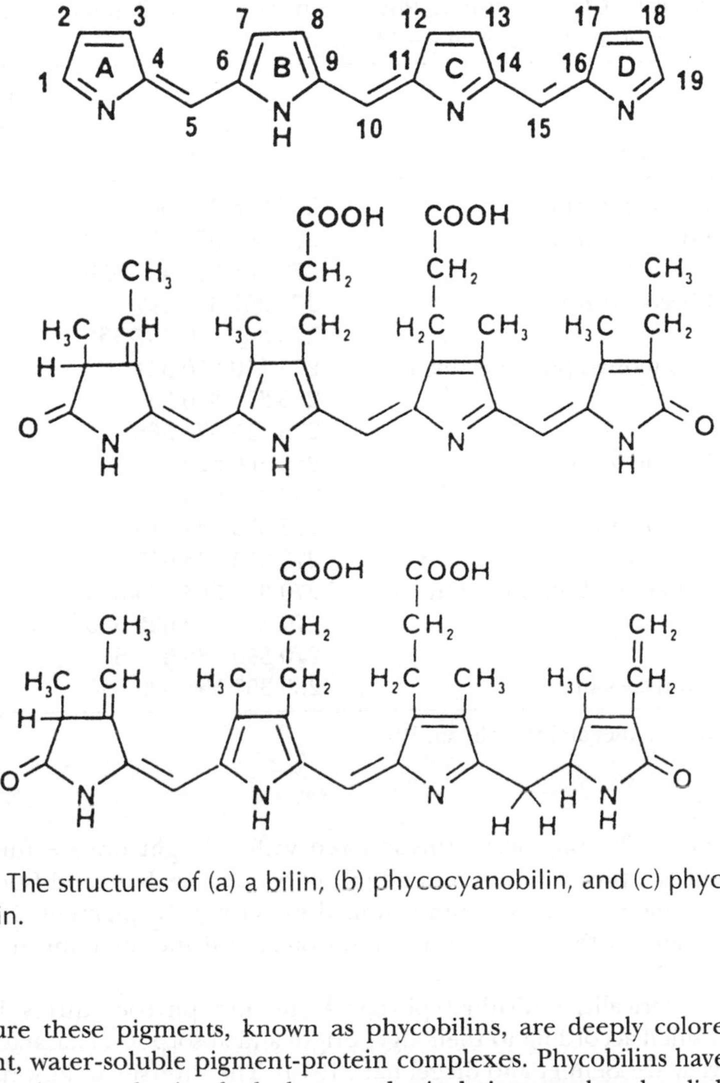

FIG. 8.2 The structures of (a) a bilin, (b) phycocyanobilin, and (c) phyco-erythrobilin.

In nature these pigments, known as phycobilins, are deeply colored, fluorescent, water-soluble pigment-protein complexes. Phycobilins have a central function to play in algal photosynthesis, being employed as light-harvesting pigments within their photosynthetic apparatus. They are found in all blue-green, red, and cryptomonad algae and are major biochemical constituents of these organisms, forming up to 40% of their soluble protein content. The range of algae in which phycobilins are found covers simple unicellular prokaryotic organisms like the blue-green algae as well as the familiar red seaweeds of our marine environment and includes the inhabitants of the famous acid springs in Yellowstone National Park . . . you will all have bumped into one somewhere. The pigments they contain can be classified on the basis of their spectral characteristics into three major groups: phycoerythrins (PEs), phycocyanins (PCs), and allophycocyanins

TABLE 8.1   Classification of Phycobilins and Their Characteristic
Absorption and Fluorescence Maxima

Phycobiliprotein	Absorption λ_{max} (nm)	Fluorescence emission λ_{max} (nm)
C-Phycoerythrin	280 308 380 565	577
R-Phycoerythrin	278 308 370 497 538	578
	278 308 370 497 556	
B-Phycoerythrin	278 307 370 497 545	578
	278 307 370 497 563	
Cryptomonad phycoerythrin	275 310 370 544	NR
	275 310 370 556	580
	275 310 370 568	NR
C-Phycocyanin	280 360 620	654
	280 360 615	647
R-Phycocyanin	275 355 522 610	637
	275 355 533 615	637 565
Cryptomonad phycocyanin	270 350 583 625 643	600
	270 350 588 625 630	NR
	270 350 588 625 615	637
Allophycocyanin	280 350 598 629 650	663

Source: Bennet and Siegelman, 1979.

(APCs). The phycoerythrins are red with a bright orange fluorescence, while the phycocyanins and allophycocyanins are blue and fluoresce red. The range of colors found within these groups of pigments is large, depending on the source of the biliprotein and the medium in which it is isolated.

Historically, individual phycocyanins and phycoerythrins have been classified according to their algal origin and absorption characteristics (for review see Bennet and Siegelman, 1979). The diversity of absorption maxima observed for the biliproteins in general is a function of the different apoproteins to which the bilins are attached, and their environment within the organism. Table 8.1 shows the absorption characteristics of phycocyanins and phycoerythrins from the three algal classes in which they occur and illustrates the variation between species. It is also possibile to identify species that preferentially synthesize individual phycobilins. For example, *Cyanidium*, *Nostoc*, and *Anabaena* species contain only phycocyanin, while a number of *Phormidium* species have predominantly phycoerythrin with only trace amounts of phycocyanin.

All of the algae described require large quantities of bilins for use in photosynthesis, yet most algae are capable of using simple carbon compounds as their energy source. How is this achieved?

Biosynthesis

The biosynthetic pathway of heme in all organisms has the substituted amino acid 5-aminolevulinate (ALA) as its first committed intermediate. In algae ALA is produced from the intact 5-carbon skeleton of the amino acid glutamate in an intriguing set of reactions involving glutamyl tRNA as an intermediate (Kannangara et al., 1988). This process has been the subject of considerable research interest, because the formation of ALA is likely to be a key control point in the biosynthesis of both chlorophyll and bilins in these photosynthetic organisms.

The next phase in the biosynthesis of phycobilins is the formation of heme. The process by which this occurs is essentially identical to that known for the biosynthesis of heme in mammalian tissues and illustrates how remarkably the process of tetrapyrrole biosynthesis has been conserved over such a range of organisms.

The formation of the chromophores of phycocyanin and phycoerythrin following the ring cleavage of heme, occurs through biliverdin IXα and probably one other intermediate (Brown et al., 1990). The final step in the biosynthesis of all bile pigments is likely to be the attachment of the bilin chromophore to its apoprotein. The biosynthesis of chlorophyll occurs in parallel to that of bile pigments, so that in organisms using both chlorophyll and biliproteins for photosynthesis, a single branched pathway is responsible for the biosynthesis of both pigments.

In red and blue-green algae phycobilins are present within the cell in the form of large multimeric aggregates called phycobilisomes. These structures are present in the chloroplast of the algae and can be seen as discrete features on its outer face when viewed under an electron microscope. The phycobilisomes of red and blue-green algae vary in their composition according to several factors. These include the organism from which they are isolated and the metabolic and spectral conditions in which they were grown (Gantt 1980, 1981; Grossman et al., 1986).

Metabolism

Chromatic Adaptation One of the characteristic properties of many algal species is their ability to produce different proportions of individual light-absorbing pigments in response to the different qualities of light in which the organism is growing. This is known as chromatic adaptation. Typically red light at wavelengths above 600 nm promotes the production of phycocyanin and allophycocyanin but not phycoerythrin, while green light at 500–600 nm promotes formation of phycoerythrin alone (Bogorad, 1975). In nature this allows algae to make the most efficient use of light at the available wavelengths, with the minimum expenditure of metabolic energy in synthesizing pigments. This characteristic has been known for some time (Hatiori and Fujita, 1959) and has been analyzed in detail in several different species. It is thus possible to maximize production of individual bile pigments by using the organism's natural mechanisms for adapting to light quality.

Sun-to-Shade Adaptation   Just as algae are able to adapt to the quality of light available for photosynthesis, synthesis of phycobiliproteins is also affected by light intensity. Increasing the light intensity causes an overall decrease in phycobiliprotein concentration, as part of an adaptation mechanism in which photosynthesis is maintained at its maximum rate using the minimum pigment necessary for light harvesting and energy transduction. Reductions of up to threefold in phycobiliproteins were reported in the unicellular rhodophyte *Porphyridium cruentum* (Brody and Emerson, 1959) in response to increasing light intensity. Similar reductions were observed in the blue-green alga *Anacystis nidulans* (Myers and Kratz, 1955), in which it was concluded that light intensity was one of the primary factors affecting pigment synthesis as a whole (Jones and Myers, 1965).

Effectors of Bilin Synthesis   The accumulation of any metabolite is dependent on its net rate of biosynthesis and degradation within the cell. Degradation of phycocyanin is known to occur in several species in response to nitrogen starvation (Wood and Haselkorn, 1979). A similar effect is also known following depletion of sulfur compounds from the growth medium of cyanobacteria (Schmid et al., 1982), These characteristics may, however, be used to increase pigment synthesis under certain conditions; addition of sulfate following a period of starvation produces a rapid and preferential increase in C-phycocyanin content prior to resumed cell growth. Addition of several substrates and intermediates of tetrapyrrole metabolism are known to induce bilin synthesis (Brown, et al., 1990). However, the addition of aminolevulinic acid, as a committed intermediate in bilin synthesis, is found *not* to increase overall levels of biliprotein but instead to cause the excretion of the free bilin chromophore (Beale and Weinstein, 1990).

Pigment Mutants   One means by which pigment synthesis is controlled very strictly, if artificially, is by the use of certain mutations in which biosynthesis of an individual pigment is blocked. In the case of the unicellular red alga *Cyanidium caldarium*, several such mutants exist (Nichols and Bogorad, 1962). Like the wild-type cells, these mutants are capable of growing heterotrophically in the presence of glucose in the dark. On transfer of dark-grown cells to the light, they synthesize pigments characteristic of their specific mutation as summarized in Table 8.2. The existence of these mutants demonstrates the possibility of using the independent functions of the branched biosynthetic pathway to produce individual pigments.

TABLE 8.2 Pigmentation of *Cyanidium caldarium* Mutants Transferred to Light Following Heterotrophic Growth

Cell type	Chlorophyll content (% of wild type)	Phycocyanin content (% of wild type)
Wild type	100	100
III-D-2	145	140
III-C	55	0
GGB	0	83
GGB-Y	0	0

Source: Troxler, 1980.

EXTRACTION AND PURIFICATION OF PHYCOBILINS

Phycobilins may be obtained without purification by simple freeze-drying of algal cells. This produces a brightly colored powder retaining many of the absorption characteristics of the phycobilins in their native form without loss of stability of the intact biliprotein structure. The spectrum of phycocyanin and chlorophyll in freeze-dried cells of *C. caldarium* that have been resuspended in aqueous solution is indistinguishable from that of whole cells in vivo (J. D. Houghton, unpublished result). The overall color obtained is a mixture of the total cellular pigment content of the cell—usually bilins, chlorophyll, and carotenoids. A pure blue color could be obtained by freeze-drying cells of the mutant GGB, which synthesizes phycocyanin without any detectable synthesis of chlorophyll *a*, giving the cells a pale blue color. The concentration of phycocyanin accumulated by the mutant is >80% of that achieved by wild-type cells (Troxler, 1980).

Phycobilins have been extracted from algae and purified in a range of forms from intact phycobilisomes to the protein-free chromophore. Traditionally the problem of purity has been approached by successive separation of phycobilins from the associated cellular pigments and proteins. This process is aided by the fact that phycobilins are very readily soluble in water. Following cell breakage it is therefore possible to separate the phycobilins from the lipid-soluble chlorophylls and carotenoids by high-speed centrifugation of the homogenate. This process yields brightly colored solutions of soluble protein. Further purification of the supernatant by ammonium sulfate precipitation and ion-exchange chromatography yields almost pure phycobilin in the form of linked protein subunits (Brown and Troxler, 1977).

A new and successful single-step method of purification of bilins has been developed in Leeds using preparative isoelectric focusing (Turner et al., 1997). In this method phycocyanins are separated from other soluble proteins according to their isolelectric points and form a single discrete peak of protein with an isoelectric point of about 5.9 pH units.

PROPERTIES OF EXTRACTED PHYCOBILINS

Temperature Effects

Among the characteristics of the algae that use phycobilins as light-harvesting pigments is their ability to survive under a variety of environmental stresses. Thus algae that are classified as thermophilic (high-temperature tolerant), acidophilic (acid-pH tolerant), halophilic (high-salt tolerant), or psychrophilic (low-temperature tolerant) might be expected to show different characteristics with regard to the stability of their pigments. Published data on the phycocyanins of several blue-green and red algae (Chen and Berns, 1978) indicate that resistance to denaturation varies from one species to another. The increased stability of the different phycocyanins is proposed to arise directly from differences in the primary amino acid structure of the proteins.

Phycoerythrins as a class are more stable toward denaturation than phycocyanins. R- and B-type phycoerythrins will retain their absorption and fluorescence maxima at temperatures up to 70°C (O'Carra and O'hEocha, 1976). This remarkable conformational stability has been suggested to arise from the presence of covalent disulfide bonds between cysteine residues of the protein subunits.

Solvent Effects

Because phycobilins are freely water soluble, most studies on their absorption characteristics have been carried out in aqueous media. All phycocyanins are known to be denatured in 4 M urea, while phycoerythrins are not. Among the phycoerythrins, the R and B types show increased stability over C and Cryptomonad pigments, denaturing only slowly in 8 M urea (O'Carra and O'hEocha, 1976). It is interesting to note that renaturation of biliproteins can be facilitated by simple removal of denaturants (Murphy and O'Carra, 1970), for example, by dialysis, reflecting an inherent stability of the pigment protein complex.

Phycocyanin has been extensively studied in terms of aggregation, and a review on this subject has appeared (Berns and MacColl, 1989). The stable state of a phycobilin varies from a simple monomer to distinct hexameric aggregates, reflecting the aggregation of bilins in phycobilisomes in vivo. In the case of C-phycoerythrin and C-phycocyanin, this aggregation is dependent on ionic strength and the presence of denaturants and detergents as well as concentration and pH (Bennett and Bogorad, 1971). In contrast, native forms of R- and B-phycoerythrins appear to exist in aqueous solution as stable hexamers (van der Veldt, 1973). A change in solvent, and therefore

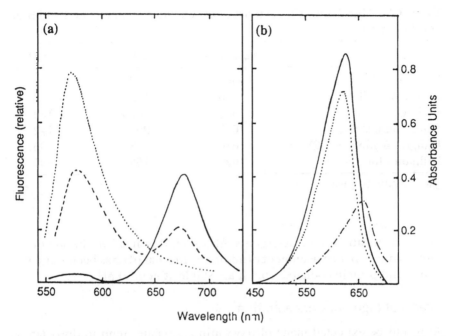

FIG. 8.3 Fluorescence emission spectra (a) and absorption spectra (b) of phycobilins of *Porphyridium* spp., showing the shifts occurring following dissociation of aggregates (——), to form partially (-----) and totally (·····) dissociated products, and finally on deraturation of the subunits (–·–·–·–).

aggregation state, of a bilin will affect both its absorption maxima and extinction coefficient. Whether the scale of these changes would be commercially significant would depend on the specific application.

Effects of pH

The visible spectral properties and fluorescence of all biliproteins are dependent on their attached apoproteins. Any change of pH affecting the interactions between bilin and apoprotein will by definition affect its absorption properties. The transition between the two stable aggregation states of R-phycocyanin is regulated by pH, with values of 6.5 or above favouring a trimeric form and values below 6.5 a hexamer. Larger pH changes bringing about complete unfolding or denaturation of the apoprotein cause a substantial decrease in the visible light absorbance (Berns and MacColl, 1989) and completely abolishes the visible fluorescence as shown in Fig. 8.3 (Rudiger, 1981). Phycocyanin from *Cyanidium caldarium* has an isoelectric point of 5.9. At this pH the positive and negative charges on the protein are equal; the protein becomes electrically neutral and will precipitate from so-

TABLE 8.3   Absorbance Characteristics of Available Synthetic Red and Blue
Food Colorants

Colorant	λ_{max} (nm)	E 1%	EmM
Ponceau 4R (E124)	505	431	71
Erythrosine (E127)	526	1154	—
Red 2G	528	620	122
Patent blue V (E131)	635	2000	172
Indigo carmine (E132)	610	489	105
Brilliant blue (FCF)	629	1637	207

Source: Walford, 1984.

lution in the absence of detergents. Although this alga lives in one of the most extreme pH environments known, its internal pH has been estimated at 6.3, and proteins isolated from it are stable at neutral pH.

Effects of Pigment Concentration

As would be expected many phycocyanin aggregates tend to dissociate at low pigment concentrations to form monomers (O'Carra and O'hEocha, 1976). In dilute solution, the extinction coefficient decreases at the absorption maxima (Neufield and Riggs, 1969; MacColl et al., 1971; Davis et al., 1986). Thus, factors affecting aggregation also affect absorbance, and it is important when making comparisons between different pigments to ensure that the physical conditions in which they are measured are comparable.

Several comparisons have been made of the colors of the phycobiliproteins with the synthetic red and blue pigments available commercially. Phycocyanin has been reported as being "between blue colour no. I (brilliant blue) and blue colour no. 2 (indigo carmine)" (Jacobson and Jolly, 1989). However, as we have seen, the exact spectral properties of all the phycobiliproteins are dependent on their algal source, state of purification and chemical environment. These spectra have been presented throughout this paper and it seems pertinent here to present for accurate comparison with the phycocyanins and phycoerythrins the absorption characteristics of the available synthetic blue and red colours (Walford, 1984) (Table 8.3).

PRESENT AND FUTURE APPLICATIONS

Although there is a long tradition in certain cultures of harvesting algae for food use, the single most promising area for further advances lies in the use of modern biotechnological methods for large-scale culture and extraction of microalgae (Cohen, 1986; Parkinson et al., 1987; Borowitzka and Borowitzka, 1988).

Following pioneering work on the extraction of β-carotene from the halophilic green alga *Dunaliella salina*, other microalgae are now beginning to be exploited for the harvesting of fine chemicals. To date, the algae most commonly cultured among the Rhodophyta and Cyanophyta are *Porphyridia* and *Spirulina* species. It is fair to say that a great many other species of algae could prove suitable for mass culture given the financial incentive to develop appropriate culture methods. This financial incentive is becoming increasingly apparent in the use of microalgae for production of vitamins, pigments, polysaccharides, and pharmaceutical compounds (Borowitzka, 1988). The processes described for the culture of *Spirulina* and *Porphyridium* are similar to those required for any number of algal species, given the biochemical knowledge to optimize their culture conditions and the economic incentive for doing so. It is reasonable to expect that, should the future demand for pigments justify it, a range of algal organisms might be employed to produce not only the C-phycocyanin and R-phycoerythrin currently available, but also the many other spectral variants of the pigments occurring in nature.

REFERENCES

Beale, S. I., and Weinstein, J. D. 1990. In *Biosynthesis of Heme and Chlorophylls*, H. A. Dailey (Ed.), pp. 287–391. McGraw Hill, New York.

Bennet, A., and Siegelman, H. W. 1979. In *The Porphyrins*, Vol. 6, D. Dolphin (Ed.), pp. 493–520. Academic Press, New York.

Bennett, A., and Bogorad, L. 1971. *Biochemistry* 10: 3625–3634.

Berns, D. S., and MacColl, R. 1989. *Chem. Rev.* 89: 807–825.

Bogorad, L. 1975. *Ann. Rev. Plant Physiol.* 26: 369–401.

Borowitzka, M. A. 1988. In *Microalgal Biotechnology*, M. A. Borowitzka, and L. J. Borowitzka (Eds.), pp. 173–175. Cambridge University Press, Cambridge, United Kingdom.

Borowitzka, M. A., and Borowitzka, L. J. 1988. *Micro-algal Biotechnology*. Cambridge University Press, Cambridge, United Kingdom.

Brody, M., and Emerson, R. 1959. *Am. J. Bot.* 46: 433–440.

Brown, A. S., and Troxler, R. F. 1977. *Biochem. J.* 163: 571–581.

Brown, S. B., Houghton, J. D., and Vernon, D. 1. 1990. *J. Photochem. Photobiol. B. Biol.* 5: 3–23.

Chen, C.-H., and Berns, D. S. 1978. *Biophys. Chem.* 8: 203–213.

Cohen, Z. 1986. In *CRC Handbook of Microalgal Mass Culture*, A. Richmond (Ed.), pp. 421–454. CRC Press, Boca Raton, FL.

Davis, L. C., Radke, G. A., Guikema, J. A. 1986. *J. Liq. Chromatogr.* 9: 1277–1295.

Fujita, Y., and Hattori, A. 1960. *Plant Cell Physiol.* 1: 293–303.

Gantt, E. 1980. *Int. Rev. Cytol.* 66: 45–80.

Gantt, E. 1981. *Ann. Rev. Plant Phjsiol.* 32: 327–347.

Grossman, A. R., Lemaux, P. G., and Conley, P. B. 1986. *Photochem. Photobiol.* 44: 827–837.

Hatiori, A., and Fujita, Y. 1959. *J. Biochem.* 46: 521–524.

Jones, L. W., and Myers, J. 1965. *J. Phycol.* 1: 7–14.

Kannangara, C. G., Gough, S. P., Bruyant, P., Hoober, J. K., Kahn, A., and Von Wettstein, D. 1988. *Trends Biochem. Sci.* 13: 139–143.

MacColl, R., Lee, J. J., and Berns, D. S. 1971. *Biochem. J.* 122: 421–426.

Murphy, R. F., and O'Carra, P. 1970. *Biochim. Biophys. Acta* 214: 371–373.

Myers, J., and Kratz, W. A. 1955. *J. Gen. Physiol.* 39: 11–22.

Neufeld, G. J., and Riggs, A. F. 1969. *Biochim. Biophys. Acta* 181: 234–243.

Nichols, K. E., and Bogorad, L. 1962. *Botan. Gaz.* 124: 85–93.

O'Carra, P., and O'hEocha, C. 1976. In *Chemistry and Biochemistry of Plant Pigments*, 2nd ed., Vol. 2, T. W. Goodwin (Ed.), pp. 328–376. Academic Press, London.

Parkinson, G., Shota, U., Hunter, D., and Sandler, N. 1987. *Chem. Eng.* 94: 19–23.

Rudiger, W. 1981. In *Pigments in Plants*, F.-C. Czygan (Ed.), pp. 314–351. Akademic-Verlag, Berlin.

Schmidt A., Erdle, I., and Kost, H.-P. 1982. *Z. Naturforsch.* 37c: 870–876.

Troxler, R. F. 1980. In *Chemistry and Physiology of Bile Pigments*, (P. D. Berk and M. I. Berlin (Eds.), pp. 431–454. Dept. of Health Education and Welfare.

Turner, L., Houghton, J. D., and Brown, S. B. 1997. *Planta* 201: 78–83.

van der Veldt, H. H. 1973. *Biochim. Biophys. Acta* 303: 246–257.

Walford, J. (Ed.) 1984. *Developments in Food Colours -2*. Elsevier Applied Science, London.

Wood, N. B., and Haselkorn. 1979. In *Limited Proteolysis in Microorganisms*, G. N. Cohen and H. Holzer (Eds.), pp. 79–1591. U.S. DHEW Publication No. N1H, Bethesda, MD.

9
Turmeric *

Ron Buescher and Luoqing Yang

University of Arkansas
Fayetteville, Arkansas

Turmeric is an important natural source of yellow pigment, spice, nutraceuticals, and herbal medicines. Turmeric is the common name used for the *Curcuma longa* L. (syn. *Curcuma domestica* val.) plant, a member of the Zingiberaceae (ginger) family and its products (American Spice Trade Association, 1982). The plant, its dried rhizomes, and products manufactured from the rhizomes are also known as curcuma, curcumin, Indian saffron, and yellow ginger. The terms turmeric and curcumin are commonly used interchangeably to refer to colorant products from turmeric rhizomes, while the pigments are called curcuminoids.

The perennial herbaceous plants are indigenous to the humid tropical climate of Southeast Asia, where the rhizomes have been used for millenia for their unique color, aroma, flavor, and health benefits. Although turmeric is cultivated in many tropical and subtropical regions of the world, most production is concentrated in the western states of India. India pro-

*Published with the approval of the Director of the Arkansas Agricultural Experiment Station, Fayetteville, AR.

TABLE 9.1   Turmeric Imported by the United States in 1997

Country of origin	Metric tons	Percent
China	1	<0.1
India	1954	96.0
Israel	1	<0.1
Other Pacific Islands (NEC)	48	2.3
Thailand	6	0.3
Turkey	9	0.4
Other[a]	25	1.1
Total	2044	

[a]Other countries, including Brazil, Haiti, Jamaica, Malaysia, Mexico, Peru.
Source: U.S. Department of Agriculture, 1998.

duces about 94% (225,000 metric tons) of the world's turmeric, which is primarily consumed domestically in the form of powdered rhizomes for flavoring and coloring foods. Since turmeric is not grown in the United States, the amount imported provides an estimate of its utilization by the United States. In 1997, 2043 metric tons of turmeric were imported by the United States in various forms, mainly for coloring (Table 9.1) (U.S. Department of Agriculture, 1998). About 96% of the imported turmeric originates from India. Import data do not specify the form of turmeric, which may be dried rhizomes, powder, oleoresins, or purified concentrates. Another consideration is that turmeric may be processed and subsequently exported by a country that does not originally produce it. Some extraction is done in the United States, but a large portion of turmeric products imported for coloring are curcumin extracts. Therefore, the actual amount of turmeric rhizomes utilized for colorant products in the United States is much greater than the amount reported as imports.

CULTIVARS AND PRODUCTION

Since turmeric is asexually propagated from its rhizomes for commercial production, genetic diversity is expected to be small, although some genotypes may be cross-pollinated to alter certain characteristics and grown from seed. The Indian Institute of Spice Research, Calicut, assumes responsibility for collecting and maintaining turmeric germplasm (Shamina et al., 1998). Currently, there are more than 50 known cultivars and 700 accessions of turmeric maintained with different names based on the locality where they were collected. A survey of isozyme polymorphism, to assess genetic variability, indicated a high degree of variability in turmeric collected from different geographical regions even though morphological features were similar (Shamina et al., 1998). In contrast, variability was low between accessions collected within a production region.

Selection of appropriate genotypes that are adapted to climatic conditions and cultural practices of a particular production region is very important for growers to attain satisfactory yields and marketable quality. Harvest yields of different turmeric cultivars vary from 10 to 43 metric tons/ha (Randhawa and Mahey, 1988). In addition, cultivars and accessions vary greatly in their dry yield and yellow pigment content (Govindarajan, 1980; Ratnambal, 1986). A survey of 184 accessions demonstrated that fresh rhizomes had 13.5–32.4% dry weight and curcuminoid content varied from 2.3 to 10.9%.

In India, six types of turmeric are marketed: Alleppey, Guddappali, Guntur, Madras, Rajpuri, and Sangali. Alleppey and Madras are the only types exported to western countries and are commonly recognized based on the name of their shipping port, even though each type may be composed of several different cultivars (Cripps, 1967). Alleppey turmeric is the type most commonly imported by the United States. The Alleppey rhizomes have a golden yellow external appearance and dark yellow to orange internal color (American Spice Trade Association, 1982). Total curcuminoids in Alleppey vary from 5 to 7%, and volatile oil content is higher than in the dark brown Madras turmeric, with less than 3% curcuminoids. Rhizomes produced in countries other than India are often classified as Madras or Alleppey turmeric, although some may assume the name of the country of origin, such as Jamaican.

Turmeric is propagated from rhizomes or pieces of rhizomes planted at a rate of 1–1.5 metric tons/ha that are usually saved by farmers from the previous year's crop. After planting, the ground is covered with leaves or straw mulch to assist with weed, moisture, and temperature control. Since the time from planting to harvest is relatively long, requiring 8–9 months, the fields are usually interplanted with faster growing crops or the turmeric is intercropped with perennial orchard crops. The plants grow to a height of about 1 m with light green oblong leaves expanding from leaf sheaths that emerge from the planted rhizome (Fig. 9.1).

During the growing season the planted rhizome, referred to as the bulb, enlarges and produces several long, cylindrical and multibranched secondary rhizomes called fingers. Foliage senescence, expressed by yellowing of leaves and stems, indicates that the rhizomes are mature and ready for harvesting. Harvesting is done by manually lifting the clusters of rhizomes from the soil. Soil, roots, and leaf scales are removed from the rhizomes, and a portion is selected and stored for planting the next crop. The bulbs and fingers are segregated and transferred to curing and drying facilities.

FIG. 9.1   Turmeric (*Curcuma longa* L.) plant containing rhizomes used for manufacturing natural colorants.

CURING, DRYING, AND POLISHING

Freshly harvested turmeric rhizomes need to be cured or dried to reduce decay during storage. Curing is usually referred to as the process of wound healing and involves callose development, suberization, and partial desiccation of epidermal tissues. This is practiced in the Philippines, where warm moist air is circulated around the rhizomes for about 2 days prior to storage (Magda, 1994). The cured turmeric has 80–85% water content and requires careful handling and storage to retard postharvest losses.

In India, the term curing is used to describe the process of cooking whole or split rhizomes in boiling water prior to drying. The rhizomes are cooked until soft. Cooking (curing) enables dehydration by disrupting the periderm and inactivates endogenous enzymes that could affect color and flavor.

Cooking also causes the curcuminoid pigments to diffuse through the tissues and reduces the rhizomes microbial population. Studies on predrying treatments and methods of drying showed that the main benefit of cooking was faster drying of whole rhizomes (Sampathu et al., 1988). Drying was also greatly accelerated by slicing the rhizomes. Curcuminoid extractability and yield were greater from noncooked rhizomes than from cooked ones, indicating that current practices involving cooking may cause losses in colorant yield or reduce extractability.

After cooking, the rhizomes are spread onto open platforms for sun-drying or transferred to hot-air dehydrators. The dried rhizomes are usually polished by physically rubbing their surfaces free of shriveled rootlets and rough, scaly materials. Turmeric powder may be applied to the rhizome surface to enhance color appearance (Sampathu et al., 1988). The smooth, bright yellow-orange product is then graded and weighed into gunny or polyethylene bags and stored until marketing or further processing (Randhawa and Mahey, 1988).

PROCESSING

Turmeric powder and oleoresins are approved for coloring foods in the United States (U.S. Food and Drug Administration, 1991). Processing of turmeric rhizomes into powder or oleoresin extracts is performed in either producing or importing countries. Powdered turmeric involves milling rhizomes to an appropriate particle size such as 60–80 mesh and removal of extraneous materials. Powdered turmeric may then be extracted to produce oleoresins by approved organic solvents (Table 9.2). The polar solvents (such as ethanol) effectively extract curcuminoids and volatile flavor substances, while nonpolar solvents (such as hexane) are used to further purify the oleoresin by removing nonpolar lipid-soluble substances. Hexane will extract lipids as well as bitter substances (Sair and Klee, 1967). The extracts are desolventized, leaving a highly viscous, dark-yellow or brownish-orange resinous product. In addition to variations occurring in turmeric powder

TABLE 9.2 Solvents Approved for Extracting and Purifying Turmeric Colorants

Acetone	Isopropanol
Ethanol	Methanol
Ethylene dichloride	Methylene chloride
Hexane	Trichloroethylene

Source: U.S. Food and Drug Administration, 1991.

composition, the amount of pigments, volatile and nonvolatile oils, and resinous substances in oleoresins will depend on the type of solvent(s) and methods used for extraction. Oleoresins usually contain about 10–12% of the total components of turmeric powder, which are mostly curcuminoids and volatile oils.

COMPOSITION

Carbohydrates, mainly starch, are the predominant components in dried turmeric rhizomes (Table 9.3) (Govindarajan, 1980). Turmeric pigments are the major component of the lipid fraction of turmeric rhizomes. Extractions with organic solvents primarily remove lipid and volatile oil substances while leaving carbohydrates, protein, and fiber as residues. About 10 to 15% of the pigments may also remain in the residue (Krishnamurthy et al., 1976). Oleoresins commonly contain 40 to 55 percent curcuminoid pigments and 15 to 20 percent volatile oil (Table 9.4). Amounts of individual curcuminoids vary considerably, although typically the amount of curcumin is more than demethoxycurcumin, which is more than bisdemethoxycurcumin. In some oleoresins demethoxycurcumin and bisdemethoxycurcumin may account for as much as 50% of the total curcuminoid content (Yang, 1990).

Sesquiterpenoids are the predominant components of the volatile oil extracted from turmeric rhizomes (Table 9.4) (Govindarajan, 1980; Hiserodt et al., 1997; He et al., 1998). While many substances have not been identified, ar-turmerone, turmerone, and curlone are known to be the major components. Further refinement of the pigments through additional processing steps may be done to eliminate the presence of the volatile oils that may influence flavor in some food products.

TABLE 9.3 Composition of Dried Turmeric Rhizomes

Component	Percent
Water	8–13
Carbohydrates	49–69
Protein	6–11
Fiber	3–6
Ash	4–9
Lipid	5–9
Volatile oil	2–6

Source: Adapted from Govindarajan, 1980.

TABLE 9.4   Composition of Turmeric Oleoresins

Component	Percent of oleoresin	Percent of main component
Total curcuminoids[a]	40–55	
Curcumin		42–81
Demethoxycurcumin		7–32
Bisdemethoxycurcumin		1–34
Volatile oil[b]	15–20	
ar-Turmerone		30–40
Turmerone		9–25
Curlone		18–25

[a]Extrapolated from Govindarajan, 1980; Yang, 1990; Tonnesen, 1992.
[b]Extrapolated from Hiserodt et al., 1997; He et al., 1998.

PRODUCTS

Natural color manufacturers market several different dry and liquid formulations of turmeric for use in food products. In general, powder formulations are ground rhizomes or purified turmeric extracts, and liquid formulations are oleoresin extracts diluted with vegetable oil, polysorbate 80, or propylene glycol (Lauro, 1991). Spray-dried products with emulsifiers such as polysorbate 60 or 80 are also manufactured. Combinations of propylene glycol with glycerides and vegetable oil with lecithin and glycerides are available to allow dispersion in both water and oil matrices of a food system (Anon., 1999). Examples of types of commercial turmeric colorant products and food applications are listed in Table 9.5.

STRUCTURE AND SYNTHESIS OF CURCUMINOIDS

The predominant color compounds of turmeric are curcumin, 1,7-bis (4-hydroxy-3-methoxyphenyl)-1,6-heptadiene-3,5-dione, and two demethoxylated curcuminoids, demethoxycurcumin and bisdemethoxycurcumin (Fig. 9.2) (Srinivasan, 1953; Tonnesen et al., 1982, 1983). Collectively these diarylheptanoids are referred to as curcumin or curcuminoids (Tonnesen, 1992a). Detailed information about the monoclinic crystal and molecular structure of curcumin has been reported based on x-ray crystallography examinations (Tonnesen et al., 1982; Gorbitz et al., 1986). The structure is essentially symmetrical with the two phenol rings deviating only 16° from coplanarity. The central diketones exist in the enol form and share a hydrogen atom, which possibly accounts for the predominately *cis*-enol configu-

TABLE 9.5 Examples of Commercial Turmeric Colorant Products and Applications

Turmeric products	Curcuminoid content (%)	Applications
Oil dispersible	10	Bakery mixes, processed cheese, margarines
Water miscible	3–8.5	Bakery mixes, beverages, breading, confections, ice cream, mustard, pickles, relish
Oil/Water dispersible	5–15	Bakery mixes, beverages, dairy products, soups, sauces, salad dressings
Powder	30–97	Cereals, dry mixes

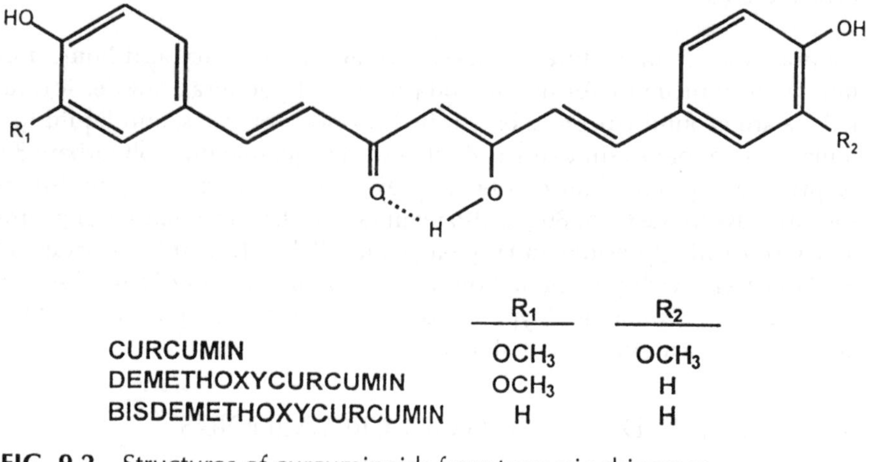

	R₁	R₂
CURCUMIN	OCH_3	OCH_3
DEMETHOXYCURCUMIN	OCH_3	H
BISDEMETHOXYCURCUMIN	H	H

FIG. 9.2 Structures of curcuminoids from turmeric rhizomes.

ration. An intermolecular interaction occurs between hydrogen and oxygen atoms (Gorbitz et al., 1986).

Curcumin can be chemically synthesized with an 80% yield from the reaction of vanillin with acetylacetone, boric anhydride, tri-isopropyl borate, and butylamine in ethylacetate (Padon, 1964). A variety of curcuminoid products can be synthesized by a similar procedure (Krackov and Bellis, 1997; Masuda et al., 1999). Since synthetic curcumin is relatively easy to produce and purify by crystallization, it has been used in many investigations to establish the structure and fundamental properties of curcumin (Gorbitz et al., 1986; Tonnesen, 1992a).

In contrast to established details on chemical procedures to synthesize curcumin, information is lacking on curcumin biosynthesis and regulatory mechanisms that influence biosynthesis within turmeric rhizomes. Based on observations that the curcumin structure is composed of two aromatic ring units, biosynthesis was suspected to involve the shikimate pathway with production of ferulates that are subsequently coupled through acetate or malonate (Roughley and Whiting, 1973). Studies adding labeled phenylalanine, cinnamate, malonate, and acetate to developing turmeric plants led to an alternative scheme whereby it was proposed that multiple additions of acetate units to a ferulate unit accounted for curcumin biosynthesis. Biosynthesis of curcuminoids through cyclization of terpenes produced from the mevalonate pathway is an alternative hypothesis. Through this scheme, biosynthesis would be similar for both curcuminoid color and sesquiterpene flavor substances in volatile oils of turmeric (Hiserodt et al., 1997; He et al., 1998). Based on the structural similarities of the sesquiterpenoids, such as ar-turmerones and curlone, to the curcuminoids, it is possible that they, or related intermediates, serve as precursors to curcuminoid synthesis (Fig. 9.3). Future studies focused on regulatory mechanisms involved with biosynthesis of both sesquiterpenoid flavor and curcuminoids should prove beneficial for the development of methods for improving turmeric production and quality.

SOLUBILITY

The curcuminoids are soluble in polar organic solvents such as acetone, ethanol, ethyl ether, ethylene chloride, glacial acetic acid, isopropanol, methanol, methylene chloride, propylene glycol, and triglyceride oils. In contrast, the curcuminoids are almost insoluble in water (pH 1–7) and in nonpolar organic solvents such as benzene, hexane, and pentane. Light absorption by the curcuminoids is dependent on the organic solvents used, although shifts in hue and chroma are usually not visually apparent. However, in ether or ethanol there is a slight green fluorescence.

The pigments are soluble in alkaline aqueous solvents above pH 7; however, in alkali the color shifts from yellow to red and brown-red. The sensitivity of curcuminoid color to alkaline pH has been used for measuring pH.

With the aid of emulsifiers, turmeric pigments are used to color aqueous foods. Liquids and surfaces of whole food tissues will be yellow; however, diffusion into aqueous tissues is often restricted, causing nonuniform coloration. Color diffusion is uniform when turmeric is used to pigment products such as relish and mustard. Another approach to accomplish solubilization or dispersion of turmeric in aqueous foods is to absorb or bond the pigments to soluble polysaccharides or proteins (Schranz, 1983; Todd,

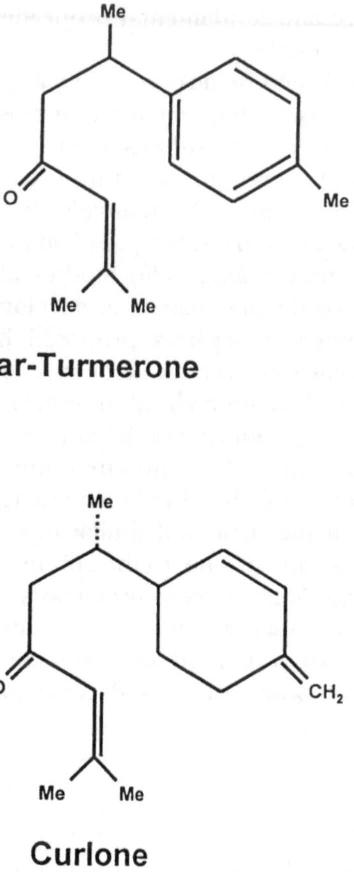

ar-Turmerone

Curlone

FIG. 9.3 Structure of common sesquiterpene flavor substances, ar-turmerone and curlone, in turmeric.

1991). Complexing with certain cations or phosphorylation also enhances water solubility of curcuminoids (Maing and Miller, 1981; Yang, 1990; Yang and Buescher, 1993).

ANALYSIS OF CURCUMINOIDS

Samples of turmeric products are commonly evaluated for their color value (CV) by dissolving and diluting in organic solvents such as ethanol or acetone and measuring absorbance at 420–430 nm. Discrepancies in the accuracy of this procedure have been reported due to observed variations in content and light absorption of individual curcuminoids (Tonnesen, 1992b). To improve standardization and accuracy, Tonnesen (1992b) recommended that the CV be determined based on the relative amount of each curcuminoid in a sample with the CV of each purified curcuminoid used as a reference for calculating the total color content. This procedure for determining CV requires chromatographic separation of the three curcuminoids.

Several procedures have been developed for separation of the curcuminoids. HPLC methods provide for rapid analysis, although thin-layer chromatography may be used, which is more time consuming but does not involve expensive equipment (Tonnesen and Karlsen, 1986; Rouseff, 1988; Tonnesen et al., 1991; Cooper et al., 1994; Price and Buescher, 1996). Using a HPLC mobile phase of THF/water buffered with citric acid at pH 3.0 allows baseline separation of the curcuminoids (Cooper et al., 1994). Citric acid is included in the mobile phase to eliminate interference of cations, especially iron, that may be complexed with the curcuminoids.

Detection is usually done spectrophometrically from 420 to 430 nm, although fluorometric methods may be used to detect very low amounts of curcuminoids. Depending on the solvent, excitation of curcuminoids is accomplished at 370–450 nm with fluorescence emission detected at 490–520 nm. Curcuminoids produce a strong fluorescence in organic solvents, which allows quantitative analysis as low as 0.3 ng/mL (Diaz and Peinado, 1992). In comparison, the limit of measurement using visible light absorbance is about 15 µg/mL (Cooper et al., 1994). For most requirements, measurement of light adsorbancy is satisfactory.

Accurate calculation of curcuminoid quantities has been a problem due to the lack of reliable molar extinction coefficients. While each curcuminoid requires a specific wavelength for absorption maxima in a given solvent, large differences in the amount of absorption between the curcuminoids have been reported resulting in differences in extinction coefficients (Tonnesen and Karlsen, 1983; Price, 1992; Cooper et al., 1994). Previous studies indicated that equal quantities of demethoxycurcumin and bisdemethoxy-curcumin had 2.4 and 3.5 times, respectively, greater absorption at 420 nm than curcumin (Tonnesen and Karlsen, 1983). More recently, it has been shown that extinction coefficients of the three curcuminoids are similar (Cooper et al., 1994). This is important since adjustments in total curcuminoid quantity based on the content of individual curcuminoids should not be required.

FACTORS AFFECTING COLOR AND STABILITY

Color and stability of curcuminoids are influenced by several factors, such as light, pH, temperature, chemical oxidants, metal ions, enzymes, and solvents. Color is most commonly altered due to degradation, although hue may also be affected by pH and metals. Instability of the curcuminoids greatly limits their utilization as a colorant. In contrast to the water-soluble synthetic yellow colorant, tartrazine, which is highly stable to light and chemical oxidation, turmeric is very unstable. However, in certain conditions the curcuminoids are highly stable. Curcuminoids in turmeric powder and oleoresin extracts are usually stable if protected from light. Also, in dry, dense, high-fat or protein foods, stability is usually not a problem if protected from exposure to light. The curcuminoids have even been shown to be stable in turmeric powder during storage for one year after being treated with 10 kGy of gamma irradiation (Chatterjee et al., 1999).

Solvents affect curcuminoid stability. In general, stability is much greater in organic solvents and declines with increasing water content (Racz et al., 1973; Tonnesen et al., 1986; Price and Buescher, 1996). In aqueous and nonaqueous solvents, oxygen has only a slight but significant influence on curcuminoid photodegradation, which is surprising because active oxygen species are involved in curcuminoid decomposition (Souza et al., 1997).

Effects of Light

Photodegradation of purified curcumin has been observed to be caused by UV (280–350 nm) as well as visible (>400 nm) radiation in both liquid and dry preparations (Tonnesen et al., 1986). Protecting against exposure to light above 500 nm by storage in brown glass containers protects curcuminoids against photodegradation. Use of clear UV or yellow filters only slightly reduces color fading (Crews, 1988). Light wavelengths from 280 to about 450 nm initiate singlet oxygen production, which contributes to sensitizing curcumin resulting in auto-oxidation, and other, yet undefined, mechanisms independent of oxygen participation (Tonnesen et al., 1986). Singlet oxygen quenchers such as β-carotene retard, but do not prevent, photodecomposition of curcumin. Exposure to triplet oxygen (ground state) has been shown to have only a minor effect on stability (Martins et al., 1994). Also, water activity ranging from 0.33 to 0.97 does not affect light stability of curcuminoids absorbed to cellulose (Souza et al., 1997).

Preventing exposure to light, especially to wavelengths <500 nm, appears to be the best protection against photodegradation. Complexing curcuminoids with aluminum ions reduces but does not prevent photodecomposition (Buescher and Yang, 1990). Other metal cations such as copper, iron, stannous, and zinc also reduce color fading, although these cations cause a change in hue (Maing and Miller, 1981; Yang, 1990).

An improved light-stable product was developed by forming a curcuminoid complex with water-soluble branched or cyclic polysaccharides or water-dispersible proteins (Todd, 1991). In addition to its resistance to photodecomposition, this curcuminoid colorant is water soluble if the polysaccharide or protein used for producing the complex is water soluble. A substantial improvement in light stability was also achieved by phosphorylating the curcuminoids (Yang and Buescher, 1993; Yang, 1990). Phosphorylated

TABLE 9.6   Decomposition (Half-Life) of Curcuminoids Affected by
Fluorescent Light, Alkaline pH, and Temperature

| | Half-life (h)[b] | | | | |
| | Light (1450 lux) | | pH | | Temperature |
Curcuminoid[a]	Water	Methanol	8	9	70°C
Curcumin (C)	12	360	330	5	3
Demethoxy C	11	210	710	14	5
Bisdemethoxy C	9	140	1300	36	11

[a]Results are from aqueous emulsions of curcuminoids except for the one indicated
to be in methanol.
[b]Calculated from first-order decomposition rate constants (Yang, 1990; Price and
Buescher 1996, 1997).

curcuminoid derivatives are water soluble and have a hue similar to unmodified curcuminoids.

Photooxidation of the three curcuminoid pigments follows first-order
kinetics (Tonnesen et al., 1986; Martins et al., 1994; Price and Buescher,
1996; Souza et al., 1997). The type of solvent has a major influence on
curcuminoid photodecomposition (Tonnesen and Karlsen, 1985b; Price
and Buescher, 1996). The curcuminoids dissolved in methanol are several times more stable in light than when emulsified in an aqueous solution (Table 9.6) (Price and Buescher, 1996). Based on first-order degradation rate constants, large differences in half-life values are observed
between the different curcuminoids in methanol with stability in the following order: curcumin > demethoxycurcumin > bisdemethoxycurcumin
(Table 9.6). The same order of stability occurs for emulsified curcuminoids;
however, differences in stability between the individual pigments are relatively small.

In decreasing order of quantity, products of curcuminoid photodecomposition are vanillin, p-hydroxybenzaldehyde, ferulic aldehyde, p-hydroxy-
benzoic acid, vanillic acid, and ferulic acid (Khurana and Ho, 1988). Exposure to light wavelengths greater than 400 nm causes curcumin to lose two
hydrogen atoms, and a yellow cyclization compound is the main degradation product (Tonnesen et al., 1986). Condensation of curcuminoids and
their degradation products would be expected since free radicals are un-
doubtedly involved in photodecomposition. Free radical–induced degradation products of curcumin are vanillin, ferulic acid, and a dimer of curcumin (Masuda et al., 1999). The dimer, bonded through the diketone
moiety of two curcumin molecules, is considered to be an unstable intermediate leading to the vanillin and ferulic acid products. Apparently some

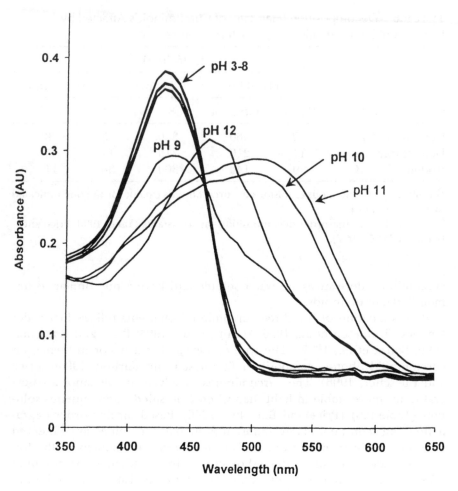

FIG. 9.4 Light absorption spectra of curcuminoids emulsified in buffers from pH 3 to 12.

of the products observed from curcumin degradation are transient, being susceptible to further oxidation by singlet oxygen and free radicals.

Effects of pH

Curcuminoids are yellow from pH 3 to pH 7 in aqueous solutions, turning brownish-red to deep red above pH 8. Major shifts in light absorption occur between pH 8 and 9, 9 and 10, and 11 and 12 that coincide with changes in curcuminoid hue (Fig. 9.4). The change in color at alkaline pH is accompanied by reduced stability and a rate of degradation several times greater above pH 8 than in acid solution. Degradation rates increase from pH 7.5 to a maximum at pH 10.2 and then gradually decline with increasing pH (Price and Buescher, 1997). The stability of each curcuminoid varies greatly, with bis-demethoxycurcumin being the most resistant to alkaline degradation (Table 9.6). Ferulic acid and feruloylmethane are the initial degradation products (Tonnesen and Karlsen, 1985a). Condensation products form, which subsequently result in a brownish-yellow hue. Microencapsulation of curcuminoids along with an acid pH buffer prevented color loss in alkaline foods (Leshik, 1981). Avoiding alkaline pH was recommended for curcuminoid use in gelatinous products such as puddings.

Effects of Temperature

The curcuminoids are considered to be relatively stable to temperature relative to their sensitivity to photooxidation; however, this does not appear to be consistent with current information. Decomposition of the curcuminoids emulsified in acidic aqueous solution at 70°C is a first-order reaction with the degradation rate of curcumin > demethoxycurcumin > bisdemethoxycurcumin (Table 9.6) (Yang, 1990; Yang and Buescher, 1996). Decomposition rate increases with temperature (Buescher and Yang, 1990; Martins et al., 1994), but aluminum ions completely inhibit thermal decomposition of the curcuminoids in acidified brine solution (Yang, 1990; Buescher and Yang, 1990).

Very large losses of curcuminoids added to foods can occur during cooking. As much as 90% of the curcuminoid content in curry powder or turmeric powder is destroyed when boiled for 15 or 30 minutes with split red gram (tur dhal) in water at pH 5.1 and 6.1 (Srinivasan et al., 1992). Less, but still large, losses (20–45%) were observed when extracted curcuminoids were used. While most studies have been conducted in model solutions, it is evident that much more research needs to be done on the thermal stability of curcuminoids in food preparations. Based on current evidence, popular acceptance that turmeric color is heat stable may need to be reconsidered.

Effect of Oxidants

Several chemical oxidants commonly present in the atmosphere or in foods cause decomposition of curcuminoids. Low levels of ozone (0.36 ± 0.11 ppm) cause decomposition of curcumin absorbed on cellulose or silica gel (Grosjean et al., 1988). Vanillin and vanillic acid are the major degradation products, which are likely formed from oxidation of ferulic acid following decomposition of curcumin. In contrast to ozone, short-time exposure of curcuminoids to hydrogen peroxide does not affect color (Yang, 1990).

Nitrites, which are present in certain processed foods or added for preservation, rapidly decompose emulsified curcuminoids (Yang, 1990). Curcuminoid color is lost at a rate of 10% per minute by the addition of

1 mM sodium nitrite. Decomposition rates are linear according to nitrite concentration.

While nitrite decomposition of curcuminoids is rapid, exposure to hypochlorite causes immediate loss of color in aqueous acidified brine solutions (Yang, 1990). Color loss increases linearly with increasing concentrate (1.5–33 μM) of sodium hypochlorite. Solutions containing 7 ppm curcuminoids lose about 10, 20, and 30% of their color from 10, 20, and 30 μM concentrations of hypochlorite, respectively. Based on these observations, relatively low concentrations of residual hypochlorite present from sanitizing hydrocooling water or washing produce and equipment can be detrimental to curcuminoid color stability. Other oxidizing agents may be similarly destructive, but their effects have not been reported.

Effects of Metal Ions

Several metal ions have been reported to enhance stability and affect other properties of curcuminoids. Boron is well known to complex with curcumin to form either rubrocurcumin or rosocyanin, depending on the type of acid solution (Dyrssen et al., 1972). The stable red pigments formed from the reaction of boron with curcuminoids provides a very sensitive method for analysis of boron. The hue of the boron-curcumin complex is pH sensitive, changing from crimson red to purple, blue, and green as the pH increases from 3 to 8. Although the pigments are not suitable for utilization in foods, the observations that boron drastically affects the color of curcumin and that the boron complex is more stable than curcumin alone is important for understanding interactions of the curcuminoids with metal ions and possibly other components in foods.

In addition to boron, other metal ions affect the solubility and stability of curcuminoids. Treatment of curcumin with potassium hydroxide forms bright orange, water-soluble crystals (Govindarajan, 1980). Treatment with potassium hydroxide causes some decomposition of curcumin resulting in potassium salts of ferulic and vanillic acids, which contribute to the product's color. Addition of curcumin to aqueous potassium sulfate solution also produces a water-soluble potassium-curcumin complex, which has improved heat and light stability (Maing and Miller, 1981). Aluminum is another metal ion that produces a water-soluble and more stable complex with curcumin without affecting hue (Maing and Miller, 1981; Yang, 1990; Yang and Buescher, 1990). Enhanced stability against light, heat, and peroxidase-induced degradation of curcuminoids can be accomplished by simply adding relatively low amounts of aluminum sulfate to emulsified aqueous solutions (Buescher and Yang, 1990; Yang, 1990). Curcuminoid complexes with stannous and zinc metal ions provide similar enhanced stability as potassium and aluminum curcuminoid complexes, although a

broad spectrum of yellow hues are created depending on the curcumi-
noid:metal ion ratio (Maing and Miller, 1981). Copper substantially reduces photodecomposition of curcuminoids; however, it causes a blue hue
(Yang, 1990).

While treatment of extracted curcuminoids with certain metal ions is
known to affect their properties, little is known about the effects of naturally
occurring metal-curcuminoid complexes in extracts on solubility, stability,
or hue. Variations in stability between batches of commercial curcuminoid
products are common and may be caused by variations in the amount of
metal ions originating naturally from the rhizomes or from other sources
such as water, organic solvents, or equipment. Treatment of curcuminoid
extracts with metal chelators has been reported to improve HPLC separations (Cooper et al., 1994) and enhance photodecomposition (Yang, 1990),
which indirectly provides evidence that a portion of extracted curcuminoids are coordinated with metal ions.

Effects of Peroxidase

Peroxidase is the only enzyme reported to catalyze curcuminoid decomposition, although it is likely that other oxidative enzymes may also affect curcuminoid stability (Buescher and Yang, 1990; Yang, 1990). Between pH 4.5
and 6.0, peroxidase from cucumbers or horseradish reduces absorbance of
an aqueous solution of curcuminoids from 0.9 to about 0.5 in one minute.
Rates of curcuminoid decomposition are reduced by a pH lower than 4.5
and increasing ionic strength. Antioxidants such as BHA and L-ascorbic
acid prevent peroxidase-catalyzed loss of color. These results indicate that
peroxidase, which is active in many food products, could play a role in the
loss of curcuminoid color and that the curcuminoids are effective substrates
for peroxidase assays. The mechanism of peroxidase-induced loss of curcuminoid color is unknown, although it is believed to be mediated through
oxidation of the phenyl hydroxyl groups instead of the conjugated ketone
chain structure. This would be similar to the oxidation of methoxyphenol
or hydroxydiphenol, both of which have been used as hydrogen donors in
peroxidase assays resulting in polymerized rather than fragmented products (Whitaker, 1994).

SUMMARY

Although turmeric rhizomes can be produced in many tropical regions of
the world, most production is concentrated in India where the greatest consumption is. The curcuminoid pigments of turmeric are important natural
colorants for foods and are gaining recognition for their antioxidant, anti-
inflammatory, anticarcinogen, and antimicrobial properties. There are

varying amounts of the three different curcuminoids in turmeric colorants that contain either 0, 1, or 2 methoxyl groups per molecule. Each of the curcuminoids has slightly different chemical properties, although the properties of a turmeric colorant preparation are usually similar to those of curcumin, which is the predominant pigment.

Both tartrazine (FD&C yellow No. 5) and the curcuminoids have similar color characteristics; however, unlike tartrazine, the curcuminoids are insoluble in water and are unstable. While water insolubility is overcome by emulsification, the problem of instability greatly affects utilization of turmeric colorants. The curcuminoids are well recognized for their sensitivity to light and alkaline pH. In addition, heat, chemical oxidants, and peroxidase can contribute to curcuminoid color loss. Progress has been made on developing an understanding of degradation mechanisms; however, information is lacking on methods to improve curcuminoid stability. A few methods have been reported to improve stability such as formation of curcuminoid complexes with metals, carbohydrates, and proteins or phosphorylation, although for various reasons these have not been widely accepted. Considering the need for a natural yellow color as an alternative to tartrazine, future development of new stable turmeric colorant products will be very important for the food industry.

ACKNOWLEDGMENTS

Financial support by the Arkansas Agricultural Experiment Station and technical assistance by Cathy Hamilton, Research Specialist, are gratefully acknowledged.

REFERENCES

American Spice Trade Association. 1982. *What You Should Know About Turmeric*, pp. 1–6. ASTA, Englewood Cliffs, NJ.

Anonymous. 1999. *Natural Colors*. Kalsec, Inc., Kalamazoo, MI.

Buescher, R., and Yang, L. 1990. Aluminum stabilizes turmeric in pickle brine against decomposition by light, heat and peroxidase. *J. Food Biochem.* 14: 263–271.

Chatterjee, S., Padwal-Desai, S., and Thomas, P. 1999. Effect of gamma irradiation on the colour power of turmeric (*Curcuma longa*) and red chillies (*Capsicum annum*) during storage. *Food Res. Int.* 31: 625–628.

Cooper, T., Clark, J., and Guzinski, J. 1994. Analysis of curcuminoids by high-performance liquid chromatography. In *Food Phytochemicals II: Teas, Spices, and Herbs*, American Chemical Society, Washington, DC, pp. 231–236.

Crews, P. C. 1988. A comparison of clear versus yellow ultraviolet filters in reducing fading of selected dyes. *Studies Conservation* 33: 87–93.

Cripps, H. D. 1967. Oleoresin turmeric applications in pickle production. *Glass Packer/Proc.* 46: 24–25.

Diaz, A., and Peinado, M. 1992. Fluorometric determination of curcumin in yogurt and mustard. *J. Agric. Food Chem.* 40: 56–59.

Dyrssen, D., Novikov, Y., and Uppstrom, L. 1972. Studies on the chemistry of the determination of boron with curcumin. *Anal. Chim Acta* 60: 139–151.

Gorbitz, C. H., Mostad, A., Pedersen, U., Rasmussen, P. B., and Lawesson, S. 1986. Structural studies of curcuminoids. V. Crystal structure of 1,7-bis (3, 4-demthoxyphenyl)-4-benzyl-1,6-heptadiene-3,5-dione (DDBHDD) and 1,7-bis (4-hydroxy-3-methoxyphenyl-4-(2-oxo-2-ethoxyethyl)-1, 6-heptadiene-3,5-dione(DHMEDD). *Acta Chem. Scand. B.* 40: 420–429.

Govindarajan, V. S. 1980. Turmeric-chemistry, technology and quality. *CRC Crit. Rev. Food Sci. Nutr.* 12: 199–301.

Grosjean, D., Whitmore, P. M., DeMoor, C., Cass, G., and Druzick, J. 1988. Ozone fading of organic colorants: Products and mechanisms of reaction of ozone with curcumin. *Environ. Sci. Technol.* 22: 1357–1361.

He, X., Lin, L., Lian, L., and Lindenmaier, M. 1998. Liquid chromatography-electrospray mass spectrometric analysis of curcuminoids and sesquiterpenoids in turmeric (*Curcuma longa*). *J. Chromatogr.* 818: 127–132.

Hiserodt, R. D., Ho, C., and Rosen, R. T. 1997. The characterization of volatile and semi-volatile components in powdered turmeric by direct thermal extraction gas chromatography-mass spectrometry. In *Spices: Flavor, Chemistry and Antioxidant Properties*, pp. 80–97. American Chemical Society, Washington, DC.

Khurana, A., and Ho, C. 1988. High performance liquid chromatographic analysis of curcuminoids and their photo-oxidative decomposition compounds in *Curcuma longa* L. *J. Liq. Chromatogr.* 11: 2295–2304.

Krackov, M., and Bellis, H. 1997. Process for the synthesis of curcumin-related compounds. U.S. Pat. 5,679,864.

Krishnamurthy, N., Mathew, A., Namburdiri, E., Shivasankar, S., Lewis, Y., and Natarajan, C. 1976. Oil and oleoresin of turmeric. *Trop. Sci.* 18: 37–42.

Lauro, G. 1991. A primer on natural colors. *Cereal Foods World* 36: 949–953.

Leshik, R. 1981. Stabilized curcumin colorant. U.S. Pat. 4,307,117.

Magda, R. R. 1994. Turmeric: A seasoning, dye and medicine. *Food Market. Technol.* (Oct.): 9–10.

Maing, I., and Miller, I. 1981. Curcumin-metal color complexes. U.S. Pat. 4,263,333.

Martins, M., Rusig, O., and Shirose, I. 1994. Temperature and light effects on turmeric (*Curcuma longa* L.) oleoresin extracts and curcumin. *Arq. Biol. Tecnol.* 37: 723–735.

Masuda, T., Hidaka, K., Shinohara, A., Maekawa, T., Takeda, Y., and Yamaguchi, H. 1999. Chemical studies on antioxidant mechanism of curcuminoid: Analysis of radical reaction products from curcumin. *J. Agric. Food Chem.* 47: 71–77.

Pardon, H. J. J. 1964. A synthesis of curcumin and related compounds. *Recueil* 83: 379–386.

Price, L. C. 1992. Kinetics and mechanisms of decomposition of curcumin and the curcuminoids by alkali and light. Ph.D. dissertation, University of Arkansas, Fayetteville, AR.

Price, L. C., and Buescher, R. W. 1996. Decomposition of turmeric curcuminoids as affected by light, solvent and oxygen. *J. Food Biochem.* 20: 125–133.

Price, L. C., and Buescher, R. W. 1997. Kinetics of alkaline degradation of the food pigments curcumin and curcuminoids. *J. Food Sci.* 62: 267–269.

Racz, I., Spiegel, P. and Jentzsch, K. 1973. Stability in solution of some curcuma pigments. *Acta Pharm. Hung.* 43: 18–24.

Randhawa, G. S., and Mahey, R. K. 1988. *Advances in the Agronomy and Production of Turmeric in India*, pp. 71–101. Oryx Press, Phoenix, AZ.

Ratnambal, M. 1986. Evaluation of turmeric accessions for quality. *Qual. Plant Foods Human Nutr.* 36: 243–252.

Roughley, P. J., and Whiting, D. A. 1973. Experiments in the biosynthesis of curcumin. *J. Chem. Soc., Perkin Trans.* I: 2379–2388.

Rouseff, R. 1988. High performance liquid chromatographic separation and spectral characterization of the pigments in turmeric and annatto. *J. Food Sci.* 53: 1823–1826.

Sair, L., and Klee, L. 1967. Debittering of turmeric. U.S. Pat. 3,340,250.

Sampathu, S. R., Krishnamurthy, N., Sowbhogya, H., and Shankaranarayana, M. 1988. Studies on the quality of turmeric (*Curcuma longa*) in relation to curing methods. *J. Food Sci. Technol.* 3: 152–155.

Schranz, J. L. 1983. Water-soluble curcumin complex. U.S. Pat. 4,368,208.

Shamina, A., Zachariah, T. J., Sasikumar, B., and George, J. K. 1998. Biochemical variation in turmeric (*Curcuma longa* Linn.) accessions based on isozyme polymorphism. *J. Hort. Sci. Biotechnol.* 73: 479–483.

Souza, C. R. A., Osme, S. F., and Gloria, M. B. A. 1997. Stability of curcuminoids in model systems. *J. Food Proc. Pres.* 21: 353–363.

Srinivasan, K. R. 1953. Chromatographic study of curcuminoids in *Curcuma longa*. *J. Pharm. Pharmacol.* 5: 448–457.

Srinivasan, K., Sambaiah, K., and Chandrasekhara, N. 1992. Loss of active principles of common spices during domestic cooking. *Food Chem.* 43: 271–274.

Todd, P. H. 1991. Curcumin complexed on water-dispersible substrates. U.S. Pat. 4,999,205.

Tonnesen, H. 1992a. *Chemistry of Curcumin and Curcuminoids*, pp. 143–153. American Chemical Society, Washington, DC.

Tonnesen, H. 1992b. Studies on curcumin and curcuminoids. XVIII. Evaluation of *Curcuma* products by the use of standardized reference colour values. *Z. Lebensm. Unters. Forsch.* 194: 129–130.

Tonnesen, H., and Karlsen, J. 1983. High-performance liquid chromatography of curcumin and related compounds. *J. Chromatogr.* 259: 367–371.

Tonnesen, H., and Karlsen, J. 1985a. Studies on curcumin and curcuminoids. VI. Alkaline degradation of curcumin. *Z. Lebensm. Unters. Forsch.* 180: 132–134.

Tonnesen, H., and Karlsen, J. 1985b. Studies on curcumins and curcuminoids. VII. Kinetics of curcumin degradation in aqueous solution. *Z. Lebensm. Unters. Forsch.* 180: 402–404.

Tonnesen, H., and Karlsen, J. 1986. Studies on curcumin and curcuminoids. VIII. Chromatographic separation and quantitative analysis of curcumin and related compounds. *Z. Lebensm. Unters. Forsch.* 182: 215–218.

Tonnesen, H., Karlsen, J., and Mostad, A. 1982. Structural studies of curcuminoids. I. The crystal structure of curcumin. *Acta Chem. Scand.* 36: 475–479.

Tonnesen, H., Karlsen, J., Mostad, A., Pedersen, U., Rasmussen, P. B., and Lowesson, S. 1983. Structural studies of curcuminoids. II. Crystal structure of 1, 7-bis (4-hydroxyphenyl)-1,6-heptadiene-3,5-dione-methanol complex. *Acta Chem. Scand.* B37: 179–185.

Tonnesen, H., Karlsen, J., and Henegouwen, G.B. 1986. Studies on curcumin and curcuminoids. VIII. Photochemical stability of curcumin. *Z. Lebensm. Unters. Forsch.* 183: 116–122.

Tonnesen, H., Grislingaas, A., and Karlsen, J. 1991. Studies on curcumin and curcuminoids. XIX. Evaluation of thin-layer chromatography as a method for quantitation of curcumin and curcuminoids. *Z. Lebensm. Unters. Forsch.* 193: 548–550.

U.S. Department of Agriculture. 1998. United States: Imports of specified condiments, seasonings, and flavoring materials by country of origin. Horticultural and Tropical Products Div. U.S. Department of Commerce, Washington, DC.

U.S. Food and Drug Administration. 1991. Code of Federal Regulations. 21CFR, Part 73.

Whitaker, J. 1994. *Principles of Enzymology for the Food Sciences*, pp. 570–573. Marcel Dekker, New York.

Yang, L. 1990. Stabilization of curcumin as a food colorant. Ph.D. dissertation, University of Arkansas, Fayetteville, AR.

Yang, L., and Buescher, R. 1993. Dye and method for making same. U.S. Pat.
5,210,316.

Yang, L., and Buescher, R. W. 1996. Light and heat induced decomposition
of curcumin and curcuminoids in aqueous solution, pp. 393–404. The
Second International Symposium on Natural Colorants. S.I.C. Publishing Co., Hamden, CT.

10
Chlorophylls

George A. F. Hendry
University of Dundee
Dundee, Scotland

CHLOROPHYLLS AS PIGMENTS UNDER NATURAL CONDITIONS

On its voyage through the solar system, the Galileo satellite passed planet Earth and recorded three unique features—an atmosphere rich in highly flammable oxygen, a stream of radio emissions in the MHz frequency, and a red light–absorbing pigment over much of the surface, land and sea. The red-absorbing pigment was due to chlorophyll, the oxygen from past photosynthesis. The radio emissions indicated the possible presence of an intelligent life form!

Chlorophylls are almost the only natural green plant pigment and certainly the only one in superabundance on planet Earth, with an estimated net annual global production far greater than any other biologically produced pigment. Some 75% of this production occurs in the aquatic and, in particular, the marine environment of cool temperate latitudes (Table 10.1).

The natural function of chlorophyll is to trap sunlight, to intensely concentrate the energy, forcing an electron to be ejected into an electron circuit and thus driving several chemical reactions to yield storable chemical energy. The whole process—photosynthesis—sustains life on this planet.

Table 10.1 Total Net Annual Global Production of Chlorophylls

Natural environment	Annual productivity (tons $\times 10^8$)
Terrestrial	2.92
Aquatic	8.63
Total	11.55

Source: Hendry et al., 1987.

The role of chlorophylls is central to the process of photosynthesis, although other pigments, notably the xanthophylls, certain carotenes, and bile pigments, play an auxiliary, although essential, role. The remarkable feature of photosynthesis is that the chlorophyll molecule remains highly stable despite participating in high-energy photoreactions.

Photosynthesis, as a biochemical and biophysical process, is at least 3500 million years old and, in all but fine details, has probably not altered significantly in that time. Given the powerful selective forces of evolution, this is a remarkable example of biological conservatism and says much about the efficiency of the system. Today we still have no more than five known widespread types of chlorophylls present in terrestrial plants and algae, conveniently called chlorophylls *a*, *b*, *c*, *d*, and *e*, with four bacteriochlorophylls recorded in photosynthetic bacteria (Table 10.2).

Table 10.2 The Types and Distribution of Naturally Occurring Chlorophylls and Bacteriochlorophylls

Pigment	Natural distribution
Chlorophylls	
a	All plants, algae, blue-green algae, and Prochlorophyta bacteria
b	All plants, green algae, Euglenophyta, bacterial Prochlorophyta
c	Brown algae, dinoflagellates, diatoms, and other algal groups
d	Some red algae and other algal groups
e	Some yellow-green algae
Bacteriochlorophylls	
a and *b*	Photosynthetic purple bacteria
c and *d*	Photosynthetic green and purple sulfur bacteria

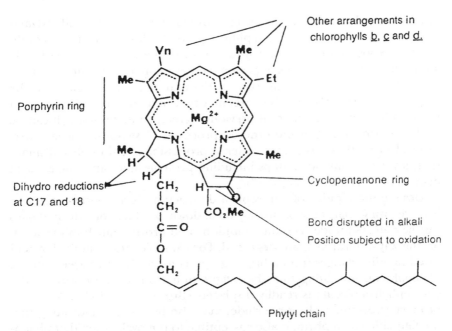

Other arrangements in
chlorophylls b, c and d.

Vn   Me

Me   Et

Porphyrin ring

Mg²⁺

Me   Me
H

Dihydro reductions   H
at C17 and 18   CH₂   H   Cyclopentanone ring

CH₂   CO₂Me

C=O   Bond disrupted in alkali

O   Position subject to oxidation

CH₂

Phytyl chain

FIG. 10.1 The structure of chlorophyll *a,* showing the location of structural differences in chlorophylls *b, c,* and *d* and the dihydro reductions present in the bacteriochlorophylls. The two positive charges on the central magnesium atom are balanced by two negative charges shared between the four pyrrole-nitrogens. The arrangement of the 10 double bonds within the macro-ring structure may also vary. (-----), Delocalized system of electrons. (See Hendry, 1996, for details of the nomenclature and numbering system of the rings and individual carbon atoms.)

The differences between the types of chlorophyll are entirely confined to the peripheral side groups (see Hendry, 1996, for a full description), and while these differences have an important functional significance in the reception of light, in terms of natural abundance and agricultural and industrial accessibility, just one type—chlorophyll *a*— dominates (Fig. 10.1).

Under natural conditions the chlorophyll molecule can be very long-lived, acquiring its stability in a lipid-rich environment in association with certain polypeptides, carotenoids, tocopherols (vitamin E), and other antioxidants located within or on the surfaces of the membranes present within the organelles of photosynthesis, the chloroplasts. The half-life of chlorophyll *a* in mature, healthy, higher plants is likely to be not much less than the half-life of the leaf itself (Hendry et al., 1987). In this sense, chlorophyll in vivo is a relatively highly stable molecule. Given that the sun is a powerful agent of photobleaching, it is remarkable that only under exceptional environmental conditions is the chlorophyll molecule destabilized and may, in susceptible species, be rapidly degraded. This occurs under conditions of exposure to high concentrations of ozone or following herbicide application or combinations of prolonged low temperatures and high light irradiance. At certain seasons, such as the end of the rainy season in the tropics or the weeks before the onset of winter in temperate latitudes, or in the days leading to ripening of previously green fruit, chlorophyll can be broken down as part of the ordered processes of season-induced senescence. During these events the ordered processing of chlorophyll dismantling involves many other steps including protein, lipid, and nucleic acid degradation (Hendry et al., 1987).

Despite the rapidity of chlorophyll destruction during senescence and fruit ripening, naturally occurring breakdown products or intermediates can be detected under certain conditions and provide much information on the way chlorophylls are degraded. For example, grass cut for hay and dried rapidly changes from a bright green color to the grey-green hue of phaeophytin, indicating that the magnesium atom at the centre of the chlorophyll molecule is readily displaced (Brown et al., 1991). A second product, grey-green phaeophorbide, may also be present, indicating that the fatty side chain phytol is also susceptible to removal. Given that phaeophytin and phaeophorbide and, to a lesser extent, the intermediate chlorophyllide are present in senescing leaves and ripening fruits, these products are, by definition, naturally occurring pigments (Fig. 10.2). And in case there is any doubt, these products do indeed form part of the natural human diet. An all too familiar example illustrating the widespread occurrence of one or more of these products is the gray-yellow hue of less-than-fresh cabbage on the supermarket shelf due to the presence of increasingly large concentrations of phaeophorbide and phaeophytin! Beyond this, the molecular processes breaking down phaeophorbide to colorless products have been described (Hortensteiner et al., 1998).

CHLOROPHYLLS IN ISOLATION

In contrast to chlorophylls under natural conditions in healthy leaves, chlorophylls in isolation are potentially highly unstable. Much effort has been spent over many years in an attempt to describe the ways and sequence in which the isolated chlorophyll molecule is broken down, carbon atom by carbon atom, and these have been described (Hendry et al., 1987; Brown

FIG. 10.2 (Opposite) Scheme showing the chemical relationship between chlorophyll a and three naturally occurring pigmented breakdown products: chlorophyllide, phaeophytin, and phaeophorbide.

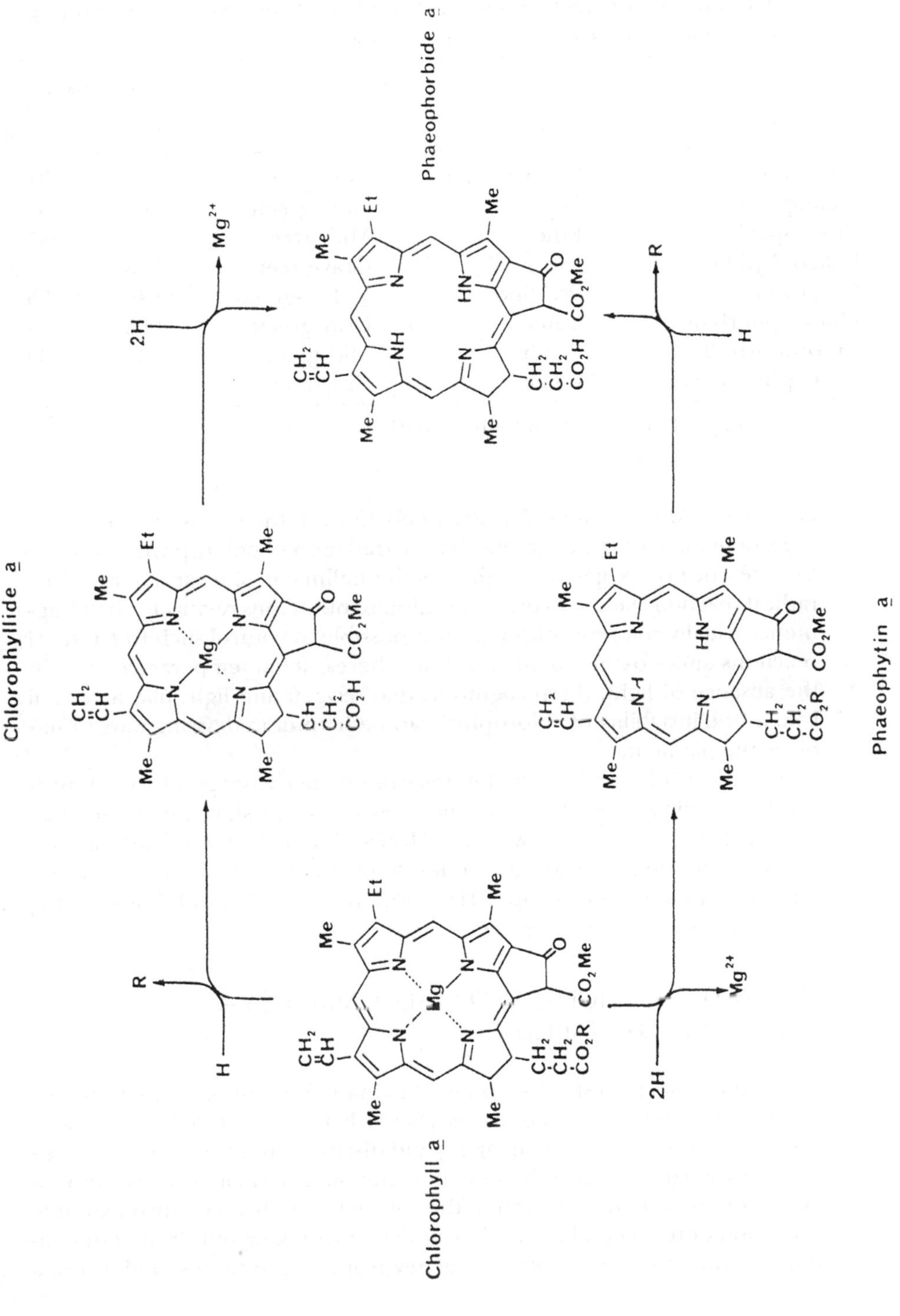
Chlorophyll a
Chlorophyllide a
Phaeophorbide a
Phaeophytin a

TABLE 10.3 Spectral Characteristics of the Most Abundant Naturally Occurring Chlorophylls and Their Respective Phaeophorbides

Pigment	Solvent	Color	Spectral absorbance λ_{max}	εmM
Chlorophyll *a*	Diethyl ether	Blue-green	662	91
Phaeophytin *a*	Ether	Gray-green	667	56
Chlorophyll *b*	Ether	Mid-green	644	51.5
Phaeophytin *b*	Ether	Gray-green	655	37.3
Chlorophyll *c*1	Pyridine	Yellow-green	640	35
Phaoeophorbide *c*1	Ether	Gray-green	666	—
Chlorophyll *c*2	Pyridine	Mid-green	642	32
Phaeophorbide *c*2	Ether	Gray-green	665	—

Source: Dawson et al., 1986; Hoff and Amesz, 1991.

et al., 1991; Merzlyak and Hendry, 1993). Only in the last decade or so has it become increasingly apparent that degradation of chlorophyll in vitro in the presence of oxygen and light has the hallmarks of a randomized free radical reaction leading to the generation of numerous relatively small fragments, usually colorless. Although it is possible to control such free radical reactions under oxygen-depleted atmospheres, at low temperatures, and in the absence of light, the recognition that oxygen and light play a central role in the instability of chlorophyll can be helpful in devising ways to stabilize the pigment.

Technical information on the absorption spectroscopy of the chlorophylls including their phaeophorbide derivative are shown in Table 10.3, with further information provided by Hoff and Amesz (1991). The table emphasizes one notable feature frequently overlooked in the industrial literature—the various chlorophylls are different in color, with hues varying from deep green to yellow-green.

FREE RADICALS, LIGHT, AND STALE CABBAGES— THE QUEST FOR STABILITY

Free radical biochemistry has its own fascination, if only because it underlies our own bodily destruction—particularly the role played by free radicals in the processes of human aging and disease—and first formulated using mammalian systems as the model by Denham Harman in 1956. Perhaps even more profoundly apparent is the role of free radicals in the inexorable operation of the second law of thermodynamics taking our dead bodily remains through the final earthly journey from corpse to dust and ashes, a process initiated by oxygen and propagated from molecule to molecule by free radical reactions. These free radical reactions are commonplace—the lipids of vegetable oils, meat, milk, and butter being readily degraded in air at room temperature to form lipid peroxides, which on further decomposition form free radicals. This is the basis of the chemistry of rancid butter and souring milk. What drives the generation of these destructive free radical reactions is usually oxygen, atmospheric oxygen.

Oxygen itself is a free radical—simply defined as an atom or molecule with an unpaired electron. A brief account of the role of oxygen and free radical reactions in the context of plants and plant products is available (Hendry, 1993). What makes oxygen so important here is its relative superabundance. Healthy chlorophyll-containing plants, in contrast to higher vertebrates, including humans, synthesize a complete complement of molecular defenses against oxidative attack in the form of antioxidants, including the tocopherols (vitamin E), ascorbic acid (vitamin C), and carotenes (pro-vitamin A) (see Salin, 1987, for a more full discussion on the defense sytems). What industrial isolation of chlorophyll in effect achieves is the separation of the chlorophyll molecule from its natural defenses, as a consequence exposing the pigment to the ravages of oxygen. The origin of almost all of the industrial chlorophyll derivatives lies in attempts to restructure the natural pigment to form a molecule relatively stable in isolation, which mimics the green hues of nature.

Because extracted chlorophylls without further modification have relatively poor stability, particularly on exposure to light, most unaltered chlorophylls are used in cosmetics and toiletries. In foods, in a limited range of commercial applications, unaltered chlorophylls can be stabilized as the oil-soluble product usually in combination with added carotenoids and phospholipids. The final color will depend on the techniques used during chlorophyll extraction and the presence of other colored compounds present—notably carotenes and xanthophyll. End use is partly determined by pH. Under acidic conditions, particularly, chlorophyll loses its magnesium atom to yield phaeophytin, so acquiring the yellow-green hue as seen in less-than-fresh cabbages. Chlorophylls are permitted as food colorants throughout western Europe, although their use appears to be limited. A trade in oil-based chlorophylls for use in coloring vegetable oils has existed for many years.

In terms of usage, however, more widespread and with a much wider range of applications, real and potential, is the practice of stabilization substituting the magnesium atom in natural chlorophyll with copper. These copper-substituted chlorophyll derivatives are, to a significant degree, less responsive to light and to oxidation and are relatively stable under alkaline conditions offering the food processor greater tinctorial strength. Removal of the fat-soluble phytol side chain by hydrolysis under alkaline conditions produces the water-soluble chlorophyllin in the form of a sodium or potassium salt. Available in powder or liquid form, copper-substituted chlorophylls have found a market in Europe as colorants in sugar confectionary, ice creams, cucumber relishes, dessert mixes, and cheeses, often in combination with carotenoids or curcumin. A fuller description of these applications is available in Henry (1996). Chlorophyllins are permitted as a food colorant in western Europe, but their use in the United State is limited to dentifrice.

The industrial processes underlying these modifications to the original chlorophyll molecule have been described in some detail by Hendry (1996). The substitution of Cu for Mg, however, does not provide complete stability; the chlorophyllin molecule itself is still subject to free radical attack (Pentilla et al., 1996). To that extent the attempts to improve stability have become a compromise between the ready availability and cheapness of chlorophyll, the ease of removing Mg and substituting Cu, and the demands of the food colorant industry for a product that is, in other respects, structurally close to the natural product chlorophyll. Almost certainly chlorophyll derivatives based on phaeophorbide could be made even more stable, but only at the price of restructuring the molecule to a degree that would make it less and less close to the original natural compound.

FUTURE PROSPECTS

Advances in our understanding of the chlorophylls and chlorophyll stability have been, in recent years, directed toward pinpointing the biophysical events associated with processing light energy, the biochemical events associated with the orderly dismantling of the chlorophyll molecule under natural conditions, and the highly disordered molecular processes surrounding the generation of destructive free radicals in vivo and in vitro. Taken together—and particularly in the present context of natural food colorants—these advances, in the author's opinion, point more and more to the need to develop techniques for the reconstitution of extracted chlorophyll in an environment more approximating to the natural environment of the chlorophyll molecule in chloroplast membranes. At the laboratory rather than the industrial scale, the techniques are well established in, for example, molecular photochemistry and biophysics. One other area of perhaps more immediate practical promise is biotechnology (O'Callaghan, 1996). The analogy of harvesting chlorophyll from an alga rather than land plants (as at present with the cultivation of the carotenoid-rich microalga *Dunaliella salina*) is intriguing. Single-celled algae grown under appropriate conditions to produce optimum concentrations of chlorophyll *a*, α-tocopherol, and ascorbic acid, harvested, extracted, and embedded in an acceptable medium could provide an abundant source of stable natural (that is unaltered) chlorophyll. It would be difficult to argue that this is not a wholly

CHLOROPHYLLS235

natural colorant. Stability would then depend on the investment needed to counter oxidative attack through natural antioxidants. The potential to develop chlorophyll—particularly natural, structurally unaltered, chlorophyll—is considerable.

The European Commission—whose writ runs now through 15 western European countries (with more to follow)—bases its food safety directives substantially on the work of the Joint UN Food and Agriculture Organization—World Health Organization Expert Committee on Food Additives (JECFA). Chlorophyll is classified by the JECFA as not requiring a toxicological limitation. Francis (1996) has covered the regulations on safety of chlorophyll more fully.

REFERENCES

Brown, S. B., Houghton, J. D., and Hendry, G. A. F. 1991. Chlorophyll breakdown. In *Chlorophylls*, H. Scheer (Ed.), pp. 465–489. CRC Press, Boca Raton, FL.

Dawson, R. M. C., Elliott, D. C., Elliott, W. H., and Jones, K. M. 1986. *Data for Biochemical Research*, 3rd ed. Clarendon Press, Oxford.

Francis, F. J. 1996. Safety of food colorants. In *Natural Food Colorants*, G. A. F. Hendry and J. D. Houghton (Eds.), pp. 112–130. Blackie Academic, London.

Harman, D. 1956. Aging: A theory based on free radical and radiation chemistry. *J. Gerontol.* 11: 298–313.

Hendry, G. A. F. 1993. Oxygen, free radical processes and seed longevity. *Seed Sci. Res.* 3: 141–153.

Hendry, G. A. F. 1996. Chlorophylls and chlorophyll derivatives. In *Natural Food Colorants*, G. A. F. Hendry and J. D. Houghton (Eds.), pp. 131–156. Blackie Academic, London.

Hendry, G. A. F., Houghton, J. D., and Brown, S. B. 1987. The degradation of chlorophyll—a biological enigma. *New Phytol.* 107: 255–302.

Henry, B. S. 1996. Natural food colours. In *Natural Food Colorants*, G. A. F. Hendry and J. D. Houghton (Eds.), pp. 40–79. Blackie Academic, London.

Hoff, A. J., and Amesz, J. 1991. Visible absorption spectroscopy of chlorophylls. In *Chlorophylls*, H. Scheer (Ed.), pp. 723–736. CRC Press, Boca Raton, FL.

Hortensteiner, S., Wuthrich, K. L., Matile, P., Ongania K-H., and Krautler, B. 1998. The key step in chlorophyll breakdown in higher plants: Cleavage of phaeophorbide *a* macrocycle by a monooxygenase. *J. Biol Chem.* 273: 15335–15339.

Merzlyak, M. N., and Hendry, G. A. F. 1993. Free radical metabolism, pigment degradation and lipid peroxidation in leaves during senescence. *Proc. Roy. Soc. Edinburgh.* 102B: 459–471.

O'Callaghan, M. C. 1996. Biotechnology in natural food colours: The role of bioprocessing. In *Natural Food Colorants*, G. A. F. Hendry and J. D. Houghton (Eds.), pp. 80–111. Blackie Academic, London.

Pentilla, A., Boyle, C. R., and Salin, M. L. 1996. Active oxygen intermediates and chlorophyllin bleaching. *Biochem. Biophys. Res. Commun.* 226: 135–139.

Salin, M. L. 1987. Toxic oxygen species and protective systems of the chloroplast. *Physiol. Plant.* 72: 681–689.

Scheer, H. 1991. *Chlorophylls*. CRC Press, Boca Raton, FL.

11
Anthocyanins

Ronald E. Wrolstad

Oregon State University
Corvallis, Oregon

INTRODUCTION

Anthocyanin pigments are responsible for the red to purple to blue colors of many fruits, vegetables, cereal grains, and flowers. There is a substantial body of knowledge on anthocyanin pigment chemistry in the scientific literature. Perhaps because of their beauty, but more certainly because of their roles as pollination attractants and phytoprotective agents, they have long been investigated by botanists and plant physiologists. As secondary plant metabolites they serve as systematic markers in taxonomic studies. Because of their role in the color quality of fresh and processed fruits and vegetables, they continue to be studied by food scientists and horticulturists.

ANTHOCYANIN-DERIVED COLORANTS
IN THE WORLD MARKET

In the United States, 26 colorants approved for food use are exempt from certification. Four of these are anthocyanin derived: grape skin extract, grape color extract, fruit juice, and vegetable juice. Grapes are the world's largest fruit crop (Nelson, 1990), and they are the source of three anthocyanin-based colorants. Grape skin extract (Enocianina) has been used as a colorant for 120 years (Francis, 1999), its first application being to enhance the color of wines. The anthocyanins are concentrated in the grape skins; hence, extraction of the grape skins as opposed to the whole fruit allows for pigment concentration. Fermentation to remove sugars further concentrates the pigment. Selections of *Vitis vinifera* that are rich in pigment are grown especially for this purpose. Grape color extract is obtained as a by-product from juice processing of Concord grapes, *V. labruscana*. The original petition for use of this colorant limited its application to nonbeverage food use, and that restriction is still in place. The third anthocyanin colorant derived from grapes is grape juice. Certain anthocyanin-rich cultivars of *V. vinifera* such as Rubired yield highly pigmented juices. Red grape juice concentrate is used to intensify the red color of many blended juice beverages.

Most of the anthocyanin-derived colorants approved for food use in the United States fall into the categories of fruit juices and vegetable juices. This classification is extended to includes aqueous extracts of dried fruits and vegetables. Examples include black raspberry and blackberry juice concentrate as well as such newer additions as red cabbage extract, chokeberry (*Aronia melanocarpa*) juice concentrate, elderberry extract, and black carrot extract. The extraction and concentration processes are to employ physical means such as evaporative distillation and membrane filtration. Use of processes regarded as chemical for isolation and concentration will not be allowed, hence resin extraction with organic solvents will not meet approval.

The European Union (EU) recognizes all anthocyanin-derived colorants as "natural colorants" under classification number E 163 (Francis, 1999). Each European country is allowed considerable choice in interpreting which sources and what processes can be permitted. Hence, there is considerable variation from one European country to another as to which anthocyanin colorants are permitted. Norway, while not being a member of the EU, does not permit any synthetic colorants. Anthocyanin-derived colorants are recognized as being appropriate for food use in Asian, Central American, and South American countries, but there are differences with respect to what sources and degree of purity is permitted. In general, other countries are not as restrictive as the United States regarding food use of anthocyanin colorants.

A QUANTITATIVE PERSPECTIVE

Clearly, materials that possess high concentrations of anthocyanins will be more suitable colorant sources. Table 11.1 lists the anthocyanin content of some fruits and vegetables taken from the literature. Typically a wide range exists for different cultivars; hence a goal of plant breeders is to develop selections that are rich in anthocyanin. Anthocyanins and polyphenolics are

TABLE 11.1   Anthocyanin Content of Selected Fruits and Vegetables

Source	Pigment content (mg/100 g fresh weight)	Ref.
Blackberries	83–326	Mazza and Miniati, 1993
Black currants	130–400	Mazza and Miniati, 1993
Black raspberries	300–400	Timberlake, 1988
Blueberries	25–497	Mazza and Miniati, 1993
Chokeberries	560	Kraemer-Schafhalter et al., 1996
Cranberries	60–200	Timberlake, 1988
Elderberries	450	Kraemer-Schafhalter et al., 1996
Grapes	6–600	Mazza and Miniati, 1993
Radishes	11–60	Giusti et al., 1998a
Red cabbage	25	Timberlake, 1988
Red-fleshed potatoes	2–40	Rodriguez-Saona et al., 1998
Red raspberries	20–60	Mazza and Miniati, 1993
Red onions	7–21	Mazza and Miniati, 1993
Strawberries	15–35	Timberlake, 1988

present in highest amounts in the epidermal tissue. Thus, extraction of skin and peels of fruits and vegetables and milling cereal grains such as corn, where anthocyanin pigments are concentrated in the seed coat, offers methods for obtaining anthocyanin-rich preparations. To be an economically viable source, additional agronomic factors need to be considered (cost and availability of raw product, ease of crop production, yield per acre, harvesting methods, etc.). Processing wastes can be an inexpensive source of pigment, the press-cake and lees from grape juice and wine processing being good examples. In some recent blueberry juice pilot-plant trials, Skrede et al. (2000) found that the press-cake was actually richer in anthocyanin pigment (mg/100 g basis) than the raw material or juice. The anthocyanins were not extracted as efficiently as sugars, acids, and other water solubles, leaving an anthocyanin-rich residue.

A QUALITATIVE PERSPECTIVE

Complete reviews on anthocyanin chemistry have appeared (Macheix et al., 1990; Mazza and Miniati, 1993; von Elbe and Schwartz, 1996; Francis, 1999; Wrolstad, 2000). Figure 11.1 shows a generalized structure for anthocyanin pigments. More than 300 anthocyanin pigments have been identified in nature (Strack and Wray, 1994). Structural variation of H, OH, and OCH_3 groups in the B-ring result in six different aglycons or anthocyanidins: pe-

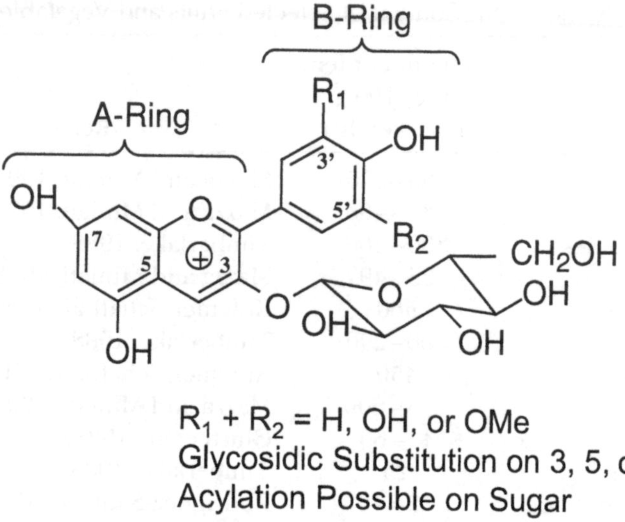

$R_1 + R_2$ = H, OH, or OMe
Glycosidic Substitution on 3, 5, or 7
Acylation Possible on Sugar

FIG. 11.1 Generalized structure for anthocyanin pigments.

largonidin, cyanidin, delphinidin, peonidin, petunidin, and malvidin. An increased number of OH and OCH_3 groups gives a bathychromic spectral shift, resulting in hue changes from red to purple to blue. With the exception of pelargonidin, grapes contain all of the anthocyanidins, with malvidin and delphinidin predominating. Cyanidin is the most common anthocyanidin in nature, and red cabbage, black carrot, and elderberry anthocyanin extracts are wholly cyanidin-based. Anthocyanins are reactive compounds, of which those containing *ortho* phenolic groups (cyanidin, petunidin, and delphinidin) are more susceptible to oxidation and will also complex with metal ions.

Anthocyanins always exist in nature as glycosides. The aglycons that can be formed by acid or enzymatic hydrolysis are extremely unstable. Glycosidic substitution increases both stability and water solubility. Substitution occurs at the 3 and 5 positions of the A-ring, rarely at the 7 position. The most common glycosidic sugars are glucose, galactose, xylose, arabinose, and rhamnose; substitution with disaccharides also occurs, with rutinose, sophorose, sambubiose, and gentiobiose being common substituting sugars. Increasing the number of sugar residues tends to increase pigment stability. The position of glycosylation can influence the structural transformations that take place with change in pH. Glycosidic substitution accounts for much of the anthocyanin diversification that occurs in nature, making them useful as taxonomic markers. Concord grapes (*Vitis labruscana*) contain substantial amounts of anthocyanin 3,5-diglucosides, whereas these are absent from *V. vinifera*.

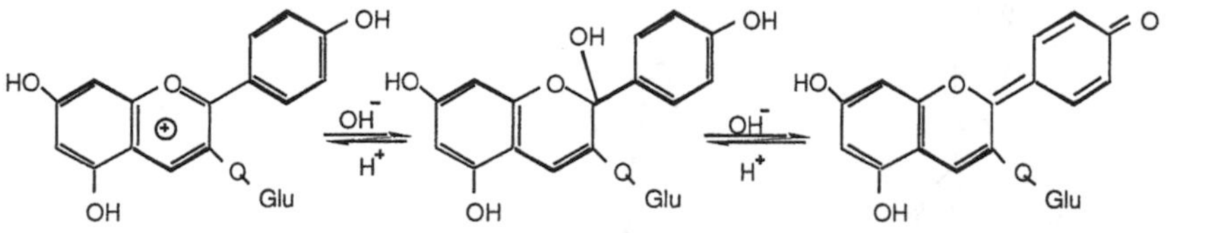

FIG. 11.2 Structural transformations of anthocyanin pigments with pH change.

A third element of structural variation is the possibility of acylation of sugar residues with cinnamic (*p*-coumaric, caffeic, ferulic) and aliphatic (acetic, malonic, succinic) acids. Acylation, particularly with the aromatic acids, has a marked influence on pigment stability. Acylation with aliphatic acids often goes undetected since these esters are vary labile to acid hydrolysis, which can occur during pigment extraction and isolation. Acylation can also affect the color shifts and structural transformations that occur with pH changes. Grapes contain some acylated pigments, and both red cabbage and black carrot anthocyanins are almost entirely acylated. This feature accounts for much of the enhanced stability for these colorants.

Saponification of acylated anthocyanins will produce the glycosidic pigment plus the free acylating acids; acid hydrolysis of anthocyanins gives the aglycons and free sugars. These techniques used in conjunction with high-performance liquid chromatography (HPLC) permit identification of the component parts of anthocyanins (Hong and Wrolstad, 1990a,b; Gao and Mazza, 1994). Identification, however, is not complete without identification of the site of sugar substitution, the nature of the glycosidic linkage, and the site of acylation. Nuclear magnetic resonance (NMR) is a powerful tool for identifying these structural features, which influence molecular conformation, the possibility of intramolecular co-pigmentation, and stacking density. All of these phenomena can affect both stability and color.

CHEMICAL AND PHYSICAL PROPERTIES OF ANTHOCYANINS

Structural Transformations with Change in pH

Figure 11.2 illustrates the reversible transformations that anthocyanin pigments undergo with pH change. The oxonium form is colored and more stable than the colorless hemiketal form. The quinoidal form actually exists

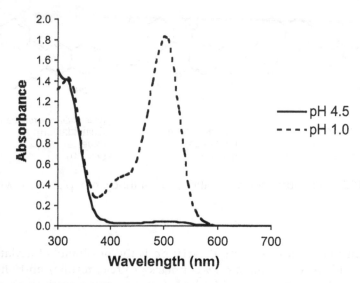

FIG. 11.3 Visible absorption spectra of radish anthocyanin extract at pH 1.0 and 4.5.

in several different resonance forms, and it shows a broad absorbance at longer wavelengths. The quinoidal forms are quite unstable, so that the color of anthocyanin-containing foods at pH values greater than 4.5 tends to be bluish and susceptible to degradation. A major limitation of anthocyanin colorants is the reduction in color intensity, change in hue, and instability that occurs in the upper pH range of many food applications. The pronounced spectral differences that occur between the oxonium and hemiketal forms do offer an analytical advantage, illustrated in Figure 11.3 for radish anthocyanin extract. By measuring the difference in absorbance of anthocyanin colorants, extracts, or beverages at pH 1 and 4.5, the amount of monomeric anthocyanin can be accurately and easily determined (Wrolstad et al., 1982). Acylation and the position of glycosidic substitution will affect the pH at which the structural and color changes occur. Table 11.2 lists the pK_A values for selected anthocyanin colorants as reported by Collins et al. (1998). Black carrot and red cabbage colorants with their higher pK_As will have a color advantage for many food applications.

Concentration Effects

Anthocyanin pigments exhibit greater stability when present in higher concentrations. Skrede et al. (1992) investigated the pigment and color stabilities of black currant and strawberry syrups. Strawberry syrup when fortified with purified strawberry anthocyanins to the same pigment level as black

TABLE 11.2 pKA Values of Selected
Anthocyanin Colorants

Anthocyanin source	pK_A
Black carrot	4.38
Red cabbage	4.35
Grape skin	3.68
Elderberry	3.32
Black currant	3.37

Source: Collins et al., 1996.

currant syrup had similar stability. They concluded that color stability was more affected by total anthocyanin concentration than by qualitative pigment composition. This concentration effect can explain why it is a difficult challenge to formulate a pink, rose-colored beverage using anthocyanin colorants that has a reasonable shelf life at room temperature. This is particularly true if the product is to be marketed in a clear container.

Influence of Light

Carlsen and Stapelfeldt (1997) studied the light sensitivity of elderberry extract and concluded that light bleaching was a major destabilizing factor for anthocyanin-containing products and that exclusion of ultraviolet light would greatly improve color stability. Giusti and Wrolstad (1996b) compared the pigment stabilities of dark versus light exposed glass-packed maraschino cherries colored with radish anthocyanin extract. Over a one-year storage period at 25°C there was a small but statistically significant improvement in stability for samples stored in the dark. A number of factors could contribute to the apparent difference between these two studies regarding the degree that light affects pigment stability: qualitative pigment composition, pigment concentration, and intensity and wavelength of light exposure.

Influence of Water Activity (a_w)

Anthocyanin chemists have observed that anthocyanins on dried paper and thin-layer chromatograms as well as purified crystalline anthocyanins showed good stability. The effect of a_w on anthocyanin stability has been investigated by exposing anthocyanin materials to different levels of relative humidity (Erlandson and Wrolstad, 1972; Brønnum-Hansen and Flink, 1985) and by dissolving anthocyanins in different proportions of glycerol and water (Kearsley and Rodriguez, 1981; Thakur and Arya, 1989; Sian and Soleha, 1991; Garzon, 1998). All of these investigations showed that anthocyanin stability increased with decreasing water activity. This feature is advantageous with respect to use of anthocyanin colorants in intermediate and low-moisture foods.

Inter- and Intramolecular Copigmentation Effects

It is well established that anthocyanins can form complexes with other polyphenolics and metal ions, which may result in spectral shifts, increased color intensity, and increased stability (Davies and Mazza, 1993). Copigmentation can occur in situ in plant materials (Yoshitama et al., 1992) as well as in solution. Hauser Chemical Research has patented a water-soluble rosemary extract, which through copigmentation intensifies color and improves the stability of various anthocyanin colorants (Lenoble et al., 1996, 1999). Intramolecular copigmentation can occur where acyl cinnamic acid residues fold over and interact with the planar pyrilium ring (Dangles et al., 1993), which can affect both color and stability. The shielding effect protects the chromophore against hydration and from coming into contact with other reactants.

Reaction of Anthocyanins with Other Food Constituents

Ascorbic acid, metal ions, and oxygen can all accelerate anthocyanin pigment degradation, as can polyphenoloxidase, peroxidase, and glycosidase enzymes, whether their source is the native fruit or vegetable, mold contamination, or side activities of processing enzymes (Wrolstad et al., 1994; von Elbe and Schwartz, 1996). Thus the color instability of anthocyanin-containing foods may be caused by interactions of anthocyanins with other food components rather than inherent instability of the anthocyanin pigment itself. Garzon (1998) studied the degradative rates of purified pelargonidin-based anthocyanins in model systems and in fortified strawberry juice and concentrate systems. Anthocyanin degradation followed first-order reaction kinetics with half-lives in the order of from only 4–9 days for natural systems to more than 6 months in model systems. Clearly the pronounced anthocyanin instability in processed strawberry products must be caused by factors other than inherent instability of pelargonidin pigments. Figure 11.4 summarizes the various factors that influence the color quality of anthocyanin-containing products.

RADISH ANTHOCYANIN EXTRACT, AN ALTERNATIVE TO FD&C RED NO. 40

Our laboratory has been investigating radish anthocyanin extract as a "natural" alternative to coloring maraschino cherries with FD&C Red No. 40 (Giusti and Wrolstad, 1996a,b; Giusti et al., 1998a,b; Rodriguez-Saona et al.,

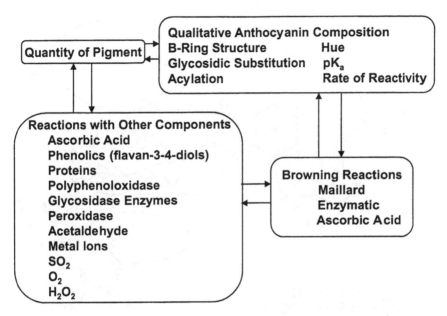

FIG. 11.4 Complexity diagram of anthocyanin color quality. (From Wrolstad, 2000.)

1999). A number of the previously discussed principles concerning anthocyanin structure and its relationship to color and stability properties are illustrated in this series of publications. They also show applications of analytical methods for determining anthocyanin structure and measuring color and stability properties. Coloring maraschino cherries with an anthocyanin colorant presents a number of challenges since processors would like a colorant with the same hue as that imparted by FD&C Red No. 40 along with a reasonable shelf life for the product, which is packed in clear glass jars and stored for long periods at ambient temperature. Our reason for investigating radish anthocyanins was not the red color of salad radishes, rather it was the anthocyanin structure as reported in the literature (Ishikura and Hayashi, 1962; Harborne, 1963). The pelargonidin chromophore should provide a red hue, and the cinnamic acid acylation and high degree of sugar substitution should enhance stability. We succeeded in matching the hue imparted by FD&C Red No. 40 with radish anthocyanin extract, and we obtained an acceptable shelf life of approximately 6 months for maraschino cherries colored with radish anthocyanin extract (Giusti and Wrolstad, 1996b). Anthocyanin pigment degradation followed first-order kinetics and had a half-life of 33 weeks. Acid hydrolysis and saponification of radish anthocyanins in combination with HPLC analyses confirmed pelargonidin-3-sophoroside-5-glucoside with p-coumaric and ferulic acid acylation as

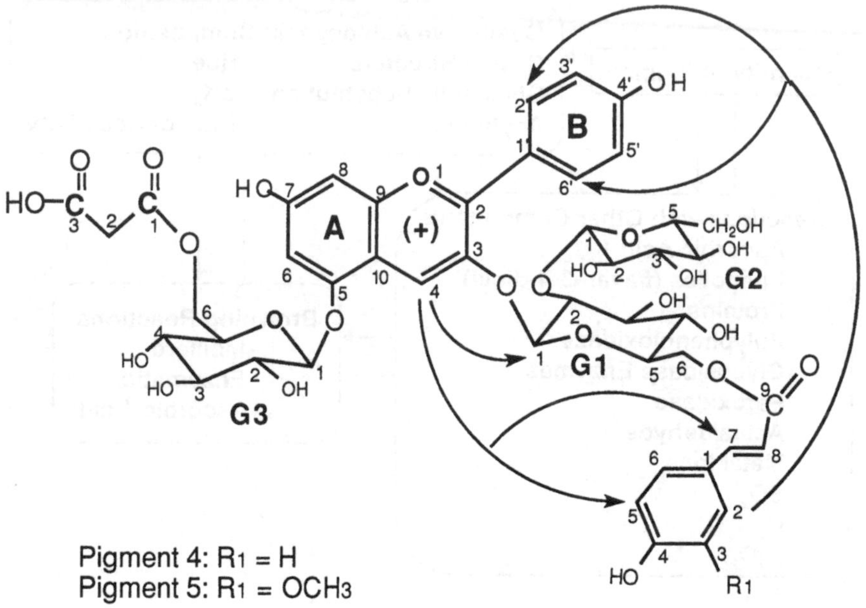

FIG. 11.5 Structure of radish anthocyanin pigments as determined by ^{1}H-NMR ^{13}C-NMR. (From Giusti et al., 1998b.)

major pigments. Electro-spray mass spectroscopy (ESMS) was critical in determining that malonic acid was also present as an acylating acid (Giusti and Wrolstad, 1996a). More complete identification was achieved with ^{1}H-NMR and ^{13}C-NMR (Giusti et al., 1998b). The first sugar of sophorose is acylated with *p*-coumaric or ferulic acid on the 6C atom, and the glucose at the 5 position is acylated with malonic acid on the 6 carbon (Fig. 11.5). There was evidence that the acyl aromatic ring was folded over the planar pyrylium ring. Intramolecular copigmentation is believed to contribute to the color and pigment stability of radish anthocyanin extract. Presently firms in both Japan and Taiwan are producing radish anthocyanin extract for use as a food colorant.

COLOR CHARACTERISTICS OF SOME ANTHOCYANIN-BASED COLORANTS

Kucza (1995) measured the color properties and spectral characteristics of a model beverage (10°Brix, pH 3.5) colored with several different anthocyanin colorants. Effort was made to have all solutions at a uniform tinctorial strength. Table 11.3 compares the color and spectral properties of some of these colorants. All colorants were from commercial sources except for

TABLE 11.3   Color and Spectral Characteristics of Selected Anthocyanin
Colorants

Colorant	Tinctorial Strength[a]	% Polymeric color[b]	λ_{max} (nm)	Hue angle[c]
Grape skin extract	1.49	63	523	24°
Welch's grape colorant	1.43	11	524	11°
Mega-red grape juice	1.46	26	525	7°
Red cabbage extract	1.67	8	536	330°
Radish extract	1.52	9	511	23°
Elderberry	1.81	9	517	24°
Black carrot	2.00	7	524	19°
FD&C Red No. 40				25°

Source: Kucza, 1995.
[a]Absorbance at λ_{max} times dilution factor.
[b]Percent pigment resistant to sodium metabisulfite bleaching.
[c]arctan (b*/a*).

radish extract, which was prepared in our pilot plant. Of particular interest
is hue angle where 0° represents a reddish blue color and FD&C Red No. 40
was measured to be 27°. In this system radish anthocyanin extract, black carrot, and elderberry had hue angles close to that of FD&C Red No. 40. There
was considerable variation in the hues of the different grape colorants, with
grape skin extract possessing a redder hue. Red cabbage was in the purple
region with a hue angle of 330°. The longer wavelength of maximum absorption for red cabbage (536 nm) is consistent with its different color. The
grape colorants had higher amounts of percent polymeric color, particularly grape skin extract. In grape skin extract manufacturing, color can be
effectively concentrated by removing the sugars through fermentation.
During fermentation, anthocyanin pigments tend to undergo polymerization and condensation reactions with other polyphenolics.

TINCTORIAL STRENGTH AND EXTINCTION
COEFFICIENTS OF ANTHOCYANIN COLORANTS

A common criticism of most natural colorants is their low tinctorial strength
as compared to synthetic colorants. Table 11.4 lists the absorptivity values
for some anthocyanin colorants and pigments with comparison to FD&C
Red No. 40. The value for the synthetic colorant is 40–250 times greater
than those for the anthocyanin extracts. Absorptivity values for purified anthocyanin pigments, however, are of the same order as that for FD&C Red

TABLE 11.4 Absorptivity Coefficients of Selected Colorants

Source	$E_{1\%}{}^a$ (pH)	ε^{2b}	Ref.
Grape skin extract	13.0 (1.0)		Hong and Wrolstad, 1990b
Welch's grape colorant	2.1 (1.0)		Hong and Wrolstad, 1990b
Red cabbage extract	10.6 (1.0)		Hong and Wrolstad, 1990b
Black carrot	5.0 (3.0)		GNT, 1998
Cyanidin-3-glucoside	573.3 (1.0)	25,700	McClure, 1967
Malvidin-3-5-diglucoside	520.4 (1.0)	37,700	Niketic-Aleksic and Hrazdina, 1972
FD&C Red No. 40	540	26,807	Marmion, 1991

^{a}Absorbance of a 1% solution in a 1 cm path length cell.
$^b\varepsilon$=Molar extinction coefficient.

No. 40. The lower tinctorial strength comes from the presence of sugars, acids, polyphenolics, and other colorless materials. Note that the absorptivity values for most of these colorants was determined at pH 1.0. At the pH of food use, the absorptivity values will be much lower because of the oxonium-hemiketal reversible transformation shown in Figure 11.2. It is technologically possible to produce more highly purified anthocyanin colorants, which should also have greater stability and be free from unwanted flavors. Regulatory constraints in the United States, however, severely restrict the purification and concentration processes that can be used for colorants classified as exempt from certification.

CURRENT AND FUTURE PERSPECTIVES

In the quest for improved anthocyanin colorants, efforts are being directed towards obtaining increased tinctorial strength, improved stability, an expanded range of hues, greater uniformity, and absence of objectionable flavors. The approaches include traditional plant breeding as well as biotechnology and genetic engineering. Technologies of nanofiltration and membrane filtration, which are physical processes, hold promise, as does supercritical fluid extraction with liquid CO_2 to remove unwanted flavors.

There is intense interest in the possible health benefits of anthocyanin pigments, which include reduced risk of coronary heart disease and stroke, improved visual activity, and antiviral and anticancer activities (Mazza and Miniati, 1993). Today some highly concentrated anthocyanin extracts are being marketed as dietary supplements in the United States, but these same extracts could not legally be used as food colorants. It seems reasonable that the possible health benefits of anthocyanin colorants should be given consideration in the protocol for colorant approval. It is highly unlikely that anthocyanin pigments per se are harmful; after all, humankind has been consuming anthocyanins from evolutionary times. It is possible, however, that toxicants and harmful substances could be concentrated or formed in the processes for pigment isolation and purification. The complete composition of highly purified anthocyanin colorants would need to be carefully evaluated. One advantage of increasing anthocyanin consumption through natural colorant usage is that the dosage should be self-limiting, i.e., consumption of extreme amounts should be unlikely since product color will be the determining factor for the amount of anthocyanin in a given formulation.

REFERENCES

Brønnum-Hansen, K., and Flink, J. M. 1985. Anthocyanin colourants from elderberry *Sambucus nigra* L.) III. Storage stability of the freeze-dried product. *J. Food Technol.* 20: 725–733.

Carlsen, C., and Stapelfeldt, H. 1997. Light sensitivity of elderberry extract. Quantum yields for photodegradation in aqueous solution. *Food Chem.* 60: 383–387.

Collins, P., Plumbly, J., and Stich, E. 1998. Black carrots. Established raw materials for the manufacture of food colours. In *Proceedings of the Third International Symposium on Natural Colorants*, Rutgers, NJ, 4/19–22/98. S.I.C. Publishing Co., Hamden, CT.

Dangles, O., Saito, N., and Brouillard, R. 1993. Anthocyanin intramolecular copigment effect. *Phytochemistry* 34: 119–124.

Davies, A. J., and Mazza, G. 1993. Copigmentation of simple and acylated anthocyanins with colorless phenolic compounds. *J. Agric. Food Chem.* 41: 716–720.

Erlandson, J. A., and Wrolstad, R. E. 1972. Degradation of anthocyanins in limited water concentration. *J. Food Sci.* 37: 592–595.

Francis, F. J. 1999. *Colorants.* Eagen Press, St. Paul, MN.

Gao, L., and Mazza, G. 1994. A rapid method for complete characterization of simple and acylated anthocyanins by high performance liquid chromatography and capillary gas liquid chromatography. *J. Agric. Food Chem.* 42: 118–125.

Garzon, G. A. 1998. The stability of pelargonidin-based anthocyanins in natural and model systems. Ph.D. dissertation, Oregon State University, Corvallis, OR.

Giusti, M., and Wrolstad, R. E. 1996a. Characterization of radish anthocyanins. *J. Food Sci.* 61: 322–326.

Giusti, M., and Wrolstad, R. E. 1996b. Radish anthocyanin extract as a natural red colorant for maraschino cherries. *J. Food Sci.* 61: 688–694.

Giusti, M. M., Rodriguez-Saona, L. E., Baggett, J. R., Reed, G. L., Durst, R. W., and Wrolstad, R. E. 1998a. Anthocyanin pigment composition of red radish cultivars as potential food colorants. *J. Food Sci.* 63: 219–224.

Giusti, M. M., Ghanadan, H., and Wrolstad, R. E. 1998b. Elucidation of the structure and conformation of red radish (*Raphanus sativus*) anthocyanins using one and two dimensional nuclear magnetic resonance techniques. *J. Agric. Food Chem.* 46: 4858–4863.

GNT. 1998. Measurement of black carrot color intensity. GNT International B.V., Aachen, Germany.

Harborne, J. B. 1963. Plant polyphenols IX. The glycosidic pattern of anthocyanin pigments. *Phytochemistry* 2: 85–97.

Hong, V., and Wrolstad, R. E. 1990a. Use of HPLC separation/photodiode array detection for characterization of anthocyanin. *J. Agric. Food Chem.* 38: 708–715.

Hong, V., and Wrolstad, R. E. 1990b. Characterization of anthocyanin containing colorants and fruit juices by HPLC/photodiode array detection. *J. Agric. Food Chem.* 38: 698–708.

Ishikura, N., and Hayashi, K. 1962. Anthocyanins in red root of a radish. Studies on anthocyanins. XXXVI. *Bot. Mag. Tokyo* 75: 28–36.

Kearsley, M. W., and Rodriguez, N. 1981. The stability and use of natural colours in foods: Anthocyanin, β-carotene and riboflavin. *J. Food Technol.* 16: 421–431.

Kraemer-Schafhalter, A., Fuchs, H., Strigl, A., Silhar, S., Kova, M., and Pfannhauser, W. 1996. Process consideration for anthocyanin extraction from black chokeberry (*Aronia melanocarpa* ELL). In *Proceedings of the Second International Symposium on Natural Colorants*, Acapulco, Mexico, 1/23–26/96. S.I.C. Publishing Co., Hamden, CT.

Kucza, M. 1995. Color and pigment stability of red natural colorants in a model beverage. B. Ingr. dissertation. I.S.A.R.A., Lyons, France.

Lenoble, R., Richheimer, S. L., Bank, V. R., and Bailey, D. T. 1996. Enhanced stability and improved color of anthocyanins by the use of natural extracts. In *Proceedings of the Second International Symposium on Natural Colorants*, Acapulco, Mexico, 1/23–26/96. S.I.C. Publishing Co., Hamden, CT.

Lenoble, R., Richheimer, S., Bank, V. R., and Bailey, D. T. 1999. Pigment composition containing anthocyanins stabilized by plant extracts. *U.S. Pat.* 5,908,650.

Macheix, J. J., Fleuriet, A., and Billot, J. 1990. *Fruit Phenolics*. CRC Press, Boca Raton, FL.

Marmion, D. L. 1991. *Handbook of U.S. Colorants*, 3rd ed. John Wiley & Sons, New York.

Mazza, G., and Miniati, E. 1993. *Anthocyanins in Fruits, Vegetables, and Grains.* CRC Press, Boca Raton, FL.

McClure, J. W. 1967. Photocontrol of *Spirodela intermedia* flavonoids. *Plant Phys.* 43: 193–200.

Nelson, K. E. 1990. The grape. In *Quality and Preservation of Fruits*, N. A. M. Eskin (Ed.), pp. 125–167. CRC Press, Boca Raton, FL.

Niketic-Aleksic, G., and Hrazdina, G. 1972. Quantitative analysis of the anthocyanin content in grape juices and wines. *Lebensm. Wiss. U. Technol.* 5: 163–165.

Rodriguez-Saona, L. E., Giusti, M. M., and Wrolstad, R. E. 1998. Anthocyanin pigment composition of red-fleshed potatoes. *J. Food Sci.* 63: 458–465.

Rodriguez-Saona, L. E., Giusti,M. M., and Wrolstad, R. E. 1999. Color and pigment stability of red radish and red-fleshed potato anthocyanins in juice model systems. *J. Food Sci.* 64: 451–456.

Sian, N. K., and Soleha, I. 1991. Carotenoid and anthocyanin contents of papaya and pineapple: Influence of blanching and predrying treatments. *Food Chem.* 39: 175–185.

Skrede, G., Wrolstad, R. E., Lea, P., and Enersen, G. 1992. Color stability of strawberry and blackcurrant syrups. *J. Food Sci.* 57: 172–177.

Skrede, G., Wrolstad, R. E., and Durst, R. W. 2000. Changes in anthocyanins and polyphenolics during processing of highbush blueberries (*Vaccinium corymbosum* L.). *J. Food Sci.* 65.

Strack, D., and Wray, V. 1994. The anthocyanins. In *The Flavonoids, Advances in Research Since 1986*, J. B. Harborne (Ed.), pp. 1–22. Chapman and Hall, New York.

Thakur, B. R., and Arya, S. S. 1989. Studies on stability of blue grape anthocyanins. *Int. J. Food Sci. Technol.* 24: 321–326.

Timberlake, C. F. 1988. The biological properties of anthocyanin compounds. *NATCOL Q. Bull.* 1: 4–15.

Von Elbe, J. H., and Schwartz, S. J. 1996. Colorants. In *Food Chemistry*, 3rd ed., O. W. Fennema (Ed.), pp. 651–722. Marcel Dekker, New York.

Wrolstad, R. E. 2000. Colorants. In *Food Chemistry: Principles and Applications*, G. L. Christen and J. S. Smith (Eds.). Science and Technology System, West Sacramento, CA.

Wrolstad, R. E., Culbertson, J. D., Cornwell, C. J., and Mattick, L. R. 1982. Detection of adulteration in blackberry juice concentrates and wines by sugar, sorbitol, nonvolatile acid and pigment analyses. *J. Assoc. Off. Anal. Chem.* 65: 1417–1423.

Wrolstad, R. E., Wightman, J. D., and Durst, R. W. 1994. Glycosidase activity of enzyme preparations used in fruit juice processing. *Food Technol.* 48: 90, 92–94, 96, 98.

Yoshitama, K., Ishikura, N., Fuleki, T., and Nakamura, S. 1992. Effect of anthocyanin, flavonol co-pigmentation and pH on the color of the berries of *Ampelopsis brevipedunculata. J. Plant Physiol* 139: 513–518.

12
Caramel Color

William Kamuf, Alexander Nixon, and Owen Parker

D. D. Williamson and Co., Inc.
Louisville, Kentucky

There are about 6000 additives available to the food industry, with more than half of these being flavors, both natural and synthetic; the remainder includes colors, preservatives, antioxidants, emulsifiers, thickeners, acids, bases, anticaking agents, flavor enhancers, glazing agents, improvers, bleaching agents, sweeteners, solvents, and a miscellaneous category (Millstone, 1984). Of these, color has always played a vitally important role in food selection and acceptance. Colorants are added to foods to make up for color that may be lost during processing, to color products that are colorless themselves but are made more appealing to the consumer when color is added (e.g., an orange-flavored beverage), to allow consumers to identify what taste to expect from a product (e.g., licorice or lime sweets), and to protect sensitive flavors from light. The colorants added to foods must be proven safe, stable, legally permitted, and effective in the particular application.

Caramel color, from the palest yellow to the deepest brown, accounts for more than 90% by weight of all the colors added to the foods we eat and drink, with consumption of more than 200,000 tons per year worldwide. In North America, approximately 80% of the caramel colors consumed are in soft drinks, while in Europe, beer, gravies, and sauces are the major uses

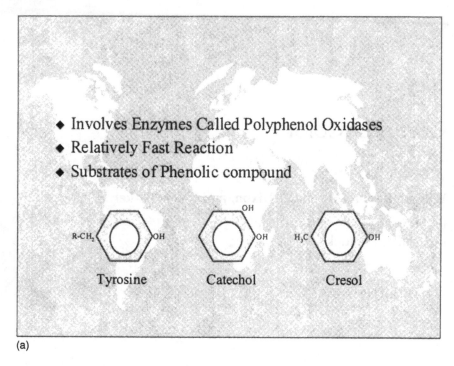

(a)

FIG. 12.1 Enzymatic browning.

(with soft drinks increasing in importance). In Asia soy and other sauces are the largest applications.

CHEMISTRY

There are two types of caramelization reactions in food products: enzymatic browning (Fig. 12.1), illustrated when damaged or cut fruit darkens at the exposed surface, and nonenzymatic browning, which occurs when food products such as coffee beans, meat, bread, or sugars are heated.

Nonenzymatic browning in foods proceeds in several ways, with two of the most important being:

1. The well-known Maillard reaction (Fig. 12.2), in which sugars, alde-hydes, and ketones react with naturally occurring nitrogen-containing compounds, such as amines and proteins, to form brown pigments known as melanins

2. Caramelization reactions, in which sugars are heated in the absence of nitrogen-containing compounds

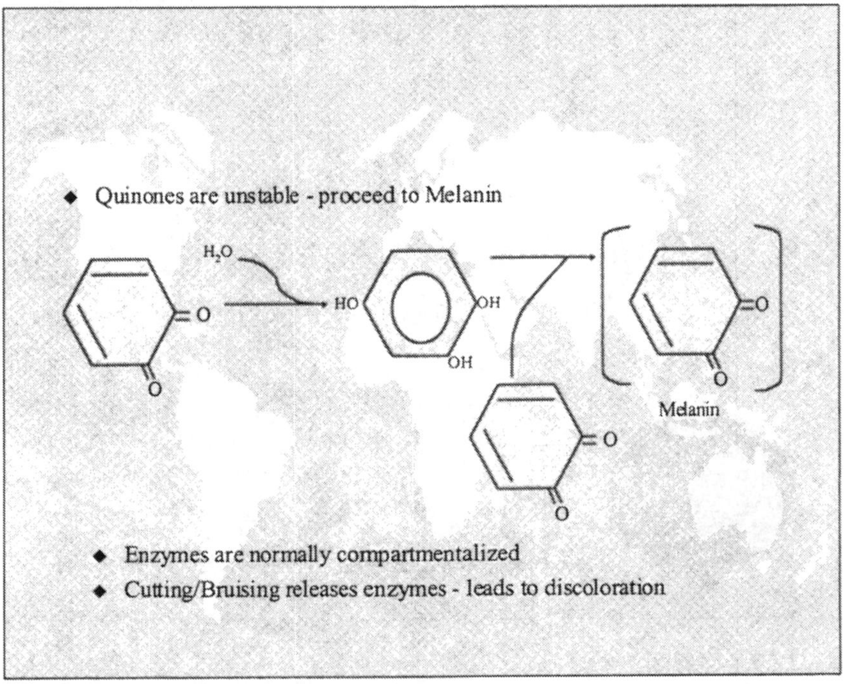

(b)

FIG. 12.1 Continued

In the latter reaction the sugars initially undergo dehydration and then condensation or polymerization into complex molecules of varying molecular weight. Lightly colored, pleasant-tasting caramel flavors are produced in the initial stages, but as the reaction continues, more high molecular weight color bodies are produced and the flavor characteristics become more bitter.

BACKGROUND

Caramel color first gained commercial importance as an additive in brewery products (e.g., porter, stout, dark beers, and ales) and as a colorant for brandy. In 1858 the first known published technical study of caramel color was authored by the French chemist M. A. Gelis. Gelis's work indicated that caramelized sucrose contained three main products: a dehydration product, caramelan ($C_{12}H_{18}O_9$), and two polymers, caramelen ($C_{36}H_{50}O_{25}$) and caramelin ($C_{96}H_{102}O_{51}$). Greenshields (1973) indicated that it is common

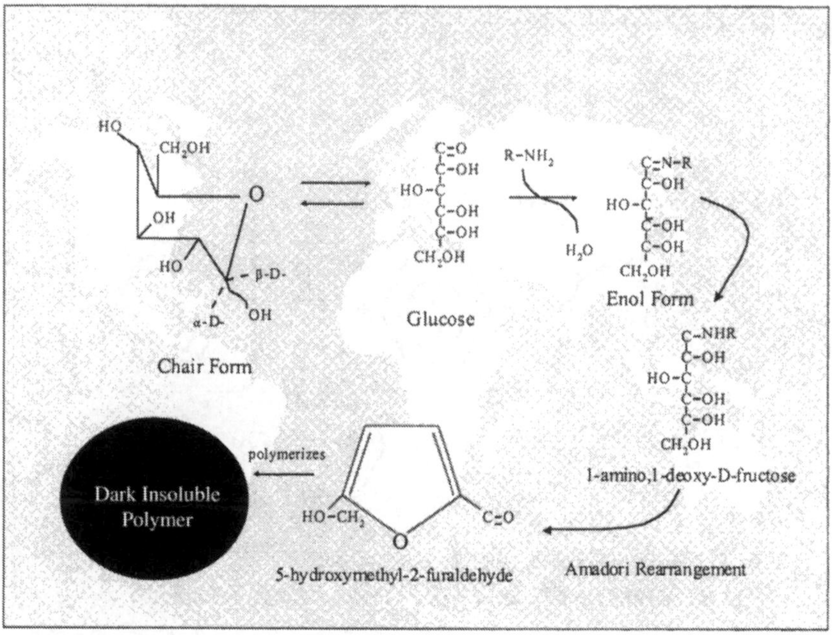

FIG. 12.2   Maillard reaction.

for both Maillard and caramelization reactions to give aldehydes and dicar-
bonyl compounds, but the former type incorporates nitrogen-containing
components. For this case Hodge and Greenshields (1953) grouped the reaction mechanisms as follows:

1. Starting reactions.

   Sugar-amino condensation

   Amadori or Heyns rearrangement

2. Degradative reactions causing the formation of colorless or yellow
   products with strong ultraviolet absorbance and the release of carbon
   dioxide.

   Sugar dehydration

   Ring splitting (Stekker degradation)

3. Polymerizing or condensing reactions forming the strongly colored
   components of relatively high molecular weight.

   Aldol condensations

   Aldehyde/Amino polymerization and formation of heterocyclic nitrogen compounds

Caramel colors have been used for so long and in such a wide variety of food products that consumers tend to think of them as a single substance, when in reality they are a family of similar materials with slightly differing properties. There are, in fact, four distinct types of caramel color to satisfy the requirements of differing food and beverage systems.

When the FAO/WHO Joint Expert Committee on Food Additives (JECFA) first examined caramel colors in 1969, they were thought to be a large number of ill-defined and complex products for which no adequate specifications existed. All that was known at the time was they were formed from various carbohydrates when heated with a range of acids, bases, and salts. In addition, an interim report by the United Kingdom Food Additives and Contaminants Committee on the Review of Coloring Matter in Food Regulation 1973, published in 1979, reported caramels to be "a multiplicity of ill-defined products geared to meet the special needs of particular users."

Many studies of caramel color were undertaken thereafter, with the largest and most comprehensive being initiated by the International Technical Caramel Association (ITCA). ITCA is the industry group composed of major users and manufacturers throughout the world formed to sponsor studies to further assure consumers and government agencies of the safety and suitability of caramel color as a food and beverage colorant. ITCA commissioned the Ontario Research Foundation (ORF) of Mississauga, Ontario, Canada to do characterization work on caramel color to prove its homogeneity. The project began in November 1979 and was completed in May 1985. The main objective of the program was to develop an analytical procedure for the characterization of caramel color that was quantitative, reproducible, and comprehensive. Caramel color was separated into size fractions using ultrafiltration, which were in turn resolved into subfractions by techniques based on charge, polarity, solubility, and functional groups. The subfractions were then profiled using chromatographic procedures along with other physical and chemical techniques. The studies defined four distinct types of caramel color and showed that while each of the four types gave differing chemical profiles, the profiles of colors varying in color intensity within a type were essentially the same. These data were submitted to JECFA and other interested regulatory agencies throughout the world in June 1985.

At this time, ITCA also submitted the proposed specifications and classification scheme, which were based on the ORF data base and on consultations with regulatory agencies and industry groups worldwide, including the European Technical Caramel Association, the British Caramel Manufacturers Association, the Canadian Caramel Association, and the Japanese Caramel Industrial Association. The specifications and classification, which are under review by JECFA and have been accepted as tentative by the European Economic Community (EEC), are described in the following sections.

REGULATORY SPECIFICATIONS

Caramel color is a dark-brown to black liquid or solid having an odor of burnt sugar and a pleasant, somewhat bitter taste. It is totally miscible with water and contains colloidal aggregates that account for most of its coloring properties and its characteristic behavior toward acids, electrolytes, and tannins.

There are four types of caramel color that are of commercial importance and that have distinctive applications in foods and beverages. Each type of caramel color has specific functional properties that ensure compatibility with a product and eliminate undesirable effects such as haze, flocculation, and separation. The four types of caramel color are Caramel Color I (also known as plain or spirit caramel or CP caramel), Caramel Color II (caustic sulfite caramel or CSC caramel), Caramel Color III (ammonia or beer caramel, bakers and confectioners caramel, or AC caramel), and Caramel Color IV (known as sulfite-ammonia or soft drink caramel, acid-proof caramel, SAC caramel, or SD caramel) (Table 12.1).

Caramel color is prepared by the controlled heat treatment of carbohydrates. The carbohydrate raw materials are commercially available, food-grade nutritive sweeteners, which are the monomers, glucose and fructose, and/or polymers thereof (e.g., glucose syrups, sucrose and/or invert sugars, and dextrose). To promote caramelization, food-grade acids, alkalis, and salts may be employed in amounts consistent with Good Manufacturing Practice (GMP) and subject to the following stipulations. Ammonium and sulfite compounds cannot be used as reactants for Caramel Color I. Sulfite compounds must be used and ammonium compounds cannot be used as reactants for Caramel Color II. Ammonium compounds must be used and sulfite compounds cannot be used as reactants for Caramel Color III. Both ammonium and sulfite compounds must be used as reactants for Caramel Color IV. The ammonium compounds that are employed are ammonium hydroxide, carbonate, bicarbonate, phosphate, sulfate, sulfite, and bisulfite. The sulfite compounds are sulfurous acid and potassium, sodium, and ammonium sulfites and bisulfites. The compounds that can be used for all four

TABLE 12.1 Analytical Requirements for Caramel Color

	Class I, E150 a	Class II, E150 b	Class III, E150 c	Class IV, E150 d
Color intensity	0.01–0.14	0.05–0.13	0.08–0.36	0.10–0.60
Total nitrogen[a]	<0.1%	<0.2%	0.7–3.3%	0.3–1.7%
Total sulfur[a]	<0.2%	1.0–3.5%	<0.2%	0.8–2.5%

[a]Expressed on an equivalent color basis.

types of caramel color are sulfuric and citric acid and sodium, potassium, and calcium hydroxide. Food-grade polyglycerol esters of fatty acids may be used as processing aids (antifoam) in amounts not greater than that required to produce the intended effect.

For the purpose of this specification, color intensity is defined as the absorbance of a 0.1% (w/v) solution of caramel color solids in water in a 1 cm cell at 610 nm. To express a parameter on a color equivalent basis, the parameter is determined on the caramel color as is, then calculated on a basis of caramel color solids, and finally expressed in terms of a product having a color intensity of 0.10 absorbance units. For example, if a caramel color contains 50% solids, 0.70% total nitrogen as is, and has a color intensity of 0.15, then the %N expressed on an equivalent color basis is: (0.70%/0.50) / (0.15/0.10) = 1.4%/1.5 = 0.93%. The general requirements for all four types of caramel color are arsenic, not more than 1 mg/kg; heavy metals, not more than 25 mg/kg; and lead, not more than 2 mg/kg. The complete analytical methodology for each of the above specifications will be found in references (FAO, 1977; ITCA, 1984; OJEC, 1995).

The Eighth Amendment to the Colors Directive of the EEC (Hodge, 1953) made it clear that these four classes of caramel color are intended for coloring and are to be distinguished from and do not correspond to the sugary aromatic product obtained from heating sugar, which is used for flavoring food and drinks such as confectionery, pastry, and some aperitif-type drinks. These are known as burnt sugars or caramelized sugars.

In the United States the permitted reactants for caramel color are to be found in the Code of Federal Regulations (CFR), Title 21, Section 73.85 and are very similar to those above (Table 12.2). Caramel color is listed as Generally Recognized As Safe (GRAS) as a general purpose food additive in CFR 21, section 182.1235, and is permanently listed and exempt from certification for use in coloring cosmetics, including cosmetics applied to the area of the eye in CFR 21, section 73.2085.

LABELING

In the United States, the responsibility for food safety and labeling rests with the U.S. Department of Agriculture (USDA) and the U.S. Food and Drug Administration of the Department of Health, Education and Welfare (FDA). Finished food labeling of caramel color is and has been relatively straightforward. The NLEA Act of 1992 indicates that color additives not subject to certification may continue to be declared as "artificial color," "color added," or by an equally informative term that makes clear that a color additive is present in the food. Alternatively, such color additives may be declared as "colored with _____" or "_____ color" with the blank filled with the name of the color additive. The *Federal Register* of Jan. 6, 1993, in-

TABLE 12.2   Purity Specifications for Caramel Color in the United States and Canada

---

Ammoniacal nitrogen: Not more than 0.6%[a]
Arsenic: Not more than 1 mg/kg[b]
Color intensity: 0.010–0.600[c]
Heavy metals (as Pb): Not more than 25 mg/kg[b]
Lead: Not more than 2 mg/kg[b]
Mercury: Not more than 0.1 mg/kg[b]
4-Methylimidazole: Not more than 0.025%[a]
Sulfur dioxide: Not more than 0.2%[a]
Total nitrogen: Not more than 3.3%[a]
Total sulfur: Not more than 3.5%[a]

---

[a]Calculated on an equivalent color basis that permits the values to be expressed in terms of a caramel having a color intensity standard.
[b]Measured as is.
[c]Color intensity is defined as the absorbance of a 0.1% w/v solution of caramel in water measured in a 1 cm cell at 610 nm expressed on a total solids basis.

dicates that full label disclosure requirements for caramel are met with the words "caramel color."

In the United States in 1986 (Canada 1992), labeling of sulfites in finished foods became mandatory if the residual sulfites from all sources by the Monier-Williams analytical method exceeded 10 ppm. Because Class II and Class IV caramel colors contain residual sulfites, products with these two types of caramel colors should be checked to ensure that they comply with the regulations.

On the issue of natural versus artificial, in the United States caramel color is defined as a noncertified color additive. According to FDA, all added colors, regardless of source, result in an artificially colored food. For example, the addition of beet juice to lemonade to make it pink would preclude the product being called natural.

While in the past it has not been necessary in all member states of the EEC to label additives, the EEC does have a Harmonization Program for additive labeling, and the European Scientific Community for Foods (ESCF) in 1962 developed a simple labeling system to identify the additives used in food products by either the additive's proper name or a number assigned to it. This was the origin of the E numbering system, in which each additive is assigned a number within one of four groups so that consumers needing to avoid a certain additive could easily find the additive on the label without having to remember the exact chemical name. The four groupings are Colors, E100–E180, (Caramel Color is in this group and is known as E150); Preservatives, E200–E290; Antioxidants, E300–E321; and Emulsifiers, stabilizers, and other additives, E322–E494. Before an additive is given an E number, it must be proven safe to the regulatory agencies. The permitted colors are governed by the European Parliament and Council Directive 94/36/EC of June 30, 1994, on colors for use in foodstuffs. Everywhere that Caramel Colors are listed, they are permitted at *quantum satis* levels.

The Labeling Directive requires that coloring matters for food must be indicated on the list of ingredients according to their function. This functionality must be expressed by the legal class name "color" followed by either the E number or the full common name of the additive.

SAFETY

Along with the specifications, in June 1985 ITCA submitted the toxicological and safety data to JECFA, the EEC, FDA, and other interested regulatory agencies throughout the world. The work was extensive with mutagenicity, short-term, and range-finding studies being performed, and in January 1985, 24-month long-term studies, feeding Fisher 344 rats and $B_6C_3F_1$ mice with 0.0, 2.5, 5.0, 7.5, and 10.0 g of caramel color per kg of body weight in the drinking water, were completed. Three laboratories carried out the bulk of the studies: Hazelton Laboratories America (Madison, WI), BioResearch Laboratories of Canada, and the Central Institute for Nutrition Research (TNO/CIVO) in Holland. As a result of these studies, the Joint FAO/WHO Expert Committee (WHO, 1986) established Acceptable Daily Intakes (ADI) for each type of caramel color (Table 12.3). These permanent ADIs are sufficient to cover all the normal usage levels currently employed by the food industry.

In the United States, FDA, after careful review of the submitted toxicological and safety data, again reaffirmed the GRAS status of Caramel Color. FDA decided to retain the specifications for Caramel Color as found in the third edition of the Food Chemicals Codex.

MANUFACTURE

Commercial caramel colors are manufactured in stainless steel reactors of 50–5000 gallon capacity under either atmospheric (open) or pressure (closed) conditions. Class I colors are usually manufactured under open

TABLE 12.3   ADIs for Caramel Color

Class	ADI
I	Unlimited
II	0–200 mg/kg body weight
III	0–200 mg/kg body weight
IV	0–200 mg/kg body weight

conditions, while Class II, III, and IV colors can be manufactured under open or closed conditions. The carbohydrate source would be added to the reactor, followed by the reactants; if a pressure reaction, the vessel would be closed; heat would be applied (normally by means of a steam coil or jacket) to 130–170°C; color formation would be monitored; and when the color target is reached, the reaction mass is cooled. Cooling is accomplished in one of three ways: if a pressure reaction, by cooling water through either the steam coils or the jacket; by "flash" cooling where the bottom valve of the reactor is opened and the pressure forces the color into an atmospheric vessel; or if an open process, by adding quench water to the reaction mass. If necessary, the specific gravity and pH would be adjusted to customer specifications and the product would be filtered through a diatomaceous earth filter for medium- and low-viscosity products or through a 100-mesh screen for high-viscosity products. The carbohydrates used in the United States and Europe are primarily glucose syrups from corn (maize) or wheat with dextrose equivalents (DE) of 40–95, with smaller amounts of sucrose also being used. In addition to sucrose and glucose from corn, glucose syrups from other carbohydrate sources, such as rice, molasses, and potato, are used in different parts of the world depending on availability. All carbohydrates used must meet the specifications of the Codex Alimentarius.

A typical Class I production process using liquid sugar would be as follows: the syrup would be added to the reactor, the pH, normally 5.0 on receipt, would be neutralized to a pH of about 7.0 with sodium hydroxide, heat would be applied with constant mixing until the temperature reaches 140–170°C, at which point the sugar is a dehydrated molten mass and color body formation is quite rapid. The reaction would be allowed to continue until the color target is reached, usually 3–5 hours, at which point quench water would be slowly and carefully added to the reaction mass. When sufficiently cooled, the color and specific gravity would be adjusted to specification with water, pH adjustment not normally being necessary. In this process the color target would be above the final product color as quench water is added, thus diluting the color, while in a pressure reaction the color target for cooling can be below the final product color, depending on the efficiency of the cooling system, because color development will continue during the cooling process.

COMMERCIAL SPECIFICATIONS

The beginnings of standardization of caramel color specifications can be traced to 1938, when W. R. Fetzer's article on analysis of caramel color was published (Fetzer, 1938). Fetzer listed methods not only for color, pH, specific gravity, etc., but also for acid-fastness, alcohol stability, tannin resistance, and compatibility. Many of these methods are still in use.

Commercial specifications will vary according to the vendor but will usually include color intensity, pH, specific gravity or Baumé, viscosity, colloidal charge, shelf life, microbiology, and, if a powder, granulation. Specialized tests would be listed depending upon the application (e.g., salt tolerance if the caramel color is to be used in soy sauce or stability and clarity in beer).

Every manufacturer and user has their preferred method of determining coloring power, and in many cases it is difficult if not impossible to correlate one method with another. The analysis becoming the reference method (listed by many regulatory agencies including FCC, JECFA, and now the EEC, several manufacturers and ITCA) is known as color intensity and is defined as the absorbance of a 0.1% w/v solution at 610 nm through a 1 cm^2 cell, expressed either on an as-is or solids basis.

A common method of color determination in Europe is the use of a colorimeter or comparator to match a solution of caramel color to a series of standardized colored glasses and, by using the appropriate multiplier, determining the color strength in European Brewery Convention (EBC) units (1950). The color of the glass standards used have a hue that is the same as beer, and solutions of Caramel Color III, the color used in beer, exactly match the hue of the glass, making it easy to determine their EBC value. The other three classes of caramel color have varying hues, and the match between solution and glass can be very difficult. JECFA and industry in the 1970s came up with a tentative correlation that relates a certain amount of absorbance of a 0.1% w/v solution at 610 nm for each class of caramel to 20,000 EBC units. For example, for Caramel Color I, 0.053 absorbance units correlates to 20,000 EBC, while the correlation for Caramel Color III is 0.076 absorbance units and for Caramel Color IV it is 0.085 absorbance units. This system works well, but the analyst has to know the type of color being checked, and experience has shown that the correlation value for a double-strength Caramel Color IV (color intensity, 0.200–0.270) should be 0.104. Dr. Satish Chandra at the University of Louisville in Kentucky, using regression analysis on a data base of 106 samples, produced a mathematical model to correlate the absorbance of a 0.1% w/v solution at 510 nm to EBC values. The regression equation, EBC = 2145 + (65,132 × abs.@510), gave R^2 values of 98.1 and is an excellent predictor of EBC values using spectrophotometric analysis.

Linner (1970) developed an equation based on spectrophotometric readings at 510 nm and 610 nm to determine the Hue Index or the "redness" of a particular caramel color.

$$\text{Hue index} = \log \frac{\text{Color at 510 nm}}{\text{Color 610 nm}} \times 10$$

For example, a solution of caramel color has an absorbance at 610 nm of 0.123 and an absorbance of 0.434 at 510 nm. The hue index would be 5.48. The range of hue index is about 3.5–7.5, with, in general, the higher the

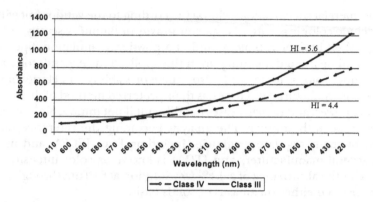

FIG. 12.3 Hue index.

value, the redder or more yellow the color. Figure 12.3 illustrates hue index by showing the amount of variation in absorbance for a Caramel Color IV (color intensity, 0.104; hue index, 4.40) and a Caramel Color III (color intensity, 0.104; hue index, 5.6). The colors have been equalized at 610 nm. It can be seen that the plots follow each other closely from 610 to 580 nm. The plots then diverge rapidly, with Caramel Color III absorbing more rapidly at the lower or blue wavelengths, thus giving a redder visual impression. A caution, however: in Class I Caramels with color intensities of about 0.015, the hue index is around 7, which would indicate an extremely red product. These colors at dilute solutions are in fact quite yellow and are sometimes used as replacements for FD&C yellow (tartrazine) to combine with FD&C blue (brilliant blue FCF) for making green.

pH can be listed on a specification either on an as-is basis or as any number of dilutions from 1 to 50%, depending on the preference of the producer. The pH of caramel color at the end of the production cycle can be quite acidic and may or may not be adjusted to a higher pH with hydroxides, depending on customer requirements.

Specific gravity is read with a hydrometer or a specific gravity balance, usually at 15.5 or 20°C. Even though there are many exceptions, generally the specific gravity range is 1.320–1.360 for Class I, 1.250–1.380 for Class III, 1.310–1.360 for single-strength Class IV, and 1.250–1.280 for double-strength Class IV. Many customers prefer the Baumé reading, which can be calculated using the specific gravity reading and the following formula:

$$Be' = 145 - \frac{145}{\text{Specific gravity at } 60°F}$$

The specific gravity is used to calculate the weight per gallon by multiplying, in the United States, by 8.328 (the weight of a U.S. gallon of water), and is also used to estimate the solids content using the Circular of the National Bureau of Standards, *Polarimetry, Saccharimetry, and the Sugars*, Table 109 on degrees Brix, specific gravity, and degrees Baumé. The values obtained from the table compare very favorably with the official method of vacuum-drying on quartz sand at 70°C or less.

Viscosities are normally read using a Brookfield or similar type viscometer at either 20 or 30°C. Viscosities will range from less than 100 cps for a double-strength Caramel Color IV to a high of 10,000 cps for some high-specific-gravity Class IIIs. Viscosity is an indicator of the quality of the product and of its age. Caramel colors that are substantially out of the stated viscosity specification on the high side when received would be suspected of (a) being abused during manufacture either by being shorted on catalyst or by being taken darker than normal during the cooking process or (b) being stored for a long period of time. Caramel color is a semi-perishable product, with both color and viscosity increasing with age. The viscosity will finally increase to where the product has reached what is known as the gel state, at which point the color will have polymerized into an insoluble rubbery mass and will be unusable.

Colloidal charge is an important feature of caramel color and determines in many applications which product must be used. Each caramel molecule carries an electrical charge formed during processing. Caramel Color I, which has the least reactants, carries a slightly negative charge. Classes II and IV, which have sulfites in the catalyst, are strongly negative, and Caramel Color III, which has only ammonium compounds in the catalyst, is strongly positive, all of these charges being as manufactured. Colloidal charge is strongly influenced by pH. By changing the pH of caramel solutions, the isoelectric point (where the charge is neutralized) can be reached. Further pH adjustment will cause the charge to switch to the opposite polarity. The charge of Caramel Color III is usually positive up to around 5; the isoelectric point will be between 5 and 7 depending upon the product and will be negative above that. Caramel Color IV has a negative charge above pH 2; the isoelectric point will usually be between 0.5 and 2 and will be positive below that. Isoelectric points below pH 4.7 can be determined using the gelatin test. The test consists of making a 0.4% solution of a caramel color and adjusting the pH of portions of the solution from 1.0 to 4.0 in 0.5 pH increments and filling test tubes about two thirds full. A series of 0.5% gelatins are made to the same pH. Add two drops of the gelatin solution of the same pH to the tube of color, and observe for the presence of haze or a cloud at the interface of the two solutions. Gelatin has a known isoelectric point of 4.7 and is positively charged below this value. A haze at the interface indicates the caramel color is negatively charged at that pH, while a positively charged color will remain clear. Above pH 4 the isoelectric point of caramel colors can be determined using a known negatively charged color such as a double-strength Class IV, making a series of solutions with pH from 4 to 8 in 0.5 increments, both of the known and the sample, and proceed to add and observe as in the gelatin test. In this instance haze at the interface indicates positive, while clear is negative.

Shelf life at ambient storage conditions for a caramel will normally be stated as either 1 or 2 years depending upon the class. The caramelization reaction continues at a slow rate during ambient temperature storage with both color and viscosity increasing with time. On a double-strength Class IV stored at ambient temperature the color increased from a color intensity of 0.235 to 0.282 over 33 months for a gain of 0.6% per month. A Caramel Color III increased from 0.111 to 0.143 over 36 months, for a gain of 0.8% per month.

Caramel color as produced is essentially a sterile product and with its relatively high solids and low pH is usually not subject to microbial attack until it is diluted. As a consequence, the micro specs for caramel tend to be tight, with a typical specification as follows:

Total plate count:	<200 per gram
Yeast:	<10 per gram
Mold:	<10 per gram
Coliform:	Negative
Salmonella:	Negative

APPLICATIONS

Although caramel is used in a wide variety of food products, in general Caramel Color I is used in high-proof spirits, Caramel Color II in high-proof spirits containing certain vegetable extracts, Caramel Color III in beer, gravies, mixes, and sauces, and Caramel Color IV in soft drinks and other food products. Use levels, of course, depend upon the shade desired, but Table 12.4 gives examples of current use levels in a variety of products for different types of caramel color of a given color intensity. If the color under test has a higher or lower color intensity, adjust accordingly. As with any food ingredient of a complex nature, application questions and problems concerning caramel color often arise. Below is a discussion of some of these issues.

Soft drinks account for a large part of the caramel market, and the necessary specifications for a product of suitable quality are well known. Caramel Color is used in colas for a variety of reasons, first and foremost to provide "eye appeal." Caramel color provides the color that looks refreshing and entices consumers to purchase the product. Among other reasons, caramel color is an emulsifier in its own right. Anwar and Calderon of the Pepsi-Cola Company patented caramel color as an emulsifying agent in 1971. They state in the patent that one of the objects of the patent is

TABLE 12.4 Current Use Levels for Different Types of Caramel Color

	Use level (%)				
	Class I	Class III	Class IV, single	Class IV, double	Class IV, powder
	0.035	0.108	0.108	0.240	0.255
Beer, light		0.02			
Beer, dark		0.3			
Brandy	0.15		0.05		
Bread, light		1.0^a			
Bread, rye		2.0^a			
Bread, pumpernin.		3.0^a			
Cake mix, spice					2.0^a
Cake mix, dark					5.0^a
Chocolate milk					0.2
Cocoa extender					40.0
Cola			0.4	0.2	
Ginger ale			0.01	0.005	
Gravy		0.6			
Gravy dry mix					3.0
Gravy browning liquid		25.0			
Ice cream wafer		4.5			
Licorice		3.6			
Maple syrup			0.08		
Pet food			0.2	0.1	
Root beer			0.3	0.15	
Rum	0.1				
Vinegar, malt		0.2			

^{a}Based on flour weight

to provide an aqueous emulsion of flavoring oils that do not contain natural gums. The water insoluble flavoring agent may be added to the caramel or the caramel may be added to the flavoring agent. The amount of solids required to emulsify the flavoring agent will depend upon the particular type of flavoring agent used. Thus an agent which contains a high percentage of terpenes, such as orange oil, is more difficult to emulsify and will require a greater proportion of caramel solids to emulsify it than an agent containing a small concentration of terpenes such as distilled lime oil. Generally, all of the water which is necessary to serve as the aqueous phase for the emulsification of the flavoring agent is present in the volume of caramel used. The aqueous admixture of caramel and flavoring agent is then emulsified. The emulsification may be accomplished by means of a homogenizer, a colloid mill or other such apparatus. It is preferred that the average diameter of the emulsified particles be less than 3 microns.

Caramel color also helps protect flavors in clear bottles from deterioration from sunlight, and while sugar and gums provide most of this function, caramel color, with its specific gravity of 1.250–1.360, contributes "body" to the mouthfeel of the finished product.

Cola beverages have a negative colloidal charge and contain acidulants, so the color used must be compatible in this environment (i.e., a negative caramel); if not, a floc-type sediment will occur. On the other hand, a "plug" in the neck of the concentrate bottle or a ring in the neck of the beverage bottle is usually a flavor oil emulsion breakdown, caused by either the average flavor particle size being too large >1 μm if caramel color is the emulsifier), or a problem with the flavor oil–gum emulsion. Caramel may sometimes cause this problem, but usually what appears to be caramel are the flavor oils that have come out of dispersion and are colored brown by the inclusion of caramel.

Ginger ale concentrates generally have a high alcohol content, and the caramel color used, in addition to being a negative type, must be able to withstand the alcohol concentration, or precipitation will occur. Precipitations of caramel due to alcohol concentration alone are often reversible with the addition of a small amount of water, but too much water will cloud a ginger ale extract.

Both positive and negative caramel colors are effective in soy sauce as long as the type chosen has the necessary salt stability (some soy sauces are preserved with up to 20% salt). Usually the positive types have inherent salt stability: in the soft drink (negative) types special formulas are used to provide this stability. It has been found that the positive types of caramel give hues to the finished soy sauce that more closely resemble the hues of naturally fermented sauces.

Chocolate milk and sometimes very dark cookies (biscuits) do not have true chocolate shades when using caramel alone. In cookies, very pleasing dark shades can be achieved by combining caramel color and alkali-processed cocoa. In milk, caramel colors have almost a "muddy" appearance, but this can be overcome by the addition of approximately 0.01% by weight of FD&C Red #40 or Amaranth (formerly FD&C Red #2), giving "Dutch" chocolate shades. By the addition to the caramel and red mix of a small amount of blue and yellow, a more chocolate brown shade can be achieved.

An attractive eggnog shade in milk can be achieved with the use of a light, yellow, Class I color. This color also does a good job giving baked or microwaved poultry an "oven-roasted" appearance, including light and dark highlights. Another use of this color is to make a variety of greens by mixing in FD&C Blue #1 (Brilliant Blue FCF). Greens as found in some of the sugars sprinkled on cookies (biscuits) at Christmas can be achieved with this combination.

Soups and gravies containing meat products and colored with caramel often give a consistent but slightly different shade before and after retorting. Both positive and negative caramels work well in this application, depending upon the shade desired. Positive caramels generally contribute more red tones.

Vinegar is the product that has probably caused the most problems for caramel manufacturers. Malt vinegars are no problem, as they consistently work well with a positively charged caramel. Cider and distilled vinegars are another story. Negative caramels from the same drum that worked perfectly last week may have a problem with a new batch of vinegar this week. The exact cause of the problem is not known, but changes in alcohol and *Acetobacter* nutrient sources may be involved. Vinegar bottlers are encouraged to make a 24-hour lab test with a new batch of vinegar whenever any change is made.

Caramel colors for beer must have a positive colloidal charge and alcohol stability. The addition of a negatively charged caramel to beer, which contains positively charged proteins, causes an immediate cloud, which will agglomerate into particles large enough to precipitate out within a short time. For wines, whiskeys, and liqueurs, either negatively charged soft drink types or specifically formulated spirit types are recommended, depending on the application.

If a wine is clarified using gelatin and tannic acid, sufficient tannic acid must be added to remove all of the gelatin. If not, the remaining gelatin (positively charged) and the caramel will precipitate and be removed in the filtration process, causing a noticeable lightening of the product.

For improved stability in liqueurs, especially the creme types, it is important to premix the caramel with alcohol before adding the other ingredients. If using dairy ingredients, it is necessary to control the pasteurizer temperatures to prevent scorching the creme, because the scorched particles, brown in appearance, tend to rise, giving the impression of caramel failure.

High-proof rums (151°) are best colored with a sucrose-based Class I color. When you compare Class I caramel processed exactly the same way except for the syrup used (sucrose or corn syrup), the sucrose-based product will tend to have the higher alcohol tolerance.

A continuing problem with alcoholic beverages in clear glass bottles is that of fade. A product such as rum or a blend containing caramel will experience moderate fade under florescent light, but the same product in direct sunlight will experience a 10-fold increase in fade rate. Current research indicates the least fade in direct sunlight is encountered with the light Class I caramels, while under florescent light the double-strength soft drink type performs best. Alcoholic solutions of soft drink Class IV caramels subjected to direct sunlight tend to fade at a faster rate in the upper visible region (610 u was used) than in the lower visible region (430 u was used); the Class I under study, in direct sunlight, tended to fade at an equal rate at both wavelengths, and gave the visual impression of less fade.

OTHER CONSIDERATIONS

The percentage of caramel needed to impart the desired color is normally so low that caramel would have no measurable impact on the nutritional profile of a product, and even though caramel is made from edible carbohydrates, the metabolic calorie content of a double strength Class IV <1kcal/g. The reason for this is that the starting carbohydrates are converted by caramelization to high molecular weight color bodies, which are not readily absorbed or metabolized. The residual sugars as profiled by ORF for a typical single-strength Class IV, color intensity 0.094, are 23.4% glucose, 11.2% maltose and isomaltose, and 4% of other di- and trisaccharides; for a double-strength Class IV, color intensity 0.240, 3% glucose, 1% maltose and isomaltose, and 1% other di- and trisaccharides. A typical nutritional profile for two products similar to these is shown in Table 12.5.

The Kjeldahl methods will give total nitrogen on the order of 0.7% and 2.5%, respectively, but again the nitrogen is tied up in the higher molecular weight substances.

TABLE 12.5   Nutritional Profile for Two Class IV Products

	Class IV Single	Class IV Double
Ash	2.0	2.0
Protein	Nil	Nil
Moisture	29.6	43.5
Fats	Nil	Nil
Fiber	Nil	Nil
Carbohydrates	68.4	54.5
kcal/g	1.5	0.2
Vitamins	Nil	Nil
Copper	1 ppm	1 ppm
Iron	10 ppm	8 ppm
Calcium	30 ppm	25 ppm
Sodium	3300 ppm	5000 ppm

Caramel colors made under the ITCA specifications contain no substances forbidden by the Jewish Dietary Code and are kosher, but because most caramel colors are made from corn syrup, they are not acceptable for use during Passover. To be certified as "Kosher for Passover," the caramel color needs to be made from cane or beet sugar under rabbinical supervision.

The minimum temperature that caramel color should be subjected to so that it will still flow depends a great deal on the specific gravity of the product. Two caramel colors, one with an S.G. of 1.320 and viscosity of 300 cps @15.5°C and the other with an S.G. of 1.270 and viscosity of 80 cps, were subjected to −5.5 and −16°C temperatures for 3 days. At −5.5°C both samples were fluid, with viscosities of 1525 and 775 cps, respectively. At −16°C the 1.270 S.G. sample was frozen, although a glass thermometer could be inserted into the mass, while the 1.320 S.G. sample was thick but pourable with a viscosity of 100,000 cps. It did not appear that freezing damaged the 1.270 sample because all analysis appeared normal after thawing.

REFERENCES

Anwar, and Calderon. 1971. Emulsions of flavoring oils and process for making same. U.S. Pat. 3,622,343.

EBC method. 1950. *J. Inst. Brewing* 56: 373.

Fetzer, W. R. 1938. Analysis of caramel color. *Ind. Eng. Chem.*

Gelis, M. A. 1858. *Ann. Chim. Phys.* 352: 352.

Greenshields, R. N. 1973. Caramel—part 2. Manufacture, composition and property. *Proc. Biochem.* (April).

Hodge, J. E. 1953. *Agric. Food Chem.* 1(15): 928.

ITCA. 1984. *Specifications for Caramel Color.* International Technical Caramel Association, Washington, DC.

JECFA. 1992. *Compendium of Food Additive Specifications.* Joint Expert Committee on Food Additives, FAO/WHO.

Kroplien, U. 1986. Quantitative analysis of 2-acetyl-4(5)-tetrahydroxybutylimidizole *J Chromatogr.* 362. 286–290.

Linner, R. T. 1970. Caramel coloring—a new method of determining its color hue and tinctorcal power. Proceedings of the Society of Soft Drink Technologists, Washington, DC.

Mathewson, P. 1999. *Enzymes in Food Processing.* Food Technology Resource Group, Salt Lake City.

Millstone, E. 1984. *New Scientist* (Oct.): 20–24.

National Academy of Science. 1996. *Food Chemicals Codex*, 4th ed. National Academy Press, Washington, DC.

Official Journal of the European Communities, Commission Directive 95/45/EC, July 1995.

Richardson, T., and Hyslop, D. B. 1985. In *Food Chemistry*, 2nd ed., O. R. Fennema (Ed.), pp. 50, 98, 133, 144, 448. Marcel Dekker, New York.

WHO. 1977. Technical Report Series No. 617. Food and Agriculture Organization of the United Nations, World Health Organization, Rome.

WHO. 1986. Twenty-ninth report of the Joint FAO/WHO Expert Committee on Food Additives. World Health Organization, Geneva.

13

The Measurement of Color

Kevin Loughrey

Minolta Corporation
Ramsey, New Jersey

INTRODUCTION

Appearance has a strong influence on the consumer's opinion about food quality. Color is the primary attribute of appearance along with geometric attributes such as texture and gloss. The interest in natural "uncertified" food colors (colors exempt from certification) is growing as the consumer demand for more healthful foods increases. The industry use of natural colorants gives rise to a need for objective measurement of these colorants and the foods containing them.

Color has been measured in the food industry by subjective visual inspection, including the use of visual color standards. There has been a movement towards using color instruments to objectively measure the color of food. Color instruments are tools that assist the eye and, with proper usage, can provide repeatable, meaningful color data that agree with visual measurement. Effective color measurement requires an essential understanding of the subjective visual experience of color and its direct relationship with the objective color instrument.

When the fundamentals of visual and instrumental color measurement are grasped, they can be applied to natural colors using standardized methods. Color data resulting from correct measurements can be communicated throughout industry. This chapter will highlight the basic principles of color and color measurement as they apply to natural food colors. Hopefully, this will promote a better understanding of color and ultimately a working relationship between the visual and instrument measurement.

THE EXPERIENCE OF COLOR

The human eye is wonderful in its ability to combine with the brain and produce the sensation of color. However, in everyday use, the eye has limitations, including loss of color memory, color blindness, eye fatigue, differences in viewing conditions, and other factors that affect the eye's ability to distinguish color. Color perception needs to be further examined to understand the relationship between the subjective eye of an observer and the objective numbers from a color instrument. Normal color observers have three types of color receptors (cones) in their eyes that are responsible for a tristimulus response to color. How then would different people see and describe the same color? The experience of color is personal to each observer, so we can expect people to use different words when describing the same color. Observers that are tested and have "normal" color vision still vary widely in their discrimination and description of color. Standard methods to measure and communicate color are critical. We refer to color as being "psychophysical," involving:

The physics of light and objects as they interact

The physiology of the eye and brain

The psychology of the human mind

Color begins with light, and simply put, *without light there is no color*! The conditions under which a color is viewed directly determine the experience of color. The light source used by an observer to view a product is the most important single factor affecting appearance. Many products are evaluated for color under commercial lighting. Commercial light sources do not have balanced color output and should not be used for the critical appraisal of color.

Proper color measurement should begin with a standard visual method that would address the many factors that affect our visual experience, including:

The spectral quality and illumination level of the light source

The difference in observer response

The angle of illumination and view

The background and surround

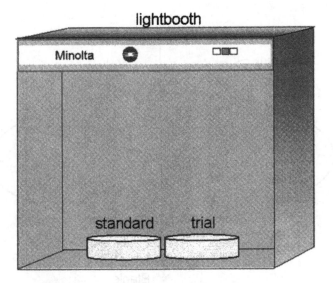

FIG. 13.1   Lightbooth.

The size of the colored sample

Directional differences in samples

A proper lightbooth (Fig. 13.1) is designed to standardize viewing conditions for controlled visual measurement and to agree with the objective color measurement of an instrument. ASTM test method D 1729 provides guidelines for selecting the appropriate conditions to be used by a person making visual judgments. The foundation for all our color work is based on the eye being the final judge of color, and the numbers from the instrument must agree with the visual measurement.

DESCRIBING COLOR

Color as we visually experience and describe it has three dimensions. Systems to organize color were developed to better identify and communicate colors. When colors are classified they can be represented in terms of their hue (color), lightness (brightness or value), and chroma (saturation).

The systems developed for measuring color have represented color based on three dimensions:

1. Hue is the name of the color, such as red, green, and so forth. Hues form what we know as the color wheel. The human eye has the ability to identify more than a million different hues.

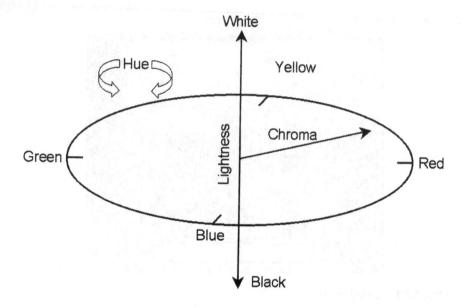

FIG. 13.2   Color model.

2. Lightness is easy to understand when classifying colors and separating those that are bright, midtone, or dark. Lightness can be represented as a vertical scale, with white at the top, gray in the middle, and black at the bottom. It can be measured independent of the hue.

3. Chroma or saturation is completely separate from hue and lightness. As a color moves away from a central neutral gray, it increases in chroma and is moving from dull toward vivid.

Hue, lightness, and chroma form a three-dimensional model (Fig. 13.2) that can be represented as a visual color system using physical colored samples. It is interesting to note that the ability of a color observer to perceive color differences between a target color (product standard) and a trial color is not equal for each of the three dimensions. The average, normal observer will first perceive small differences in hue, followed by chroma and, lastly, lightness. It is important to keep this in mind when studying the visual observing situation and relating it to the color instrument.

The Munsell system, still in use today, is just such a color model and defines the three dimensions as hue, value (lightness), and chroma. Visual color standards such as Munsell may be used when comparing colors under standard viewing conditions. Object colors can be identified with letters and numbers using such a system. However, a visual system is still subjective and at times difficult to communicate. The early color scientists began explor-

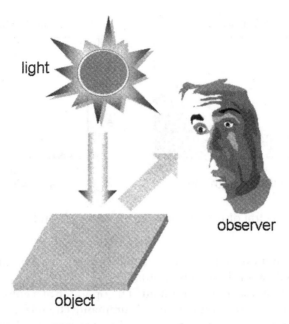

light

observer

object

FIG. 13.3 Observer situation.

ing methods for more accurate color communication, and their work led to the development of numerical color scales based on instruments.

OBSERVING COLOR

The sensation of color experienced by an observer occurs when viewing an object illuminated by a light source. This interaction of light, object, and observer is known as the observer situation. A simple arrangement, it is at the heart of understanding color and color measurement. The subjective response of the human observer to a color should relate directly to the objective measurement of the color instrument. Fig. 13.3 is an example of an observer situation with a light source, object, and observer. Accurate color measurement depends upon reproducing this relationship when measuring color.

Different light sources will make the same color appear different. The light source as the first element in the observer situation can be represented as a standard illuminant. The International Commission on Illumination (CIE) first defined standard illuminants in 1931, and their work has continued with the development of many standard illuminants.

Standard illuminants widely used in color measurement today are D_{65} (average daylight), F_2 (cool white fluorescent), and A (incandescent). The

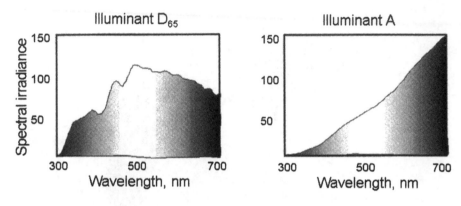

FIG. 13.4 Spectral distributions of illuminants D65 and A.

spectral distributions of D_{65} and A are very different, as seen in Figure 13.4. The spectral data for these illuminants is stored so color measurements can be calculated for a specific illuminant. The operator of a color spectrophotometer has only to select the desired illuminant. The selection of standard illuminants raises an important distinction between a colorimetric spectrophotometer and a tristimulus filter colorimeter. The traditional colorimeter is designed to measure color under daylight conditions only based on the response of the filters. The colorimeter design is a light source filter and photodetector combination that reproduces the eye's response. The spectrophotometer uses the spectral distribution of the standard illuminant selected and can calculate color under daylight conditions as well as incandescent and fluorescent illumination.

Differences between the colorimeter and spectrophotometer arise again when considering the object or product to be measured. The object, as the second element of the observer situation, needs to be evaluated as to how it interacts with the incident light. Most products are less than ideal, so one must consider opacity, uniformity, flatness, and other factors when preparing to measure a sample for color. Natural food colors are available in different forms such as powder, liquid, or slurry. Light illuminating a sample could cause a number of interactions to occur (Fig. 13.5), including:

1. Surface reflectance (specular reflectance or gloss)
2. Refraction into the object
3. Selective absorption
4. Diffuse reflectance (scattered light)
5. Transmittance (light passing through the product)

Any color measurement must consider the nature of the sample and how it interacts with the incident light:

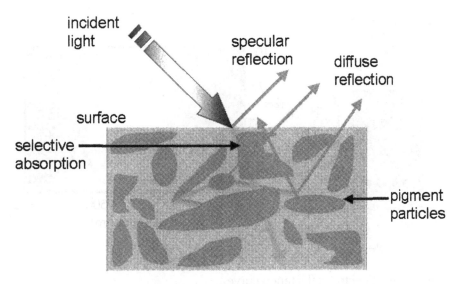

FIG. 13.5   Interaction of light and object.

Opaque samples reflect light.

Translucent samples reflect and transmit light.

Transparent samples primarily transmit light.

These interactions must be examined when measuring the color of a product. A powder sample would be measured for reflected light. It is this scattered light, known as diffuse reflectance, which is responsible for the color of the powder. (It is important to note that the light reflected (or transmitted) by an opaque product is responsible for the color we perceive. The light absorbed by the product never reaches the eye-brain to cause our sensation of color. Thus, analytical color measurements of absorbed light that are made in food applications are not based on this model of how we perceive color.)

Using a spectrophotometer the reflected light can be measured at wavelength intervals across the visible spectrum. The spectral data for a natural color (Fig. 13.6) are a physical measurement of the reflectance of this object. The spectrum is like a fingerprint, being unique for every object measured. The data can be displayed as a table of reflectance values or plotted to form a spectral curve of the object. It is important to remember that the color of an object depends upon three critical factors:

1. The nature of the incident light
2. The reflectance of the object
3. The response of the observe

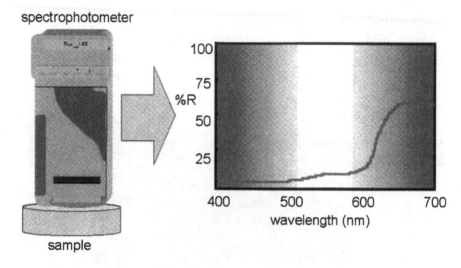

spectrophotometer

sample

FIG. 13.6   Spectral reflectance curve.

The observer is the last of the three elements that combine to form the observer situation. The "standard observer" is based on the spectral sensitivity of the normal eye, and its development is the basis for all instrumental color measurement. In 1931 the CIE first defined the 2 Degree Standard Color Observer. Experiments with real observers were conducted to collect the data that was then transformed mathematically to represent the average observer having normal color vision (Fig. 13.7).

The CIE defined a second color observer in 1964 known as the CIE 1964 10 Degree Supplementary Color Observer. Both observers are in use today, but most color applications use the 10 Degree Observer. Filter colorimeters are fixed for one observer and utilize the 2 Degree Observer. Color spectrophotometers provide the operator a choice of either observer along with a choice of various illuminants.

COLOR SCALES

CIE XYZ numbers, known as the tristimulus values, can be calculated for a color as it appears for an observer under a defined illuminant. The three elements of the observer situation, light object, and observer are measured numbers that can be combined to calculate a set of XYZ tristimulus values:

Spectral energy		Reflectance		Observer		X
of the light	×	of the object	×	response	=	Y
						Z

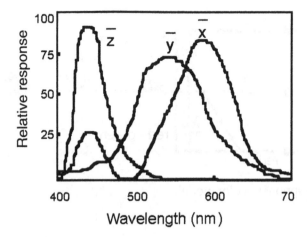

FIG. 13.7   The CIE Standard Color Observer.

The spectrophotometer measures the spectral reflectance of the object and integrates these data with the standardized data for the selected illuminant and observer. The resulting tristimulus values identify the object color for a given illuminant and observer. However, XYZ numbers do not easily relate to observed color and color difference. Considerable work has been done to develop alternate color scales based on the basic principles of normal color vision.

The opponent theory of color explained the mechanics of normal color vision. There are three types of color receptors in the normal human eye. Light reflected by an object reaches the receptors, which are excited. Three sets of signals are coded and sent to the brain of the observer via the optic nerve. The signals correspond to light versus dark (opposing), red versus green, and yellow versus blue. Color scales were developed that supported the accepted theory and agreed with the experience of normal observers observing color and color difference. Work continued to develop color scales based on the human vision system. The L,a,b model was founded on the opponent theory of color perception.

Fig. 13.8 outlines the opponent signals sent to the brain of a normal observer describing a color. The signals are a brightness signal and two hue signals. One hue signal describes the amount of red or green. The other hue signal describes the amount of blue or yellow. Improvements to develop a more visually uniform color space were made in the original L,a,b system. In 1976 the CIE published the CIELAB color scale known as L*a*b*. CIELAB numbers are widely accepted in many food and packaging applications. L*a*b* is a rectangular coordinate system that defines a color in terms of lightness, red versus green, and yellow versus blue (Fig. 13.9). The L* value indicates lightness from 100 = white to 0 = black. A +a* num-

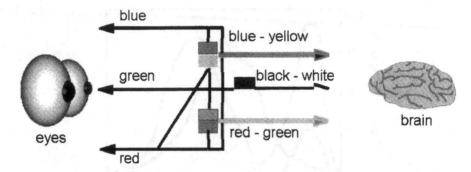

FIG. 13.8 Opponent color model.

ber indicates redness and −a* indicates greenness. A +b* indicates yellow-ness and −b* indicates blueness. The CIELAB (L*a*b*) system was intro-duced with a second color scale known as CIE L*C*h. The L*C*h color scale is a cylindrical or polar system. The L* indicates lightness and is the same L* as in the CIELAB system, the C* is the chroma, and the h is the hue angle. The hue is expressed as an angle in degrees; moving counterclock-wise 0° = red, 90° = yellow, 180° = green, and 270° = blue.

MEASURING COLOR WITH THE INSTRUMENT

The color scale numbers provide an accurate method based on interna-tional standards to objectively measure color and color difference using a color instrument. The perception of color by an observer can change when the viewing conditions change, including the light source and background. The colorimeter is designed to reproduce the sensitivity of the human eye, but it uses a fixed set of conditions for the light source and observer. The visual measurement process is based on the light reflected by the object reaching the eye and then being sent to the brain to identify the color. The colorimeter illuminates the object, captures the reflected light, passes it through three filters (red, green, blue), and then calculates L*a*b*. The spectrophotometer also measures L*a*b* color for the object, but by a dif-ferent method (Fig. 13.10). A closer look reveals that the colorimeter uses filters (red, green, and blue) designed to have the spectral sensitivity of the human eye.

The key to a spectrophotometer is how it measures the object and uses the data collected. The reflectance of any natural color can be measured and will be different for each color. The spectral reflectance curve for any product indicates the amount of light reflected across the visible spectrum. This spectral curve is like a fingerprint being unique for a given product. The spectrophotometric method measures the spectral reflectance data of

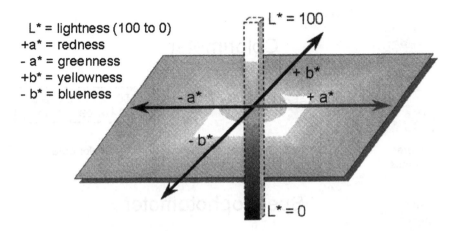

L* = lightness (100 to 0)
+a* = redness
- a* = greenness
+b* = yellowness
- b* = blueness

FIG. 13.9 CIELAB color space.

the product at individual wavelengths (10 or 20 nm) using a spectral sensor. The microcomputer then converts the reflectance data into the tristimulus values X, Y, and Z. The X, Y, Z values can be used to calculate values in the color scale selected such as L*a*b.

Colorimeters have been widely used in food applications because they are simple to use, provide fast data, and the measurements are always based on the same observer and illumination method. Spectrophotometers are more complex, using the actual reflectance data across the visible spectrum. The reflectance data can be integrated with a choice of illuminant and a choice of observer to calculate color numbers under both observers and a variety of light sources.

THE PREPARATION AND PRESENTATION OF SAMPLES

At the heart of any successful color-measurement program is a consistent method of specimen (standards and trials) preparation and measurement. Color numbers first need to be repeatable before they can be correlated to visual estimates. Unfortunately, this area is often overlooked and may only be noticed when disagreement results from comparing color data. The person performing color measurements must be aware of how "less than ideal" the specimens are and what effect any preparation procedure may have on the optical properties of the specimens (standards and trials).

The method by which all products are prepared and presented for visual and instrument color measurement is critical to the success of any color program. If all products were flat, smooth, uniform, totally opaque, nondirectional, nonfluorescent, and not sensitive to heat or light, we would still require a defined method. A natural color may be in the form of a powder

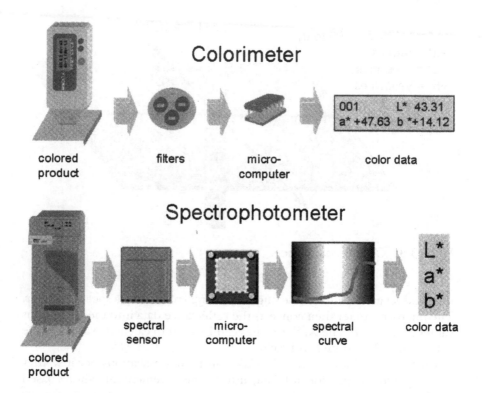

FIG. 13.10   Colorimeter and spectrophotometer.

and there is more than one method to prepare such a sample. The powder can be placed in an optically clear cell (glass or plastic) and measured through the bottom of the cell (Fig. 13.11). The instrument measures the light reflected light by the uniform surface of the powder, which was created by the powder pressing against the glass or plastic interface.

A second method of measuring the powder would be to present the surface of the powder directly to the instrument (Fig. 13.12). The powder would need to be prepared so the instrument can measure a flat, smooth surface. The consideration of how to prepare and present a sample for color measurement is an area overlooked even by experienced colorists. This really comes back to the basic properties of objects and how they reflect and/or transmit light. Some of the factors to be considered are:

1. Opacity, translucency, or transparency
2. Uniformity, texture, smoothness, flatness, curvature, etc.
3. Thickness, backing, surround
4. Directionality

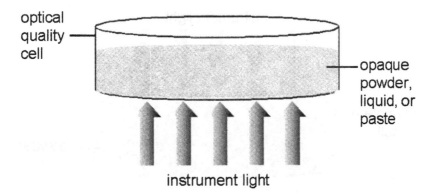

optical
quality
cell

opaque
powder,
liquid, or
paste

instrument light

FIG. 13.11 Measuring powder through a cell interface.

5. Thermochromism (sensitive to heat) or photochromism (sensitive to light)

6. Pressure, tension

7. Size, shape, etc.

8. Number of color measurements per specimen (average)

The list is not complete and is to serve only as encouragement to investigate the nature of colored products prior to visual and instrumental color measurement. The number of color measurements per sample is particularly important because in many color applications a single instrument reading will not adequately represent a product. An average of two or more readings is required to ensure repeatability between samples (standards and trials) and instrument operators. This is just one of a number of steps that need to be addressed in order to obtain meaningful numbers that agree with the eye.

THE KEY TO SUCCESS

The color data collected is very valuable information, provided the necessary steps have been taken to ensure the numbers are "repeatable." The color numbers will be repeatable and meaningful if the variables within the color application are identified and controlled. A sound color program is built on a consistent method that has been defined and documented.

Some of the more important factors are as follows:

Viewing conditions—level of illumination, spectral quality of light source, incident/viewing angle, background/surround, color response functions of observers, etc.

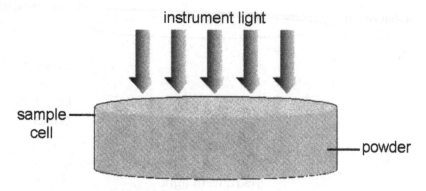

FIG. 13.12 Measuring a powder surface directly.

Instrument conditions—color scale, primary/secondary illuminant, standard observer, etc.

Selection and use of standards and trials—stability, availability, uniformity, etc.

Preparation and measurement of standards and trials—size, backing, number of readings

An example of a simple written color specification could include:

1. Viewing

Standard lightbooth	Manufacturer and Model
Angle of illumination/viewing	0°/45°
Primary illuminant	Cool white fluorescent
Secondary illuminant	North sky daylight

2. Specimen Preparation and Measurement

Thickness	Fold sample to create four-ply
Size (desired)	8″ × 10″
Measurement	Three random areas and average

3. Color Calculations

Color scale	CIELAB (L*a*b*)
Primary illuminant	F2 (cool white fluorescent)
Secondary illuminant	D65 (daylight)
Color observer	CIE 10 Degree Standard Observer
Color tolerance	Delta L*, delta a*, and delta b*

Not all of the answers to color and appearance problems are contained in this brief introduction to color. The real "key to success" is the effort made to better understand and apply control, consistency, and communication in all color work! This begins with the manner in which color is observed (observer situation) and carries through to the color instrument. This is the secret to taking the first steps toward the goal of agreement between the eye and the numbers from the color instrument. The final goal of all color analysis, using a colorimeter or spectrophotometer, is to achieve repeatable color numbers that agree with visual perception. The eye will maintain its position of importance and may be assisted by the color instrument to achieve consistent results from person to person and from day to day. People with normal color vision highlight the basic principles of color and color measurement as they apply to natural food colors. Hopefully, this will promote a better understanding of color and ultimately a working relationship between visual and instrument measurement.

REFERENCES

Minolta Co., Ltd. 1998. *Precise Color Communication*, Minolta Co., Ltd.

Loughrey, K. 1994. The basics of color and appearance. Paper presented at Spinneybeck National Sales Meeting, June 23, Rochester, NY.

Loughrey, K. 1993. An introduction to color. Seminar material, TransMir Associates, South Deerfield, MA.

14
Health Aspects of Natural Colors

G. Mazza

Pacific Agri-Food Research Centre
Agriculture and Agri-Food Canada
Summerland, British Columbia, Canada

INTRODUCTION

Natural colors are generally defined as materials extracted, isolated, or otherwise derived from plants, animals, or minerals that are capable of imparting a distinguishing color when added to foodstuffs. The most important natural colorants include anthocyanins (flavonoids), carotenoids, chlorophylls (porphyrins), curcuminoids (turmeric), betalaines (beetroot), and quinonoids (carmines). There is now a considerable amount of evidence suggesting that some natural colors may be important nutritional antioxidants and that their presence in the diet may reduce the risk of cardiovascular disease, cancer, and other diseases associated with aging. The evidence for nutritional prevention of diseases by anthocyanins, carotenoids, and curcuminoids has been obtained from a variety of interdisciplinary studies including epidemiological studies, distribution of the pigments in human organs and tissues, supplementation studies, studies with small animals, and studies with transformed and nontransformed animal and human cells. This chapter will present information from studies that show that some natural colors have beneficial physiological effects including the reduction of the risk of diseases such as coronary heart disease and specific types of cancer via a number of different pathways. Other health-related properties of natural colors, including antimicrobial and anti-inflammatory activities, will also be discussed.

ANTHOCYANINS

In recent years numerous studies have shown that flavonoids and other phenolics, present in fruit and vegetable products, possess anticarcinogenic, anti-inflammatory, antihepatotoxic, antibacterial, antiviral, antiallergic, antithrombotic, and antioxidant effects (Wagner, 1979; Macheix et al., 1991; Middleton and Kandaswami, 1992; Tsuda et al., 1996; Wang et al., 1999).

Antioxidant Activity

Antioxidation is one of the most important mechanisms for preventing or delaying the onset of major degenerative diseases of aging, including cancer, heart disease, cataracts, and cognitive disfunction. The antioxidants are believed to exert their effects by blocking oxidative processes and free radicals that contribute to the causation of these chronic diseases (Ames, 1983; Block, 1992). Several plant phenolics, especially catechins, flavonols, anthocyanins, and tannins, have been shown to perform these functions (St. Léger et al., 1979; Ames, 1983; Block, 1992; Decker, 1995; Mackerras, 1995; Bors et al., 1996; Mazza, 1997).

Recently, several authors reported that anthocyanins show strong antioxidative activity (Tsuda et al., 1994a,b, 1996; Mazza, 1997; Wang et al., 1997; Narayan et al., 1999). Tsuda et al. (1996) investigated the reaction products of cyanidin 3-O-β-D-glucoside with 2,2′-azobis(2,4-dimethylvaleronitrile) and concluded that the antioxidative mechanism of cyanidin 3-O-β-D-glucoside may be different from that of α-tocopherol; cyanidin 3-O-β-D-glucoside would produce another radical scavenger, as it would break down the structure and scavenge the radicals.

Narayan et al. (1999) investigated the antioxidant effect of anthocyanin on enzymatic and nonenzymatic lipid peroxidation. Their results show that in vitro enzymatic and nonenzymatic polyunsaturated fatty acid peroxidation was significantly inhibited in a dose-dependent manner by purified anthocyanin pigment from carrot cell culture. The kinetics shows that anthocyanin is a noncompetitive inhibitor of lipid peroxidation. Anthocyanin was found to be a potent antioxidant compared to classical antioxidants such as butylated hydroxyanisole (BHA), butylated hydroxytoluene (BHT), and α-tocopherol.

Anti-inflammatory Activity

Reports of anthocyanins possessing anti-inflammatory activity were first published in the 1970s (Lietti et al., 1976). Recently, Wang et al. (1999) reported that cyanidin 3-glucosylrutinoside, cyanidin 3-rutinoside, cyanidin 3-glucoside, and cyanidin aglycon isolated from tart cherries exhibited in vitro antioxidant and anti-inflammatory activities comparable to commercial products. The inhibition of lipid peroxidation of the three anthocyanins and their aglycon were 39, 70, 75, and 57%, respectively, at 2 mM concentrations. The antioxidant activities of the anthocyanins and cyanidin were comparable to the antioxidant activities of *tert*-butylhydroquinone and butylated hydroxytoluene and superior to vitamin E at 2 mM concentrations. In the anti-inflammatory assay, cyanidin gave IC_{50} values of 90 and 60 mM, respectively, for prostaglandin H endoperoxide synthase-1 and prostaglandin H endoperoxide synthase-2 enzymes. Possible mechanisms for the anti-inflammatory activity of anthocyanins include inhibition of arachidonic acid metabolism (Ferrandiz and Alcaraz, 1991) and inhibition of the prostaglandin synthase cyclooxygenase activity (Hoult et al., 1994). The property of anthocyanins to decrease the fragility and permeability of blood capillaries is common to other flavonoids and was originally identified by Albert Szent-Gyorgyi, who coined the term vitamin P (Rusznyàk and Szent-Györgyi, 1936) for those compounds that reduce capillary permeability.

Several prescription and nonprescription pharmaceutical products containing anthocyanins from bilberry (*Vaccinium myrtillus*) as the active principle are used to control capillary permeability and fragility (Wagner, 1979; Boniface et al., 1986; Pizzorno and Murray, 1987; Beretz and Cazahave, 1988; Morazzoni and Bombardelli, 1996). The anti-inflammatory activity of these anthocyanin extracts accounts for their significant antiedema properties and their action on diabetic microangiopathy (Boniface et al., 1986; Morazzoni and Bombardelli, 1996).

Clinical applications for bilberry anthocyanin extracts include treatment of visual disorders related to nighttime visual acuity, to aid in adapting to low light conditions and to decrease recovery time after exposure to glare. The use of these products is recommended to pilots, truck drivers, those with poor night vision, and those using video display terminals. A recent study on the effect of anthocyanosides on night vision, however, concluded that single oral administration of 12–36 mg of anthocyanins appears to lack significant effect on militarily relevant night vision tests (Levy and Glovinsky, 1998).

Bilberry fruit anthocyanin extract is reported to be useful in treatment of inflammatory conditions of the joints due to its positive effects on collagen structure and metabolism. The same extract is touted to prevent glaucoma and to be effective in the treatment of retinopathies, including diabetic retinopathy. Small clinical trials have demonstrated the efficacy of bilberry extract in the treatment of peripheral vascular disease, i.e., reduced blood supply to lower limbs and in the pre- and postoperative treatment of varicose veins and hemorrhoids (Morazzoni and Bombardelli, 1996).

TABLE 14.1   Antioxidant and Pro-oxidant Activity of Selected Anthocyanins

Compound	DPPH · method Antiradical activity[a]	HPLC method Antioxidant activity[b] (μM of compound added)	Pro-oxidant activity[c]
Cyanidin	−7.40 ± 0.22	200–300	+
Cyanidin 3-glucoside	−6.81 ± 0.21	300–400	+
Cyanidin 3,5-diglucoside	−3.32 ± 0.08	500–1000	+
Delphinidin	−8.86 ± 0.31	500–1000	+
Malvidin	−4.42 ± 0.16	1500–2000	+
Malvidin 3-glucoside	−4.29 ± 0.19	500–1000	+
Malvidin 3,5-diglucoside	−2.56 ± 0.05	2000–2500	+
Pelargonidin	−4.63 ± 0.25	1500–2000	+
Pelargonidin 3-glucoside	−3.95 ± 0.09	2000–2500	+
Pelargonidin 3,5-diglucoside	−2.04 ± 0.06	2000–2500	+
Peonidin	−4.05 ± 0.16	1500–2000	+
Peonidin 3-glucoside	−3.38 ± 0.13	2500–3000	+
Ascorbic acid	−1.83 ± 0.03	>4000	+
α-Tocopherol	−1.88 ± 0.04	200–2500 (50%)	nd
BHA	−2.61 ± 0.04	1000–1500	nd

[a]Values are calculated coefficients of the slope by linear regression ± standard errors in μM of DPPH ·/μM of compound tested.
[b]Antioxidant activity was defined by the concentration range of added compound needed to reach 0% malonaldehyde of the control.
[c]Prooxidant activity was positive (+) if the % malonaldehyde of the control was >100% between 0 and 400 μM of added compound.
Source: G. Mazza and L. Fukumoto, unpublished.

To our knowledge, there are no apparent scientific reasons for the use of bilberry anthocyanin to the exclusion of other anthocyanin-rich plant foods such as elderberries, chokeberries, blueberries, raspberries, grapes, etc. However, the antioxidant efficacy of anthocyanins is higher in delphinidin and cyanidin than other anthocyanins (Table 14.1). Thus the high concentration of glycosides of delphinidin and cyanidin in bilberries (Baj et al., 1983; Mazza and Miniati, 1993) may account for the apparently higher antioxidant and anti-inflammatory activity attributed to this fruit.

Anticarcinogenic Effects

The anticarcinogenic activity of phenolics has been correlated with the inhibition of colon, esophagus, lung, liver, mammary, and skin cancers (Decker, 1995). Examples of plant phenolics that inhibit carcinogenesis include anthocyanins (Kamei et al., 1995), resveratrol (Jang et al., 1997), quercetin (Verma et al., 1988; Deschner et al., 1991; Verma, 1992), caffeic acid (Kuenzig et al., 1984; Hirose et al., 1993), ellagic acid (Chang et al., 1985; Mandel and Stoner, 1990; Maas et al., 1991), and flavan-3-ols (Liu and Castonguay, 1991; Yoshizawa et al., 1992). Anthocyanins have been found to significantly suppress the growth of cultured tumor cells and have been shown to have greater inhibitory effect than other flavonoids (Kamei et al., 1995). Flavonoids and polyphenols have been shown to impact on the initiation step of cancer development by protecting the cells against direct-acting carcinogens such as nitrosamines (Verma et al., 1988) or alter their metabolic activation (Chang et al., 1985). Particularly active polyphenols are the monomeric flavan-3-ols (Liu and Castonguay, 1991; Yoshizawa et al., 1992) and ellagic acid (Mandel and Stoner, 1990; Maas et al., 1991); for the latter compound, a mechanism based on the structural similarity with polycyclic aromatic compounds has been proposed (Sayer et al., 1982). Antitumor-promoting activity of flavonoids has been connected to the inhibition of various cellular DNA and RNA polymerases (Ohno and Nakane, 1990) and/or the inactivation of ornithine decarboxylase (Agarual et al., 1992). In addition, some flavonoids such as quercetin appear to be effective in assisting cell cycle progression.

Antiatherogenic Effects

The association between grape phenolics and coronary heart disease (CHD) has been ascribed in part to the presence of anthocyanins and resveratrol in red wine (Frankel et al., 1993; Kinsella et al., 1993; Kanner et al., 1994). In addition, several epidemiological studies have shown that coronary heart disease mortality can be decreased by moderate consumption of red wine (Rimon et al., 1991; Graziano et al., 1993; Klatsky, 1994). The primary mechanisms believed to be responsible for this reduced risk factor include reduced platelet coagulability (Elwood et al., 1991; Renaud et al., 1992) and higher circulatory high-density lipoprotein cholesterol (HDL), which is increased by ethanol in a dose-response manner (Graziano et al., 1993; Klatsky, 1994). Other mechanisms such as inhibition of lipoprotein oxidation, free-radical scavenging, and modulation of eicosanoid metabolism (Bors and Saran, 1987; Afanas'ev et al., 1989; Steinberg et al., 1989; Esterbauer et al., 1992) are also thought to play a role in the reduction of atherosclerosis.

Antibacterial and Antiviral Activity

Most phenolic compounds possess antibacterial or antiviral activities (Pisha and Pezzuto, 1994). Hydroxybenzoic, salicylic, gallic, and protocatechuic acids have also been reported to possess an antibacterial effect (Balansard et al., 1980; Dumenil et al., 1980). The antiviral potential of epicatechin observed in vitro has been suggested to be due to inhibition of the reverse transcriptase derived from Moloney murine leukemia virus (Pisha and Pezzuto, 1994). Reverse transcription inhibitors can be very specific for retro-viruses because the action of reverse transcriptase is different from cellular enzymes including DNA polymerase (Pisha and Pezzuto, 1994). Chloro-genic acid has been found to be active against the Epstein-Barr and HIV viruses (Pisha and Pezzuto, 1994). Most of the phenolic compounds tested for antimicrobial properties, however, have been shown to display consid-erably lower activity than products such as antibiotics (Wagner, 1985).

Absorption and Metabolism

One of the first studies to provide evidence for the uptake of anthocyanins in humans in vivo was published by Paganga and Rice-Evans (1997). An-thocyanins, rutin, and other quercetin glycosides were detected in human plasma in the glycosilated form by HPLC analysis. The results reveal that phloretin and quercetin are absorbed from the diet as glycosides. More re-cently, Miyazawa et al. (1999) reported the results of a human study involv-ing seven male and five female adults. The subjects orally ingested 2.7 mg cyanidin 3-glucoside (Cy 3G) and 0.25 mg Cy 3,5-diglycoside/kg body weight. In plasma, 30 minutes after intake the level of Cy 3G was 11 µg/L. At 60 minutes after intake, plasma Cy 3G was 13 µg/L. Only traces of Cy 3,5diG were found in plasma. These results demonstrate the direct intes-tinal absorption of anthocyanins in structurally intact forms after oral supplementation and that anthocyanins are present in the plasma after only 15 minutes. Other studies on bioavailability of anthocyanins have been re-ported by Morazzoni et al. (1991), Lapidot et al. (1998), and Cao and Prior (1999).

CAROTENOIDS

Carotenoids are a large group of fat-soluble C-40 plant pigments that are synthesized by photosynthetic microorganisms and plants (Briton and Goodwin, 1982). Fruits and vegetables of green, orange, and red color are the most important sources of carotenoids in the human diet. Over 600 carotenoids have been identified and are classified into hydrocarbon carot-enoids, with β-carotene and lycopene being the most prominent members, and oxycarotenoids (xanthophylls), to which belong β-cryptoxanthin, lu-tein, zeaxanthin, canthaxanthin, and astaxanthin (Straub, 1987).

Carotenoids are of physiological interest in human nutrition because some are vitamin A precursors and many exhibit radical or single oxygen trapping activity (Miller et al., 1996), and as such have potential antioxidant effects in vivo. Their presence in the diet may reduce the risk of cardiovas-cular disease, lung cancer, cervical dysplasia, age-related macular degener-ation, and cortical cataract (Prasad and Edwards-Prasad, 1990; Mathews-Roth, 1991; Olson, 1996; Duthie et al. 1998; Kritchevsky, 1999).

Carotene

β-Carotene is the most prevalent vitamin A precursor followed by β-crypto-xanthin and the α- and γ-carotenes. Vitamin A plays a central role in many essential biological processes. Its function as a chromophore in the visual process has been recognized for decades. It is also involved in fetal development and in the regulation of proliferation and differentiation of many types of cells throughout life. However, it has become increasingly clear that carotenoids mediate cellular functions in addition to serving as vitamin A precursors (Olson, 1996; Duthie et al. 1998). The beneficial effects of β-carotene are thought to occur through one of several modes: singlet oxygen quenching (photoprotection), antioxidant protection, and enhancement of the immune response. β-Carotene is approved for the general treatment of erythropoietic protoporphyria, a genetically inherited, light-sensitive skin disease (Mathews-Roth et al., 1970). It has been shown to be effective in protecting lipid membranes from free radical damage (Krinsky and Deneke, 1982), particularly under low oxygen partial pressure (Burton and Ingold, 1984).

Early in vitro studies of LDL oxidation showed the β-carotene carried in LDL is oxidized prior to the onset of oxidation of LDL polyunsaturated fatty acids, suggesting the possibility of β-carotene delaying the onset of LDL oxidation (Esterbauer et al., 1989). Subsequent investigations have yielded conflicting results, especially with regard to the role of β-carotene in the prevention of cardiovascular disease (Table 14.2). Thus, although several epidemiological studies have shown an inverse association between serum/adipose β-carotene levels and coronary heart disease risk, randomized clinical trials have not shown any benefit and perhaps even an adverse effect of β-carotene supplementation (Kritchevsky, 1999). Several possible confounding factors that may help explain the inconsistency between the trials and epidemiological evidence have been suggested. One such factor is the other carotenoids that are correlated with β-carotene both in the diet and in the blood might be important factors, as might other plant-derived compounds. Alternatively, low serum carotenoid levels may reflect either increased lipoprotein density or the presence of inflammation, both factors emerging as important new risk factors for coronary heart disease. Whereas the trial results do not support a preventive role for β-carotene, the epidemiological evidence does generally support the idea that a diet rich in high-carotenoid foods is associated with a reduced risk of heart disease (Kritchevsky, 1999).

In animal studies, β-carotene has also been shown to exhibit chemopreventive effects. It has been found to decrease tumorigenesis and inhibit conversion of papillomas to carcinomas in the skin of 7,12-dimethylbenz[a]anthracene (DMBA) – treated mice (Kornhauser et al., 1994). Similarly, spontaneous mammary tumor formation in C3H (Jax) mice (Bhide et al., 1994) and the incidence of preneoplastic GST liver foci in rats treated

TABLE 14.2 Epidemiological Studies of Blood/Tissue Carotenoids and Coronary Disease

Study	Outcome[a]	Carotenoid[b]	Adjusted relative risk[c]
Riemersma et al., 1991	Angina	Carotene (P)	↓[d]
Kardinaal et al., 1993	Nonfatal MI	β-Carotene (A)	↓↓[e,f]
Gey et al., 1993	CHD death	Carotene (P)	↓[e]
Morris et al., 1994	CHD	Total carotenoids (S)	↓[e]
Street et al., 1994	MI	β-Carotene (S)	↓↓↓[e,f]
		Lycopene (S)	↓↓↓[e,f]
		Lutein (S)	↓↓↓[e,f]
		Zeaxanthin (S)	↓↓[e,f]
Sahyoun et al., 1996	CHD death	Carotenoids (S)	↔
Iribarren et al., 1997	Carotid intima-media thickening > 90th percentile	α-Carotene (S)	↔
		β-Carotene (S)	↔
		β-Cryptoxanthin (S)	↔
		Lutein/Zeaxanthin (S)	↓
		Lycopene (S)	↔
Kohlmeier et al., 1997	Nonfatal MI	α-Carotene (A)	↓[e,g]
		β-Carotene (A)	↓↓[e,g]
		Lycopene (A)	↓↓[e]
Evans et al., 1998	CHD death	Total carotenoids (S)	↓
	Nonfatal MI		↔

[a]MI = myocardial infarction; CHD = coronary heart disease.
[b]Measured in serum (S), plasma (P), or adipose tissue (A).
[c]Relative risk (RR) contrasting the highest carotenoid category to the lowest adjusting for confounders except for Iribarren et al. (1997), where the relative risk is for a 1 SD difference in serum level.
[d]↓ denotes an RR < 0.8 but > 0.5; ↓↓ an RR between 0.5 and 0.33; ↓↓↓ an RR < 0.33; ↔ an RR between 0.8 and 1.25.
[e]Significant, $p < 0.05$.
[f]Results for cigarette smokers only.
[g]Not significant after adjusting for lycopene.
Source: Kritchevsky, 1999.

with the mutagen 2-amino-3-methylimidazo[4,5-f] quinoline are inhibited by β-carotene (Tsuda et al., 1994c).

β-Carotene and other carotenoids have also been shown to enhance both specific and nonspecific immune functions (Bendich, 1990). A number of modulating immunological reactions can increase the tumoricidal activity of cytotoxic T cells, macrophages, and/or natural killer cells and enhance traditional antimicrobial immunological functions (Mathews-Roth, 1991).

Lycopene

Lycopene provides the familiar red color to tomato products and is one of the major carotenoids in the Western diet. It accounts for about 50% of carotenoids in human serum (Gerster, 1997).The interest in lycopene has increased dramatically in recent years following the recent publication of epidemiological studies implicating lycopene in the prevention of cardiovascular disease and cancers of the prostate and gastrointestinal tract (Gerster, 1997; Clinton, 1998; Giovannucci, 1999).

Among the common dietary carotenoids, lycopene has the highest singlet oxygen quenching capacity in vitro. Other outstanding features are its high concentration in testes, adrenal gland, and prostate (Gerster, 1997). Remarkable inverse relationships between lycopene intake or serum values and risk have been observed in particular for cancers of the prostate, pancreas, and, to a certain extent, the stomach. In some studies lycopene was the only carotenoid associated with risk reduction (Gerster, 1997; Clinton, 1998).

A very recent review of the epidemiological literature on tomatoes, tomato-based products, lycopene, and cancer conducted by Giovannucci (1999) shows that among 72 studies identified, 57 reported inverse associations between tomato intake or blood lycopene level and the risk of cancer at a defined anatomical site; 35 of these inverse associations were statistically significant. The evidence for a benefit was strongest for cancers of the prostate, lung, and stomach. Data were also suggestive of a benefit for cancers of the pancreas, colon and rectum, esophagus, oral cavity, breast, and cervix. It should be noted, however, that although lycopene may account for or contribute to these benefits, there is the possibility that other potentially beneficial components of tomatoes and/or interactions among multiple components may contribute to the anticancer properties of tomatoes.

There are several biochemical mechanisms potentially underlying the protective effects of lycopene. These include antioxidant activity, induction of cell-cell communication, and growth control (Sies and Stahl, 1998).

Recently, Klebanov et al. (1998) compared the antioxidant properties of lycopene in three different model oxidative systems. In egg yolk liposomes in the presence of 2.5 mM $FeSO_4$ and 200 mM ascorbate, lycopene, α-tocopherol, and β-carotene inhibited the accumulation of lipid peroxidation products reacting with 2-thiobarbituric acid (TBARS) in a dose-dependent mode. In the liposomes subjected to illumination with a He-Ne laser (632.8 nm) at a dose of 10.5 J/cm^2 in the presence of 32.5 µg/mL hematoporphyrin derivatives, TBARS accumulated, and this effect was inhibited by lycopene, α-tocopherol, and dihydroquercetin with approximately equal efficiencies. In both systems studied, sodium azide at a concentration of 10 mM inhibited TBARS accumulation by no more than 20%. Apparently, the inhibitory action of not only lycopene, but also β-carotene and α-tocopherol was the result of their antiradical action, rather than quenching of the singlet oxygen in an aqueous medium. The antiradical activity of lycopene was confirmed by the method of luminol photochemiluminescence (PCL), which showed that lycopene increased the PCL lag period and decreased the PCL amplitude, implying its antiradical and SOD-like activity in this system.

In another recent study on the antioxidant properties, Agarwal and Rao (1998) provided dietary lycopene to 19 healthy human subjects using tomato juice, spaghetti sauce, and tomato oleoresin for a period of 1 week each. Blood samples were collected at the end of each treatment and analyzed for serum lycopene levels, cholesterol levels, serum lipid peroxidation, and LDL oxidation . The results show that dietary supplementation of lycopene significantly increased serum lycopene levels at least twofold. Although there was no change in serum cholesterol levels [total, low-density lipoprotein (LDL), or high-density lipoprotein], serum lipid peroxidation and LDL oxidation decreased significantly.

Inhibition of cell proliferation by lycopene was reported by Levy et al. (1995) in a study with endometrial (Ishikawa), mammary (MCF-7), and lung (NCI-H226) human cancer cells. Their results show that lycopene is a more potent inhibitor of human cancer cell proliferation than either α-carotene or β-carotene. For example, in Ishikawa cells, a 4-fold higher concentration of α-carotene or a 10-fold higher concentration of β-carotene was needed for the same order of growth suppression. The inhibitory effect of lycopene was detected after 24 hours of incubation, and it was maintained for at least 3 days. In contrast to cancer cells, human fibroblasts were less sensitive to lycopene, and the cells gradually escaped growth inhibition over time. In addition to its inhibitory effect on basal endometrial cancer cell proliferation, lycopene also suppressed insulin-like growth factor I–stimulated growth.

Lycopene is absorbed from the colon and from the small intestine, and both routes of absorption contribute to a comparative amount of lycopene accumulation in the colon mucosa after ingestion of this carotenoid (Oshima et al., 1999). Concentration of lycopene in human plasma is determined by lycopene intake, plasma cholesterol level, and, to a lesser degree, by lifestyle and demographic factors such as marital status (Mayne et al., 1999). Its bioavailability depends on various factors such as food processing or coingestion of fat (Porrini et al.,1998). Also, *cis*-lycopene is more bioavailable than *trans*-lycopene in vitro and in vivo in lymph-cannulated ferrets (Boileau et al., 1999). Little is known about the metabolism of lycopene. Potentially biologically active oxidation products of lycopene have been identified in human plasma (Sies and Stahl,1998).

Lutein

Several studies in the past few years have linked lutein to lower risk for eye, skin, and other health disorders. One proposed mechanism of protection by lutein and other carotenoids centers on their putative antioxidant activity. Thus, in a recent investigation on the antioxidant potential of lutein, Martin et al. (1996) demonstrated that lutein or its metabolites protect human HepG2 liver cells against oxidant-induced damage. The presence of metabolites of lutein were reported from a supplementation study (Granado et al., 1998) in which lutein was supplied as part of an intervention trial to test whether the consumption of this phytochemical reduces oxidative damage to human tissue components. Lutein from a natural source (15 mg/d as mixed ester forms) was supplied for 4 months to 18 nonsmoking, apparently healthy volunteers (9 men, 9 women) aged 25–45 years. The serum carotenoid profile was analyzed at baseline and monthly thereafter. On average, lutein concentrations increased fivefold after the first month of supplementation (range 0.6–3.34 μmol/L). On reviewing the results, in those volunteers whose lutein levels surpassed 1.05 μmol/L (14 of 17), the authors tentatively identified lutein monopalmitate along with another unidentified ester (possibly a monoketocarotenoid) in serum. Lutein levels returned to baseline values, and ester forms were not present 3 months after supplementation was discontinued. Their concentrations did not correlate with and represented less than 3% of the lutein levels achieved in serum.

A supplementation study with lutein (4 months) and α-tocopherol (2 months) was conducted by Olmedilla et al. (1997) to determine if the consumption of a diet rich in carotenoids reduces oxidative damage to human tissue components. Fifteen mg/day of lutein and 100 mg/day of α-tocopherol were supplied to 10 volunteers. Lutein levels increased during the first month (given alone) in all controls, but they behaved very differently during the following 3 months. Ketocarotenoids (not supplied) increased, whereas anhydroluteins (not supplied) did not change, suggesting that they are formed in vivo.

According to Khachik et al. (1995, 1997) the lutein metabolites zeaxanthin (a dietary dihydroxy carotenoid isomeric to lutein), 3'-epilutein, and 3-hydroxy-β,epsilon-caroten-3'-one in the human retina may be interconverted through a series of oxidation-reduction reactions, and the presence of the direct oxidation product of lutein and 3'-epilutein (metabolite of lutein and zeaxanthin) in human retina suggests that lutein and zeaxanthin may act as antioxidants to protect the macula against short-wavelength visible light. This oxidative-reductive process may thus play an important role in prevention of age-related macular degeneration and cataracts (Khachik et al. 1997).

Landrum et al.(1997) gave two human subjects lutein esters, equivalent to 30 mg of free lutein per day, for a period of 140 days and measured macular pigment optical density by heterochromatic flicker photometry before, during, and after the supplementation period. Twenty to 40 days after the subjects commenced taking the lutein supplement, their macular pigment optical density began to increase uniformly at an average rate of 1.13 ± 0.12 milliabsorbance units/day. During this same period, the serum concentration of lutein increased roughly 10-fold. The optical density curve eventually leveled off 40–50 days after the subjects discontinued the supplement. During the same 40–50 days, the serum concentration returned to baseline. Thereafter, little or no decrease in optical density was observed. The mean increases in the macular pigment optical density were 39% and 21% in the eyes of the two subjects, respectively. The authors concluded that lutein supplementation produced in the subjects a 30–40% reduction in blue light reaching the photoreceptors, Bruch's membrane, and the retinal pigment epithelium, the tissues affected by age-related macular degeneration (AMD).

In a very recent study on dietary supplementation of lutein and zeaxanthin with 11 moderately hypercholesterolemic men and women, Handelman et al. (1999) showed that cooked chicken egg yolk is a highly bioavailable source of lutein and zeaxanthin. However, the benefit of introducing these carotenoids into the diet with egg yolk is counterbalanced by potential LDL cholesterol elevation from the added dietary cholesterol.

Studies linking lutein to cancer have been reported by Thurnham (1989), Khachik et al. (1995), Nishino (1998), Park et al. (1999), and others. The general conclusion is that lutein or its oxidative metabolites appears to have chemoprevention activity. Further studies on cancer-prevention properties of lutein and other carotenoids are, however, needed.

CURCUMINOIDS

Turmeric has a long history of use as a natural yellow colorant due primarily to the curcuminoids curcumin, which is a diarylheptanoid, as well as to demethoxycurcumin and bisdemethoxy curcumin (Srinivasan, 1953). Fresh rhizome of turmeric contains a total of about 0.8 % of these three curcuminoids. In addition, some other curcumin related compounds such as, cyclocurcuminoids (Kiuchi et al., 1993) and diarylpentanoids (Masuda et al., 1993) have been isolated from the rhizome.

Antioxidant Activity

It has been known since the early 1950s that turmeric has strong antioxidant activity (Chipault et al., 1952; Ruby et al., 1995), and curcumin is the main compound responsible for the activity. Sreejayan and Rao (1996) compared the free radical–scavenging activity of curcumin, demethoxycurcumin, bisdemethoxycurcumin, and acetylcurcumin and found that curcumin is the most potent scavenger of superoxide radicals followed by demethoxycurcumin and bisdemethoxycurcumin. Acetylcurcumin was found to be inactive. Interaction with DPPH showed a similar activity profile. This study indicates that the phenolic group is essential for the free radical–scavenging activity, and the presence of a methoxy group further increases the activity. Toda et al. (1985) examined the antioxidant effect of three curcuminoids on linoleic acid and found that it increased in the order bisdemethoxycurcumin > demethoxycurcumin > curcumin. It was speculated that the antioxidant activity was associated with the stabilization of the radical, which resulted from the delocalization of the unpaired electron to a ketone function, because curcumin, formed by condensation of two ferulic acids, showed stronger activity than ferulic acid (Cuvelier et al., 1992).

Curcumin, demethoxycurcumin, and bisdemethoxycurcumin (curcumins, I, II, and III) were also checked for their antioxidant activity by Ruby et al. (1995), who found that the amount of curcuminoid (I, II, and III) needed for 50% inhibition of lipid peroxidation was 20, 14, and 11 µg/mL, respectively. Concentrations needed for 50% inhibition of superoxides were 6.25, 4.25, and 1.9 µg/mL and those for hydroxyl radicals were 2.3, 1.8, and 1.8 µg/mL, respectively. The ability of these compounds to suppress the superoxide production by macrophages activated with phorbol-12-myristate-13-acetate (PMA) indicated that all three curcuminoids inhibit superoxide production and bisdemethoxycurcumin produces maximum effect. Furthermore, curcumin has been found to have a positive antioxidant effect for hemolysis and lipid peroxidation of mouse erythrocytes induced by hydrogen peroxide (Toda et al., 1988), oxidation of phosphatidylcholine liposomal membranes and rat liver homogenate induced by free radicals (Noguchi et al., 1994), and oxidation of rat brain homogenate (Sharma, 1976).

Anti-inflammatory and Anti-Tumor-Promoting Effects

The most important biological activities of curcumin are anti-inflammatory and anti-tumor-promoting effects (Kikuzaki, 2000). In the late 1980s, curcumin was found to inhibit both tumor initiation and promotion (Huang et al., 1988; Nishino et al., 1988). Huang et al. (1992a) reported that topical application of curcumin inhibited benzo[a]pyrene-mediated DNA adduct formation in the epidermis of mouse skin in the initial stage. Furthermore, the anti-tumor-promoting activity of curcumin was established by its inhibition of TPA-induced increase in epidermal ornithine decarboxylase activity (Huang et al., 1992b), TPA-induced epidermal thickness, and leukocyte infiltration (Huang et al., 1992c). Anto et al. (1996) reported the results of an investigation in which three natural and five synthetic curcuminoids were assessed for their antimutagenic and antipromotional activity. The natural curcuminoids used, curcumin, demethoxycurcumin, and bisdemethoxycurcumin were isolated from *Curcuma longa* and were found to be potent inhibitors of mutagenesis and crotean oil- induced tumor promotion. Bisdemethoxycurcumin produced 87.6% inhibition to 2-acetamidofluorene (2-AAF) induced mutagenesis, at a concentration of 100 µg/plate, demethoxycurcumin and curcumin produced 70.5% and 68.3% inhibition at the same concentration. All the synthetic curcuminoids were found to inhibit 2-AAF-induced mutagenicity, among which salicyl- and anisylcurcuminoids were the most active. Bisdemethoxycurcumin was the most effective antipromoter among natural curcuminoids. While 90% of the control animals were having papillomas on the 10th week of tumor initiation, only 10% of the bisdemethoxycurcumin- treated animals, 20% of the demethoxycurcumin-treated animals, and 40% of the curcumin-treated animals were having papillomas. Salicylcurcuminoid, which was causing no papillomas by the 10th week, was the most potent anticarcinogen among the synthetic curcuminoids. Piperonal curcuminoid also exhibited antipromotional activity. A study of the clinical application of curcumin as a chemopreventive anticancer drug was carried out at the National Cancer Institute (Kelloff et al., 1996).

During the past few years, a large number of papers has been published about the anti-inflammatory and anti–tumor-promoting effects of curcumin with the objective of elucidating the molecular mechanism of their effects. Table 14.3 shows some physiological properties of curcumin concerning anti-inflammatory and anti–tumor-promoting effects.

CHLOROPHYLLS

Chlorophyll is ubiquitous in all green plant parts. Chlorophyllin is a water-soluble derivative of chlorophyll in which the central magnesium atom is replaced by other metals, such as cobalt, copper, or iron. Chlorophyllin is an efficient antimutagenic agent and has been used as a dietary supplement or to diminish the intensity of the discomforting side effects of cyclophosphamide therapy (Sarkas et al, 1994; Te et al., 1997). Cyclophosphamide is a potent antitumor agent used against many forms of cancer and against certain other diseases.

Te et al. (1997) undertook to determine the antimutagenic effectiveness of chlorophyllin against cyclophosphamide in a mouse model and to determine whether the antitumor efficacy of cyclophosphamide was compromised in vivo by chlorophyllin treatment. Experiments utilized chlorophyllin administered either in drinking water (1%) for 2 days before treatment or by gavage (200 mg/kg) 2 hours before treatment with cyclophosphamide (220 mg/kg). Urinary mutagenicity following cyclophosphamide treatment, as determined by the Salmonella/microsome assay, was decreased by both regimes of chlorophyllin cotreatment. Similarly, the increase in micronuclei in bone marrow polychromatic erythrocytes in response to cyclophosphamide was reduced by concomitant chlorophyllin treatment. In contrast, antitumor efficacy, as determined by growth delay of colon 38 adenocarcinomas, was not diminished by chlorophyllin treatment. The authors conclude that chlorophyllin may have beneficial effects when used in combination with cyclophosphamide therapy.

Recently, Kumar et al. (1999) examined chlorophyllin for its ability to protect DNA against radiation induced strand breaks using an in vitro plasmid DNA system. Gamma-radiation, up to a dose of 6 Gy (dose rate 1.25 Gy/min), induced a dose-dependent increase in single-strand breaks (ssbs)

TABLE 14.3   Anti-inflammatory and Anti–Tumor-Promoting Effects
of Curcumin

Effect	Study
Scavenging effect of superoxide anion free radicals	Kunchandy and Rao, 1990
Scavenging effect of nitric oxide	Sreejayan and Rao, 1997
Inhibition of nitric oxide synthase in RAW 264.7 macrophages activated with lipopolysaccharide and interferon-γ	Brouet and Ohshima, 1995
Reduction of inducible nitric oxide synthase—RNA expression in the livers of lipopolysaccharide-injected mice by oral treatment of curcumin	Chan et al., 1998
Inhibition of activation of the necrosis factor-kappa B (NF-kB)	Lienhard et al., 1998
Inhibition of the covalent binding of benzo[a]pyrene (B[a]P) to epidermal DNA by topical application of curcumin	Huang et al., 1992a
Suppression of the formation of skin tumor induced by B[a]P and DMBA in mice	Huang et al., 1992a
Inhibitory effect of dietary curcuminoid on skin carcinogenesis in mice induced by DMBA and TPA	Limtrakul et al., 1997
Inhibition of platelet aggregation induced by arachidonate, adrenaline, and collagen	Srivastava et al., 1995
Inhibition of in vitro lipoxygenase and cyclooxygenase activities in mouse epidermis	Huang et al., 1991
Suppression of TPA-induced skin inflammation of edema of mouse ears	Huang et al., 1988
Inhibition of TPA-induced ornithine decarboxylase activity	Huang et al., 1988
Inhibition of TPA-induced protein kinase C activity	Lin et al., 1998
Suppression of TPA-induced expression of c-fos, c-jun, and c-myc proto-oncogene messenger RNAs in mouse skin	Kakar and Roy, 1994
Inhibition of the proliferation and cell cycle progression of human umbilical vein endothelial cell	Singh et al., 1996
Inhibitory effect of dietary curcumin on azosymethane-induced colon tumorigenesis in mice or rats	Huang et al., 1994; Pereira et al., 1996

Source: Kikuzaki, 2000.

in plasmid pBR322 DNA. Chlorophyllin per se did not induce but inhibited radiation-induced ssbs in a concentration-dependent manner, 500 μM giving about 90% protection. The protection afforded by chlorophyllin was comparatively less than that of Trolox, a water-soluble analog of α-tocopherol. By studying the reaction of chlorophyllin with the radiation-derived hydroxyl radical (OH•) and deoxyribose peroxyl radical (ROO•), the authors found that chlorophyllin exhibited free radical–scavenging properties, and that chlorophyllin effectively protects plasmid DNA against ionizing radiation. Kumar et al. (1999) concluded that the ability of chlorophyllin to scavenge OH• and ROO• may contribute to its protective effects against radiation induced DNA damage.

CONCLUSIONS

From the foregoing, it is evident that anthocyanins, carotenoids, chlorophylls, and curcuminoids have a wide range of biochemical and pharmacological effects including anticarcinogenic, anti-inflammatory, and antioxidant activities. A few studies involving supplementation of the human diet have been carried out with β-carotene, lycopene, lutein, anthocyanins and curcumin, and most of these studies have demonstrated beneficial health effects, including risk reduction for some cancers. Further investigations on the prevention of disease and improvement of human health by these natural products are however required.

REFERENCES

Afanas'ev, I. B., Dorozhko, A. I., Brodskii, A. V., Kostyuk, V. A., and Potapovitch, A. I. 1989. Chelating and free radical scavenging mechanisms of inhibitory actions of rutin and quercetin in lipid peroxidation. *Biochem. Pharmacol.* 38: 1763–1769.

Agarual, R., Katiyar, S. K., Zaidi, S. I. A., and Mukthar, H. 1992. Inhibition of skin tumor promoter-caused induction of epidermal ornithine decarboxylase in SENCAR mice by polyphenolic fraction isolated from green tea and its individual epicatechin derivatives. *Cancer Res.* 52: 3582–3588.

Agarwal, S., and Rao, A. V., 1998. Tomato lycopene and low density lipoprotein oxidation: a human dietary intervention study. *Lipids* 33: 981–984.

Ames, B. N. 1983. Dietary carcinogens and anticarcinogens. Oxygen radicals and degenerative diseases. *Science* 221: 1256–1264.

Anto, R. J., George, J., Babu, K. V., Rajasekharan, K. N., and Kuttan, R. 1996. Antimutagenic and anticarcinogenic activity of natural and synthetic curcuminoids. *Mutat. Res.* 370: 127–131.

Baj, A., Bombardelli, E., Gabatta, B., and Martinelli, E. M. 1983. Qualitative and quantitative evaluation of *Vaccinium myrtillus* anthocyanins by high-resolution gas chromatography and high-performance liquid chromatography. *J. Chromatogr.* 279: 365–371.

Balansard, G., Zamble, D., Dumenil, G., and Cremieux, A. 1980. Mise en évidence des propriétés antimicrobiennes du latex obtenu par incision du tronc de *Alafia multiflora* Stapf. Identification de l'acide vanillique. *Plant. Med. Phytother.* 2: 99–104.

Bendich, A. 1990. Carotenoids. In *Chemistry and Biology*, N. I. Krinsji, M. M. Mathews-Roth, and R. F. Taylor (Eds.), pp. 323-336. Plenum Press, New York.

Beretz, A., and Cazahave, J.-P. 1988. The effect of flavonoids on blood vessel wall interactions. In *Plant Flavonoids in Biology and Medicine II*, V. Cody, E. Middleton, J. B. Harborne, and A. Beretz (Eds.), pp. 187–200, Alan R. Liss, Inc., New York.

Bhide, S. V., Azuine, M. A., Lahiri, M., and Telang, N. T. 1994. Chemoprevention of mammary tumor virus- induced and chemical carcinogen-induced rodent mammary tumors by natural plant products. *Breast Cancer Res. Treat.* 30: 233–242.

Block, G. 1992. The data support role for antioxidants in reducing cancer risk. *Nutr. Rev.* 50: 207–213.

Boileau, A. C., Merchen, N. R., Wasson, K., Atkinson, C. A., and Erdman, J. W. Jr. 1999. *cis*-Lycopene is more bioavailable than trans-lycopene in vitro and in vivo in lymph-cannulated ferrets. *J. Nutr.* 129: 1176–1181.

Boniface, R., Miskulin, M., Robert, L., and Robert, A. M. 1986. Pharmacological properties of *Myrtillus* anthocyanosides: correlation with results of treatment of diabetic microangiopathy. In *Flavonoids and Bioflavonoids*, L. Farkas, M. Gabor, and F. Kallay (Eds.), pp. 193–201. Elsevier, Amsterdam.

Bors, W., and Saran, M. 1987. Radical scavenging by flavonoid antioxidants. *Free Rad. Res. Commun.* 2: 289–294.

Bors, W., Heller, W., Michel, C., and Stettmaier, K. 1996. Flavonoids and polyphenols. In *Chemistry and Biology Handbook of Antioxidants*, pp. 409–465. Marcel Dekker, Inc., New York.

Briton, G., and Goodwin, T. W. 1982. *Carotenoid Chemistry and Biochemistry*. Pergamon Press, Elmsford, NY.

Brouet, I., and Ohshima, H. 1995. Curcumin, an anti-tumor promoter and anti-inflammatory agent, inhibits induction of nitric oxide synthase in activated macrophages. *Biochem. Biophys. Res. Commun.* 206: 533–540.

Burton, G. W., and Ingold, G. W. 1984. β-Carotene: an unusual type of lipid antioxidant. *Science* 224: 569–573.

Cao, G., and Prior, R. L. 1999. Anthocyanins are detected in human plasma after oral administration of an elderberry extract. *Clin. Chem.* 45: 574–576.

Chan, M. M.-Y., Huang, H.-I., Fenton, M. R., and Fong, D. 1998. In vivo inhibition of nitric oxide synthase gene expression by curcumin, a cancer preventive natural product with anti-inflammatory properties. *Biochem. Pharmacol.* 55: 1955–1962.

Chang, R. L., Huang, T., and Wood, A. W. 1985. Effect of ellagic acid and hydroxylated flavonoids on the tumorigenicity of benzo(a) pyrene and (±)-7 ,8-dihydroxy-9,10-epoxy, 7,8,9,10-tetrahydrobenzo(a)pyrene on mouse skin in the newborn mouse. *Carcinogenesis* 6: 1127–1133.

Chipault, J. R., Mizuno, G. R., Hawkins, J. M., and Lundberg, W. O. 1952. The antioxidant properties of natural spices. *Food Res.* 17: 46–54.

Clinton, S. K. 1998. Lycopene: chemistry, biology, and implications for human health and disease. *Nutr. Rev.* 56: 35–51.

Cuvelier, M.-E., Richard, H., and Berset, C. 1992. Comparison of the antioxidative activity of some acid-phenols: structure-activity relationship. *Biosci. Biotech. Biochem.* 56: 324–325.

Decker, E. A. 1995. The role of phenolics, conjugate linoleic acid, carinosine, and pyrrologlunolinc quinone as nonessential antioxidants. *Nutr. Rev.* 53: 49–58.

Deschner, E. E, Ruperto, J., Wong, G., and Newmark, H. L. 1991. Quercetin and rutin as inhibitors of azoxymethanol-induced colonic neoplasia. *Carcinogenesis* 7: 1193–1196.

Dumenil, G., Vasquez, M., Cremieux, A., and Balansard, G. 1980. Action antibactérienne des acides phénols de la série benzoique. *Bull. Biaison Groupe Polyphénols* 10: 302–308.

Duthie, S. J., Collins, A. R., and Duthie, G. G. 1998. The role of carotenoids in modulating DNA stability and lipid peroxidation. Importance for human health. *Subcell Biochem.* 30: 181–207.

Elwood, P. C., Renaud, S., Sharp, D. S., Beswick, A. D., O'Brien, J. R., and Yarnell, J. W. G. 1991. Ischemic heart disease and platelet aggregation. *Circulation* 83: 38–44.

Esterbauer, H., Striegl, G., Puhl, H., and Rotheneder, M. 1989. Continuous monitoring of in vitro oxidation of human low density lipoprotein. *Free Radic. Res. Commun.* 6: 67–75.

Esterbauer, H., Gebicki, J., Puhl, H., and Jurgens, G. 1992. The role of lipid peroxidation and antioxidants in oxidative modification of LDL. *Free Radic. Biol. Med.* 13: 341–390.

Evans, R. W., Shaten, B. J., Day, B. W., and Kuller, L. H. 1998. Prospective association between lipid soluble antioxidants and coronary heart disease in men: The Multiple Risk Factor Intervention Trial. *Am. J. Epidemiol.* 147: 180–186.

Ferrandiz, M. L., and Alcaraz, M. J. 1991. Anti-inflammatory and inhibition of arachidonic acid metabolism by flavonoids. *Agents Actions* 32: 283–288.

Frankel, E. N., Kanner, J., and Kinsella, J. E. 1993. Inhibition in vitro of oxidation of human low density lipoproteins by phenolic substances in wine. *Lancet* 341: 454–457.

Gerster, H. 1997. The potential role of lycopene for human health. *J. Am. Coll. Nutr.* 16: 109–126.

Gey, K. F., Stähelin, H. B., and Eichholzer, M. 1993. Poor plasma status of carotene and vitamin C is associated with higher mortality from ischemic heart disease and stroke: Basel Prospective Study. *Clin. Invest.* 71: 3–6.

Giovannucci, E. 1999. Tomatoes, tomato-based products, lycopene, and cancer: review of the epidemiologic literature. *J. Natl. Cancer Inst.* 91: 317–331.

Granado, F., Olmedilla, B., Gil-Martínez, E., and Blanco, I. 1998. Lutein ester in serum after lutein supplementation in human subjects. *Br. J. Nutr.* 80: 445–449.

Graziano, J. M., Buring, J. E., and Breslow, J. L. 1993. Moderate alcohol intake: increased levels of high-density lipoprotein and its subfractions and decreased risk of myocardial infarction. *N. Engl. J. Med.* 329: 1829–1834.

Handelman, G. J., Nightingale, Z. D., Lichtenstein, A. H., Schaefer, E. J., and Blumberg, J. B. 1999. Lutein and zeaxanthin concentrations in plasma after dietary supplementation with egg yolk. *Am. J. Clin. Nutr.* 70: 247–251.

Hirose, M., Hoshiya, T., Akagi, K., Takahashi, S., Hara, Y., and Ito, N. 1993. Effects of green tea catechins in a rat multiorgan carcinogenesis model. *Carcinogenesis* 14: 1549–1553.

Hoult, J. R., Moroney, M. A., and Paya, M. 1994. Action of flavonoids and cumarins on lipoxygenase and cyclooxygenase. *Meth. Enzymol.* 234: 443–455.

Huang, M. T., Smart, R. C., Wong, C. Q., and Conney, A. H. 1988. Inhibitory effects of curcumin, chlorogenic acid, caffeic acid, and ferulic acid on tumor promotion in mouse skin by 12-O-tetradecanoylphorbol-13-acetate. *Cancer Res.* 48: 5941–5946.

Huang, M.-T., Lysz, T., Ferraro, T., Abidi, T. F., Laskin, J. D., and Conney, A. H. 1991. Inhibitory effects of curcumin on in vitro lipoxygenase and cyclooxygenase activities in mouse epidermis. *Cancer Res.* 51: 813–819.

Huang, M.-T., Wang, Z. Y., Georgiadis, C. A., Laskin, J. D., and Conney, A. H. 1992a. Inhibitory effects of curcumin on tumor initiation by benzo[a]pyrene and 7,12-dimethylbenz[a] anthracene. *Carcinogenesis* 13: 2183–2186.

Huang, M.-T., Lysz, T., Ferraro, T., and Conney, A. H. 1992b. Inhibitory effects of curcumin on tumor promotion and arachidonic acid metabolism in mouse epidermis. In *Cancer Chemoprevention*, L. Wattenberg, M., Lipkin, C. W., Boone, and G. J. Kelloff (Eds.), pp. 375–391. CRC Press, Boca Raton, FL.

Huang, M.-T., Robertson, F. M., Lysz, T., Ferraro, T., Wang, Z. Y., Georgiadis,

C. A., Laskin, J. D., and Conney, A. J. 1992c. Inhibitory effects of cur-
cumin on carcinogenesis in mouse epidermis. In *Phenolic Compounds in
Food and Their Effects on Health II. Antioxidants and Cancer Prevention*, M.-T.
Huang, C.-T. Ho, and C. Y. Lee (Eds.), pp. 338–349. American Chemical
Society, Washington, DC.

Huang, M.-T., Lou, Y.-R., Ma, W., Newmark, H. L., Reuhl, K. R., and Conney, A. H. 1994. Inhibitory effects of dietary curcumin on forestomach,
duodenal colon carcinogenesis in mice. *Cancer Res.* 54: 5841–5847.

Iribarren, C., Folsom, A. R., Jacobs, D. R., Jr., Gross, M. D., Belcher, J. D.,
and Eckfeldt, J. H. 1997. Association of serum vitamin levels, LDL sus-
ceptibility to oxidation, and autoantibodies against MDA-LDL with
carotid atherosclerosis: a case-control study. *Arterioscler. Thromb. Vasc.
Biol.* 17: 1171–1177.

Jang, M., Cai, L., Udeani, G. O., Slowing, K. V., Thomas, C. F., Beecher,
C. W. W., Fong, H. S., Farnsworth, N. R., Kinghorn, A. D., Mehta, R. G.,
Moon, R. C., and Pezzuto, J. M. 1997. Cancer chemopreventive activity of
resveratrol, a natural product derived from grapes. *Science* 275: 218–220.

Kakar, S. S., and D. Roy. 1994. Curcumin inhibits TPA induced expression
of C-fos, C-jun and C-myc proto-oncogenes messenger RNAs in mouse
skin. *Cancer Lett.* 87: 85–89.

Kamei, H., Kojima, T., Hasegawa, M., Koide, T., Umeda, T., Yukawa, T., and
Terabe, K. 1995. Suppression of tumor cell growth by anthocyanins in
vitro. *Cancer Invest.* 13: 590–594.

Kanner, J., Frankel, E., Granit, R., German, B., and Kinsella, J. E. 1994. Natural antioxidants in grapes and wines. *J. Agric. Food Chem.* 42: 64–69.

Kardinaal, A. F., Kok, F. J., Ringstad, J., Gomez-Aracena, J., Mazaev, V. P.,
Kohlmeier, L., Martin, B. C., Aro, A., Kark, J. D., Delgado-Rodriguez, M.,
Riemersma, R. A., van't Veer, P., Huttunen, J. K., and Martin-Moreno,
J. M. 1993. Antioxidants in adipose tissue and risk of myocardial infarc-
tion: the EURAMIC study. *Lancet* 342: 1379–1384.

Kelloff, G. J., Crowell, J. A., Hawk, E. T., Steele, V. E., Lubet, R. A., Booney,
C. W., Covey, J. M., Doody, L. A., Omenn, G. S., Greenwald, P., Hong,
W. K., Parkinson, D. R., Bagheri, D., Baxter, G. T., Blunden, M., Doeltz,
M. K., Eisenhauer, K. M., Johnson, K., Knapp, G. G., Longfellow, D. G.,
Malone, W. F., Nayfield, S. G., Seifried, H. E., Swall, L. M., and Sigman,
C. C. 1996. Strategy and planning for chemopreventive drug develop-
ment: clinical development plans II. *J. Cell Biochem.* 26(Suppl.): 54–71.

Khachik, F., Beecher, G. R., and Smith, Jr., J. C. 1995. Lutein, lycopene, and
their oxidative metabolites in chemoprevention of cancer. *J. Cell Biochem.*
22(Suppl.): 236–246.

Khachik, F., Bernstein, P. S., and Garland, D. L. 1997. Identification of
lutein and zeaxanthin oxidation products in human and monkey retinas.
Invest. Ophthalmol. Vis. Sci. 38: 1802–1811.

Kikuzaki, H. 2000. Ginger for drug and spice purposes. In *Herbs, Botanicals and Teas as Functional Foods and Nutraceuticals*, G. Mazza, and B. D. Oomah (Eds.), in press. Technomic Publishing Inc., Lancaster, PA.

Kinsella, J. E., Frankel, E., German, B., and Kanner, J. 1993. Possible mechanisms for the protective role of antioxidants in wine and plant foods. *Food Technol.* 47: 85–89.

Kiuchi, F., Goto, Y., Sugimoto, N., Akao, N., Kondo, K., and Tsuda, Y. 1993. Nematocidal activity of turmeric: Synergistic action of curcuminoids. *Chem. Pharm. Bull.* 41: 1640–1643.

Klatsky, A. L. 1994. Epidemiology of coronary heart disease-influence of alcohol. *Alcohol Clin. Exp. Res.* 18: 88–96.

Klebanov, G.I., Kapitanov, A. B., Teselkin, Y. O., Babenkova, I. V., Zhambalova, B. A., Lyubitsky, O. B., Nesterova, O. A., Vasil'eva, O. V., Popov, I. N., Lewin, G., and Vladimirov, Y. A. 1998. The antioxidant properties of lycopene. *Membr. Cell Biol.* 12: 287–300.

Kohlmeier, L., Kark, J. D., Gomez-Garcia, E., Martin, B. C., Steck., S. E., Kardinaal, A. F., Ringstad, J., Thamm, M., Masaev, V., Riemersma, R., Martin-Moreno, J. M., Huttunen, J. K., and Kok, F. J. 1997. Lycopene and myocardial infarction risk in the EURAMIC Study. *Am. J. Epidemiol.* 146: 618–626.

Kornhauser, A., Wamer, W. G., Lambert, L. A., and Wei, R. R. 1994. β-Carotene inhibition of chemically induced toxicity in vivo and in vitro. *Food Chem. Toxicol.* 32: 149–154.

Krinsky, N. I., and Deneke, S. M. 1982. Interaction of oxygen and oxyradicals with carotenoids. *J. Natl. Cancer Inst.* 69: 205–210.

Kritchevsky, S. B. 1999. β-Carotene, carotenoids and the prevention of coronary heart disease. *J Nutr.* 129: 5–8.

Kuenzig, W., Chan, J., and Norkus, E. 1984. Caffeic acid and ferulic acid as blockers of nitrosoamine formation. *Carcinogenesis* 5: 309–314.

Kumar, S. S., Chaubey, R. C., Devasagayam, T. P., Priyadarsini, K. I., and Chauhan, P. S. 1999. Inhibition of radiation-induced DNA damage in plasmid pBR322 by chlorophyllin and possible mechanism(s) of action. *Mutat. Res.* 425: 171–179.

Kunchandy, E., and Rao, M. N. A. 1990. Oxygen radical scavenging activity of curcumin. *Int. J. Pharm.* 38: 237–240.

Landrum, J. T., Bone, R. A., Joa, H., Kilburn, M. D., Moore, L. L., and Sprague, K. E. 1997. A one year study of the macular pigment: The effect of 140 days of a lutein supplement. *Exp. Eye Res.* 65: 57–62.

Lapidot, T., Harel, S., Granit, R., and Kanner, J. 1998. Bioavailability of red wine anthocyanins as detected in human urine. *J. Agr. Food Chem.* 46(10): 4297–4302.

Levy, J., Bosin, E., Feldman, B., Giat, Y., Miinster, A., Danilenko, M., and Sharoni, Y. 1995. Lycopene is a more potent inhibitor of human cancer cell proliferation than either α-carotene or β-carotene. *Nutr. Cancer* 24: 257–266.

Levy, Y., and Glovinsky, Y. 1998. The effect of anthocyanosides on night vision. *Eye 1998* 12: 967–969.

Lienhard, S. M., Hehner, S. P., Bacher, S., Droege, W., and Heinrich, M. 1998. Transcription factor NF-kappa B. *Dtsch. Apoth. Ztg.* 138: 4881–4886.

Lietti, A., Cristoni, A., and Picci, M. 1976. Studies on *Vaccinium myrtillus* anthocyanins. I. Vasoprotective and anti-inflammatory activity. *Arzneim. Forsch./Drug Res.* 26: 829–831.

Limtrakul, P., Lipigorngoson, S., Namwong, O., Apisariyakul, A., and Dunn, F. W. 1997. Inhibitory effect of dietary curcumin on skin carcinogenesis in mice. *Cancer Lett.* 116: 197–203.

Lin, J. K., Chen, Y.-C., Huang, Y.-T., and Lin-Shiau, S.-Y. 1997. Suppression of protein kinase C and nuclear oncogene expression as possible molecular mechanisus of cancer chemoprevention by apigenin and curcumin. *J. Cell. Biochem.* Suppl. 28–29: 39–48.

Liu, L., and Castonguay, A. 1991. Inhibition of the metabolism and genotoxicity of 4-(methylnitrosamino)-1-(3-pyridyl)-1-butanone (NNK) in rat hepatocytes by (+)-catechin. *Carcinogenesis* 12: 1203–1208.

Maas, J. L., Galletta, G. I., and Stoner, G. D. 1991. Ellagic acid, an anticarcinogen in fruits, especially strawberries: A review. *HortScience* 26: 10–14.

Macheix, J. J., Fleuriet, A., and Billot, J. 1991. *Fruit Phenolics*, pp. 1–103. CRC Press Inc., Boca Raton, FL.

Mackerras, D. 1995. Antioxidants and health. *Food Austr.* 47(Suppl.): 1–23.

Mandel, S., and Stoner, G. D. 1990. Inhibition of N-nitrosobenzyl-methylamine-induced esophageal tumorigenesis by ellagic acid. *Carcinogenesis* 11: 55–61.

Martin, K. R., Failla, M. L., and Smith, Jr., C. R. 1996. β-carotene and lutein protect HepG2 human liver cells against oxidant-induced damage. *J. Nutr.* 126: 2098–2106.

Masuda, T., Jitoe, A., Isobe, J., Nakatani, N., and Yonemori, S. 1993. Antioxidative and anti-inflammatory curcumin-related phenolics from rhizomes of *Curcuma domestica*. *Phytochemistry* 32: 1557–1560.

Mathews-Roth, M. M. 1991. Recent progress in the medical application of carotenoids. *Pure and Appl. Chem.* 63: 147–156.

Mathews-Roth, M. M., Pathak, M. A., Fitzpatrick, T. B., Harber, L. C., and Kass, E. H. 1970. β-Carotene as a photoprotective agent in erythropoietic protoporphyria. *N. Engl. J. Med.* 282: 1231–1234.

Mayne, S. T., Cartmel, B., Silva, F., Kim, C. S., Fallon, B. G., Briskin, K., Zheng, T., Baum, M., Shor-Posner, G., and Goodwin, Jr., W. J. 1999. Plasma lycopene concentrations in humans are determined by lycopene intake, plasma cholesterol concentrations and selected demographic factors. *J. Nutr.* 129: 849–854.

Mazza, G. 1997. *Anthocyanins in Edible Plant Parts: A Qualitative and Quantitative Assessment. Antioxidant Methodology In Vivo and In Vitro Concepts*, pp. 119–140. AOCS Press, Champaign, IL.

Mazza, G., and E. Miniati. 1993. *Anthocyanins in Fruits, Vegetables and Grains*, pp. 112–114. CRC Press Inc., Boca Raton, FL.

Middleton, E., and Kandaswami, C. 1992. Effects of flavonoids on immune and inflammatory cell functions. *Biochem. Pharmacol.* 43: 1167–1179.

Miller, N. J., Sampson, J., Candeias, L. P., Bramley, P. M., and Rice-Evans, C. A. 1996. Antioxidant activities of carotenes and xanthophylls. *FEBS Lett.* 384: 240–242.

Miyazawa, T., Nakagawa, K., Kudo, M., Muraishi, K., and Someya, K. 1999. Direct intestinal absorption of red fruit anthocyanins, cyanidin-3-glucoside and cyanidin-3,5-diglucoside, into rats and humans. *J. Agric. Food Chem.* 47: 1083–1091.

Morazzoni, P., and Bombardelli, E. 1996. *Vaccinium myrtillus* L. *Fitoterapia* 68: 3–28.

Morazzoni, P., Livio, S., Scilingo, A., and Malandrino, S. 1991. *Vaccinium myrtillus* anthocyanosides pharmacokinetics in rats. *Arzneim.-Forsch./ Drug Res.* 41(1): 128–131.

Morris, D. L., Kritchevsky, S. B., and Davis, C. E. 1994. Serum carotenoids and coronary heart disease: The Lipid Research Clinics Coronary Primary Prevention Trial and Follow-up Study. *J. Am. Med. Assoc.* 272: 1439–1441.

Narayan, M. S., Naidu, K. A., Ravishankar, G. A., Srinivas, L., and Venkataraman, L. V. 1999. Antioxidant effect of anthocyanin on enzymatic and non-enzymatic lipid peroxidation. *Prostaglandins Leukot. Essent. Fatty Acids* 60: 1–14.

Nishino, H. 1998. Cancer prevention by carotenoids. *Mutat. Res.* 402: 159–163.

Nishino, H., Nishino, A., Takayasu, J., and Hasegawa, T. 1988. Antitumor-promoting activity of curcumin, a major constituent of the food additive turmeric yellow. *Kyoto-furitsu Ika Daigaku Zasshi* 96: 725–728.

Noguchi, N., Komuro, E., Niki, E., and Willson, R. L. 1994. Action of curcumin as an antioxidant against lipid peroxidation. *J. Jpn. Oil Chem. Soc.* 43: 1–7.

Ohno, K., and Nakane, H. 1990. Mechanism of inhibition of various cellular DNA, and RNA polymerases by several flavonoids. *J. Biochem.* 108: 609–613.

Olmedilla, B., Granado, F., Gil-Martinez, E., and Blanco, I. 1997. Supplementation with lutein (4 months) and α-tocopherol (2 months), in separate or combined oral doses, in control men. *Cancer Lett.* 114: 179–181.

Olson, J. A. 1996. Benefits and liabilities of vitamin A and carotenoids. *J. Nutr.* 126 (Suppl. 1208S–1212S): 4.

Oshima, S., Inakuma, T., and Narisawa, T. 1999. Absorption and distribution of lycopene in rat colon. *J. Nutr. Sci. Vitaminol. (Tokyo)* 45: 129–134.

Paganga, G., and Rice-Evans, C. A. 1997. The identification of flavonoids as glycosides in human plasma. *FEBS Lett.* 401: 78–82.

Park, J. S., Chew, P. B., Wong, T. S., Zhang, J. X., and Magnuson, M. S. 1999. Dietary lutein but not astaxanthin or β-carotene increases pim-1 gene expression in murine lymphocytes. *Nutr. Cancer* 33: 206–212.

Pereira, M. A., Grubbs, C. J., Barners, L. H., Li, H., Olson, G. R., Eto, I., Juliana, M., Whitake, L. M., and Gary, J. 1996. Effects of the phytochemicals, curcumin and quercetin, upon azosymethane-induced colon cancer and 7,12-dimethylbenz[a]anthracene-induced mammary cancer in rats. *Carcinogenesis* 17: 1305–1311.

Pisha, E., and Pezzuto, J. M. 1994. Fruits and vegetables containing compounds that demonstrate pharmacological activity in humans. *Econ. Med. Plant Res.* 6: 189–233.

Pizzorno, I. E., and Murray, M. T. 1987. *Vaccinium myrtilus*. In *A Textbook of Natural Medicines*, pp. 1–6. John Bastyr College Publications, Washington, DC.

Porrini, M., Riso, P., and Testolin, G. 1998. Absorption of lycopene from single or daily portions of raw and processed tomato. *Br. J. Nutr.* 80: 353–361.

Prasad, K. N., and Edwards-Prasad, J. 1990. Expressions of some molecular cancer risk factors and their modification by vitamins. *J. Am. Coll. Nutr.* 9: 28–34.

Renaud, S. C., Beswick, A. D., Fehily, A. M., Sharp, D. S., and Elwood, P. C. 1992. Alcohol and platelet aggregation: the Caerphilly prospective heart disease study. *Am. J. Clin. Nutr.* 55: 1012–1017.

Riemersma, R. A., Wood, D. A., Macintyre, C. C., Elton, R. A., Gey, K. F., and Oliver. M. F. 1991. Risk of angina pectoris and plasma concentrations of vitamins A, C, and E and carotene. *Lancet* 337: 1–5.

Rimon, E. B., Giovanucci, E. L., and Willett, W. C. 1991. Prospective study of alcohol consumption and risk of coronary disease in men. *Lancet* 338: 464–468.

Ruby, A. J., Kuttan, G., Babu, K. D., Rajasekhara, K. N., and Kuttan, R. 1995. Anti-tumour and antioxidant activity of natural curcuminoids. *Cancer Lett.* 94: 79–83.

Rusznyàk, S., and Szent-Györgyi, A. 1936. Vitamin P: flavonols as vitamins. *Nature* 138: 27.

Sahyoun, N. R., Jacques, P. F., and Russell, R. M. 1996. Carotenoids, vitamins C and E, and mortality in an elderly population. *Am. J. Epidemiol.* 144: 501–511.

Sarkas, D., Sharma, A., and Talukder, G. 1994. Chlorophyll and chlorophyllin as modifiers of genotoxic effects. *Mutat. Res.* 318: 239–247.

Sayer, J. M, Yagi, H., Wood, A. W., Connery, A. H., and Jerina, D. M. 1982. Extremely facile reaction between the ultimate carcinogen benzo[a]-pyrene-7,8-diol 9,10-epoxide and ellagic acid. *J. Am. Chem. Soc.* 104: 5562–5564.

Sharma, O. P. 1976. Antioxidant activity of curcumin and related compounds. *Biochem. Pharmacol.* 25: 1811–1812.

Sies, H., and Stahl, W. 1998. Lycopene: Antioxidant and biological effects and its bioavailability in the human. *Proc. Soc. Exp. Biol. Med.* 8: 121–124.

Singh, A. K., Sidhu, G. S., Deepa, T., and Maheshwari, R. K. 1996. Curcumin inhibits the proliferation and cell cycle progression of human umbilical vein endothelial cell. *Cancer Lett.* 107: 109–115.

Sreejayan, N., and Rao, M. N. 1996. Free radical scavenging activity of curcuminoids. *Arzneimittelforschung* 46: 169–171.

Sreejayan, N., and Rao, M. N. 1997. Nitric oxide scavenging by curcuminoids. *J. Pharm. Pharmacol.* 49: 105–107.

Srinivasan, K. R. 1953. Chromatographic study of the curcuminoids in *Curcuma longa. J. Pharm. Pharmacol.* 5: 448–457.

Srivastava, K. C., Bordia, A., and Verma, S. K. 1995. Curcumin, a major component of food spice turmeric (*Curcuma longa*) inhibits aggregation and alters eicosanoid metabolism in human blood platelets. *Prostaglandins, Leukotrienes Essent. Fatty Acids* 52(4): 223–227.

St. Léger, A. S., Cochrane, A. L., and Moore, F. 1979. Factors associated with cardiac mortality in developed countries with particular reference to the consumption of wine. *Lancet* May 12, 1: 8124, 1017–1020.

Steinberg, D., Parsatharathy, S., Carew, T. E., Khoo, J. C., and Witztum, J. L. 1989. Beyond cholesterol. Modifications of low-density lipoprotein that increase its atherogenicity. *N. Engl. J. Med.* 320: 915–924.

Straub, O. 1987. *Key to Carotenoids*, 2nd ed. Birkhauser Verlag, Basel.

Street, D. A., Comstock, G. W., Salkeld, R. M., Schüep, W., and Klag, M. J. 1994. Serum antioxidants and myocardial infarction: Are low levels of carotenoid and α-tocopherol risk factors for myocardial infarction? *Circulation* 90: 1154–1161.

Te, C., Gentile, J. M., Baguley, B. C., Pearson, A. E., Gregory, T., and Ferguson, L. R. 1997. In vivo effects of chlorophyllin on the antitumour agent cyclophosphamide. *Int. J. Cancer* 70: 84–89.

Thurnham, D. I. 1989. Lutein, cholesterol, and risk of cancer [letter]. *Lancet* 19: 441–442.

Toda, S., Miyase, T., Arichi, H., Tanizawa, H., and Takino, Y. 1985. Natural antioxidants. III. Antioxidative components isolated from rhizome of *Curcuma longa* L. *Chem. Pharm. Bull.* 33: 1725–1728.

Toda, S., Ohnishi, M., Kimura, M., and Nakashima, K. 1988. Action of curcuminoids on the hemolysis and lipid peroxidation of mouse erythrocytes induced by hydrogen peroxide. *J. Ethnopharmacol.* 23: 105–108.

Tsuda, T., Watanabe, M., Ohshima, K., Norinobu, S., Choi, S.-W., Kawakishi, S., and Osawa, T. 1994a. Antioxidative activity of the anthocyanin pigments cyanidin 3-O-β-D-glucoside and cyanidin. *J. Agric. Food Chem.* 42: 2407–2410.

Tsuda, T., Ohshima, K., Kawakishi, S., and Osawa, T. 1994b. Antioxidative pigments isolated from the seeds of *Phaseolus vulgaris* L. *J. Agric. Food Chem.* 42: 248–251.

Tsuda, T., Uehara, N., Iwahori, Y., Asamoto, M., Iigo, M., Nagao, M., Matsumoto, K., Ito, M., and Hirono, I. 1994c. Chemopreventive effects of β-carotene, α-tocopherol and five naturally occurring antioxidants on initiation of hepatocarcinogenesis by 2-amino-3-methylimidazo[4,5-f]quinoline in the rat. *Jpn. J. Cancer Res.* 85: 1214–1219.

Tsuda, T., Ohshima, K., Kawakishi, S., and Osawa, T. 1996. Oxidation products of cyanidin 3-O-β-D-glucoside with a free radical initiator. *Lipids* 31: 1259–1263.

Verma, A. K. 1992. Modulation of mouse skin carcinogenesis and epidermal phospholipid biosynthesis by the flavonol quercetin. In *Phenolic Compounds in Food and Their Effects on Health*, Vol. II. *Antioxidants and Cancer Prevention*, C.-T. Ho, C. Y. Lee, and M.-T. Huang (Eds.), pp. 250–264. American Chemical Society, Washington, DC.

Verma, A. K., Johnson, J. A., Gould, M. N., and Tanner, M. A. 1988. Inhibition of 7,12-dimethylbenz(a)anthracene and N-nitrosomethylurea induced rat mammary cancer by dietary flavonol quercetin. *Cancer Res.* 48: 5754–5788.

Wagner, H. 1979. Phenolic compounds in plants of pharmaceutical interest. In *Biochemistry of Plant Phenolics*, T. Swain, J. B. Harborne, and C. F. Van Sumere (Eds.)., p. 581. Plenum Press, New York.

Wagner, H. 1985. New plant phenolics of pharmaceutical interest. In *Ann. Proc. Phytochem. Soc. Eur.*, Vol. 15, C. F. Van Sumere and P. J. Lea (Eds.), pp. 409–425. Clarendon Press, Oxford, United Kingdom.

Wang, H., Cao, G., and Prior, R. L. 1997. Oxygen radical absorbing capacity of anthocyanins. *J. Agric. Food Chem.* 45: 304–309.

Wang, H., Nair, M. G., Strasburg, G. M., Chang, Y. C., Booren, A. M., Gray, J. I., and DeWitt, D. L. 1999. Antioxidant and antiinflammatory activities of anthocyanins and their aglycon, cyanidin, from tart cherries. *J. Nat. Prod.* 62: 294–296.

Yoshizawa, S., Horiuchi, T., and Suganuma, I. 1992. Penta-o-galloyl-β-d-glucose and (-)-epigallocatechin gallate: cancer preventive agents. In *Phenolic Compounds in Food and Their Effects on Health*, Vol. II. *Antioxidants and Cancer Prevention*, C.-T., Ho, C. Y. Lee, and M.-T. Huang (Eds.), pp. 316–325. American Chemical Society, Washington, DC.

15
Regulations in Europe and Japan

Bruce S. Henry

Phytone, Ltd.
Burton-on Trent, Staffordshire, England

INTRODUCTION

Of the five senses, sight is one of the most developed and important to humans. We use our ability to see in almost every aspect of our daily lives, and it continually provides us with information about our environment. Color adds a further dimension that gives us additional data as well as frequently adding to our pleasure and enjoyment. It is, therefore, not surprising that from early times food has been prepared to be more visually appealing by the use of colorful ingredients.

In 1857 a paper was published that reported on the results of the analysis for colorants found in samples of colored sweets. The list by today's standards make horrifying reading, including, as it did, such mineral compounds as lead chromate, red lead, mercuric sulfide, and copper arsenite, to name just a few. Fortunately for the long-term survival of both the sugar confectionery industry and its consumers, this is no longer the current practice, and since that time there has been a gradual evolution of regulatory control over the use of additives to foods.

EUROPEAN DIRECTIVES

In 1988 the Council of the European Communities published a directive (frequently referred to as the "framework directive") that empowered the European commission to implement rules to control the use of food additives to be used as ingredients in the manufacture or preparation of foodstuffs.

This directive divides food additives into 24 categories according to their functionality (e.g., antioxidants, preservatives, colorants, emulsifiers, humectants) and listed a number of substances excluded from the directive, including processing aids, nutrients, flavorings, and agrochemicals.

The framework directive also laid down the following definition of "food additive": "any substance not normally consumed as a food in itself and not normally used as a characteristic ingredient of food, whether or not it has nutritive value, the intentional addition of which to food for a technological purpose in the manufacture, processing, preparation, treatment, packaging, transport or storage such food results, or may be reasonably expected to result, in it or its by-products becoming directly or indirectly a component of such food." Thus, by definition any substance that is added to food should either be a foodstuff or a food additive or one of the substances excluded from the food additive definition (e.g., flavor, nutrient, processing aid, or agrochemical). Thus, a purified extract of red rice (a monascus extract) would be considered a food additive but is not listed as a permitted additive in the relevant directive.

Any food additive must be supported by data demonstrating that it presents no hazard to the health of the consumer, must be used in a way that does not deceive the consumer, and must have a proven technological need. In the European Union (EU) safety data is evaluated by the Scientific Committee for Food (SCF), an advisory expert committee of the European Commission. Another important committee, working on a global basis, which also assesses the safety of food additives, is the Joint WHO/FAO Expert Committee on Food Additives (JECFA). Following evaluation of the toxicological data, both committees allocate an Acceptable Daily Intake (ADI) value to an additive, which may have no specified numerical value or one expressed as mg/kg body weight per day, otherwise its use is not recommended (Table 15.1).

In June 1994 the Directive on Colours for Use in Foodstuffs reference 94/36/EC was published, followed in July 1995 by Directive 95/45/EC, which provided the specific purity criteria for the permitted colorants. As a consequence, the Colours in Food Regulations 1995 came into force in the United Kingdom on January 1, 1996 and in other European countries around the same time. This directive is very similar but not identical in all countries within the EU and includes the following definition of colorants: "For the purposes of this Directive, 'colourants' are substances which add or

TABLE 15.1   Acceptable Daily Intake Values for Some Permitted
Food Colors

Pigment	EC No.	ADI (JECFA) (mg/kg body weight per day)
Curcumin	E100	0.1
Tartrazine	E102	7.5
Carmines	E120	5.0
Erythrosine	E127	0.1
Brilliant blue FCF	E133	12.5
Chlorophylls	E140	NS
β-Carotene	E160(a)	5.0
Annatto	E160(b)	0.065
Paprika	E160(c)	NS
Beetroot red	E162	NS
Anthocyanins	E163	NS

NS = Not specified.

restore colour in a food, and include natural constituents of foodstuffs and natural sources which are normally not consumed as foodstuffs as such and not normally used as characteristic ingredients of food. Preparations obtained from foodstuffs and other natural source materials obtained by physical and/or chemical extraction resulting in a *selective extraction* of the pigments relative to the nutritive or aromatic constituents are colourants within the meaning of this Directive."

These regulations consist of a series of five annexes, and with their introduction have come a number of changes in both principle and detail when compared to those that were previously in place. Annex 1 (Table 15.2) lists some of those colors that are permitted, identifying them by both E number and by name, and includes E100 curcumin and E160(c) paprika extract. It is important to note that the word turmeric does not appear in the U.K. version of this document. In the EU directive itself turmeric is given as an example of exclusions from the definition of colorant where article 1, paragraph 3, states "foodstuffs, whether dried or in concentrated form and flavourings incorporated during the manufacturing of compound foodstuffs, because of their aromatic, sapid or nutritive properties together with a secondary colouring effect such as paprika, turmeric and saffron."

From the paragraphs detailing the definitions of colorants and exclusions from the scope of the directive, one can attempt to define when a food is a color additive. Thus, since turmeric is a food, its total (nonselective) extract, which retains the full flavor characteristic of turmeric (turmeric oleo-

TABLE 15.2   Some of the Colorants Listed in Annex 1

Colorant	EC No.
Curcumin	E100
Riboflavin and riboflavin-5-phosphate	E101
Carmine	E120
Chlorophylls and chlorophyllins	E140
Copper chlorophylls and chlorophyllins	E141
Caramels	E150
Vegetable carbon	E153
Carotenes	E160(a)
Annatto	E160(b)
Paprika	E160(c)
Lycopene	E160(d)
Apo-carotenal	E160(e)
Apo-carotenoic ester	E160(f)
Lutein	E161(b)
Canthaxanthin	E161(g)
Beetroot red	E162
Anthocyanins	E163

resin), is still a food, but when the coloring components (the curcumins) are selectively extracted to yield a color (Curcumin E100), then this is a food additive. This is very different from the regulations in the United States, where essentially any extract (or juice) that is used to provide color is deemed to be a colorant. Thus, when turmeric oleoresin or cherry juice is used to give color, it is declared as color.

From this example it could be assumed that spinach extract and carrot extract are both foods within the term of the EC colors directive. However, when the coloring components are selectively isolated, they would be considered color additives (chlorophylls E140 and carotenes E160(a), respectively). It is a matter of debate when a spinach extract becomes E140.

As mentioned above, all vegetable and fruit juices used in the United States to provide color are considered colorants. In the EC only one juice, that of beetroot, is listed as a color even when the betanin content is as low as 0.4%, whereas a fruit juice concentrate with, say, 1% pigment is always a juice.

Annex II of the EU directive lists those foodstuffs to which colors may not be added unless specifically provided for in subsequent annexes and includes staple foods such as milk, bread, eggs, flour, jam, fish, and meat. Annex III lists those foodstuffs to which only certain permitted colorants may be added and includes 27 categories, of which breakfast cereals is one

TABLE 15.3 Colorants Permitted in Breakfast Cereals, as Detailed in Annex III

Cereal	EC No.	Colorant	Amount
Extruded, puffed, and/or	E150(c)	Ammonia caramel	Quantum satis
fruit-flavored breakfast	E160(a)	Carotenes	Quantum satis
cereals	E160(b)	Annatto	25 mg/kg
	E160(c)	Paprika extract	Quantum satis
Fruit-flavored breakfast	E120	Cochineal, carminic	200 mg/kg
cereals		acid, carmines	
	E162	Beetroot red	(individually
			or in
	E163	Anthocyanins	combination)

(Table 15.3). As can be seen from this listing, a cinnamon-flavored breakfast cereal cannot contain any red colorants and a lime-flavored breakfast cereal cannot contain any green colorants—in fact, all breakfast cereals are expressly prohibited from containing any green colorant. Thus, in order to produce a lime-flavored variety the manufacturer has to use a food extract such as spinach. Interestingly, the only permitted caramel for this application is ammonia caramel E150(c); the use of a plain caramel E150(a) produced using sugar and water only is not permitted. This is also the only category of food where a quantitative limit exists for anthocyanins (E163); in all other applications where colorants are permitted, anthocyanins may be used quantum satis. Since this particular application refers specifically to fruit-flavored products, it is likely that a fruit juice or purée is also part of the recipe and that anthocyanins would be a naturally occurring component when a red fruit is used. It is difficult to see the logic of this type of restrictive legislation (see the Japanese regulations, discussed later).

Cheese is another food included in Annex III for which color regulation is also overly complicated (see Table 15.4). Note that curcumin is not listed in the categories of cheese described. Note also that whereas red marbled cheese is a category, there is no equivalent green marbled cheese (Sage Derby is specifically named instead.) It would seemingly have been simpler to list one category of marbled (veined) cheese, allowing a fuller range of colors. (Flavored processed cheese is listed in Annex V.)

Annex IV lists those colorants that are restricted to use in specific foods only. The colors referred to in Annex IV are generally those that have been allocated a low ADI and whose widespread use may give cause for concern. Justification for their inclusion as permitted colors is usually on the basis of their satisfying a particular functional need that cannot easily be achieved with alternative permitted colorants.

Annatto extracts have been used for many years; records of annatto ex-

TABLE 15.4 Colorants Permitted in Cheese, as Detailed in Annex III

Cheese	EC No.	Colorant	Amount
Sage Derby cheese	E140	Chlorophylls Chlorophyllins	Quantum satis
	E141	Copper complexes of chlorophylls and chlorophyllins	
Ripened orange, yellow, and broken-white cheese;	E160(a)	Carotenes	Quantum satis
unflavored processed	E160(b)	Annatto	15 mg/kg
cheese	E160(c)	Paprika	Quantum satis
Red Leicester cheese	E160(b)	Annatto	50 mg/kg
Mimolette cheese	E160(b)	Annatto	35 mg/kg
Morbier cheese	E153	Vegetable carbon	Quantum satis
Red marbled cheese	E120	Cochineal, carminic acid, carmines	125 mg/kg
	E163	Anthocyanins	Quantum satis

traction in London go back to the 1790s, since they are both relatively inexpensive as well as being extremely versatile as a consequence of being available in oil-soluble and water-soluble forms. The ADI for annatto is only 0.065 mg/kg of body weight (bw) per day and is based on studies carried out some 40 years ago on annatto extracts of low pigment content. Thus, although no adverse affects were seen, the calculated ADI was low because of the low quantities of pigment incorporated into the diet.

In order to safeguard the future status of annatto, it is necessary to attempt to increase the existing ADI. To this end an initiative has been undertaken by associations in Europe, Japan, the United States, and South America, and a new program of toxicological testing was begun in 1998.

Annex V lists colorants permitted in foodstuffs other than those mentioned in Annexes II and III.

Part 1 lists 15 colorants, principally those of natural origin (Table 15.5) that can be used in foodstuffs mentioned in Annex V Part 2 and in all other foodstuffs (other than those listed in Annex II and III) at quantum satis levels.

Part 2 lists 18 colorants, principally synthetic ones, but also including curcumin, carmine, lycopene, and lutein (Table 15.6) that can be used in a list of defined foodstuffs up to a maximum specified dose level (Table 15.7).

It is important to recognize that any levels specified relate to the pure coloring principle and not to the product defined in the purity criteria direc-

TABLE 15.5 Annex V Part I

E101	(i) Riboflavin
	(ii) Riboflavin-5′-phosphate
E140	Chlorophylls and chlorophyllins
E141	Copper complexes of chlorophylls and chlorophyllins
E150a	Plain caramel
E150b	Caustic sulfite caramel
E150c	Ammonia caramel
E150d	Sulfite ammonia caramel
E153	Vegetable carbon
E160a	Carotenes
E160c	Paprika extract, capsanthin, capsorubin
E162	Beetroot red, betanin
E163	Anthocyanins
E170	Calcium carbonate
E171	Titanium dioxide
E172	Iron oxides and hydroxides

The listed colorants may be used, in each case at quantum satis, in foods listed in Annex V Part 2 and in any food other than those listed in Annexes II and III.

tive. Many of the naturally derived colors contain more than a single pigment. For example, a typical grape skin extract colorant will contain at least 10 different anthocyanins derived from four principal anthocyanidins (petunidin, malvidin, delphinidin, and peonidin). Similarly, some of the naturally derived carotenoid colors such as mixed carotenes, paprika extract, and lutein contain a range of related substances, all of which contribute to the color strength of the extract.

It is clearly impractical and not particularly meaningful to quantitatively determine each individual component pigment in the final food product. In such cases it is more appropriate to make the calculation from spectrophotometric measurements using the conversion data laid down in the specific purity criteria directive.

The current Japanese legislation for naturally derived colorants sets no limits on their use in foodstuffs that may be colored and thus has no need to convert from a spectrophotometric measurement to a pigment content. Thus, colors are specified in terms of their absorption, expressed as a 10% solution in a named solvent.

CURCUMIN

As previously mentioned, curcumin is included in the list of colorants with a restricted use (Annex V Part 2) because of the fact that it has been allo-

TABLE 15.6 Annex V Part 2

E100	Curcumin
E102	Tartrazine
E104	Quinoline yellow
E110	Sunset yellow FCF
	Orange yellow S
E120	Cochineal, carminic acid, carmines
E122	Azorubine, carmoisine
E124	Ponceau 4R, cochineal Red A
E129	Allura red AC
E131	Patent blue V
E132	Indigotine, indigo carmine
E133	Brilliant blue FCF
E142	Green S
E151	Brilliant black BN, black PN
E155	Brown HT
E160d	Lycopene
E160e	Beta-apo-8'-carotenal (C 30)
E160f	Ethyl ester of beta-apo-8'-carotenoic acid (C 30)
E161b	Lutein

The listed colorants may be used singly or in combination in the foods listed in Table 15.7 up to the maximum level specified in the table. However, for nonalcoholic flavored drinks, edible ices, desserts, fine bakery wares, and confectionery, the colors may be used up to the limit indicated, but the quantities of each of the colors E110, E122, E124, and E155 may not exceed 50 mg/kg or mg/l.

cated a temporary low ADI value. JECFA has allocated curcumin a temporary ADI of 0–1.0 mg/kg bw/day. This ADI, increased from the previous temporary ADI of 0–0.1 mg/kg bw/day, was based primarily upon the results of a carcinogenicity study in mice. In order to establish a permanent ADI for curcumin, JECFA has requested the submission of results from a reproductive toxicity study for review in 2001.

JECFA has reviewed the toxicology of curcumin at least eight times in the past 25 years, and each time has requested further information from the colors industry. Unfortunately, the industry has been very slow to react to these requests.

In 1974 JECFA allocated curcumin a temporary ADI based upon a turmeric study. In 1978 JECFA agreed to extend the temporary ADI for curcumin until 1980 pending the results of long-term feeding studies. In 1982 the temporary ADI was maintained, but no results were submitted, thus JECFA requested a series of rodent toxicological studies. In 1986 JECFA reviewed new data relating to turmeric oleoresin.

TABLE 15.7 Annex 5 Part 2

Food	Maximum level
Nonalcoholic flavored drinks	100 mg/l
Candied fruits and vegetables, mostarda di frutta	200 mg/kg
Preserves of red fruits	200 mg/kg
Confectionery	300 mg/kg
Decorations and coatings	500 mg/kg
Fine bakery wares (e.g., viennoiserie, biscuits, cakes, and wafers)	200 mg/kg
Edible ices	150 mg/kg
Flavored processed cheese	100 mg/kg
Desserts including flavored milk products	150 mg/kg
Sauces, seasonings (e.g., curry powder, tandoori) pickles, relishes, chutney, and piccalilli	500 mg/kg
Mustard	300 mg/kg
Fish paste and crustacean paste	100 mg/kg
Precooked crustaceans	250 mg/kg
Salmon substitutes	500 mg/kg
Surimi	500 mg/kg
Fish roe	300 mg/kg
Smoked fish	100 mg/kg
"Snacks": dry, savory potato, cereal, or starch-based snack products:	
Extruded or expanded savory snack products	200 mg/kg
Other savory snack products and savory coated nuts	100 mg/kg
Edible cheese rind and edible casings	Quantum satis
Complete formulas for weight control intended to replace total daily food intake or an individual meal	50 mg/kg
Complete formulas and nutritional supplements for use under medical supervision	50 mg/kg
Liquid food supplements/dietary integrators	100 mg/l
Solid food supplements/dietary integrators	300 mg/kg
Soups	50 mg/kg
Meat and fish analogs based on vegetable proteins	100 mg/kg
Spirituous beverages (including products less than 15% alcohol by volume), except those mentioned in Annex II or III	200 mg/l
Aromatized wines, aromatized wine–based drinks, and aromatized wine–product cocktails as mentioned in Regulation (EEC) No. 1601/91, except those mentioned in Annex II or III	200 mg/l
Fruit wines (still or sparkling), cider (except cidre bouché), and perry aromatized fruit wines, cider and perry	200 mg/l

The temporary ADI was maintained and the results of carcinogenicity and reproduction/teratology studies were requested. In 1989 the temporary ADI was extended until 1992, while results from a National Toxicology Program (NTP) were awaited. In 1992 the temporary ADI was again extended. In 1995 some of the results of the NTP studies were reviewed, and the temporary ADI was increased to 1 mg/kg and was extended to 1998. However, their report included the following statement: "At its present meeting, it reconfirmed the need for the study and reiterated the view that previous studies with turmeric were not relevant to the evaluation of curcumin. If such studies are not submitted for review in 1998 it is unlikely that the temporary ADI can be further extended."

In 1996 an initiative was taken by associations in Europe, Japan, the United States, and India to jointly provide data to JECFA. To this end a report incorporating previously unpublished data was submitted in 1997 and a protocol for a multigeneration reproductive study was drawn up. This study should be undertaken in the year 2000.

It is important for manufacturers to appreciate that technical problems will not disappear and that these must be addressed, even though it is appreciated that natural extracts are less well specified than pure chemicals and that the industry itself is very diverse. While discussing curcumin it is also interesting to consider how the specifications for turmeric extracts and curcumin vary worldwide and how the method of analysis is important if it is thought necessary to impose a tight specification.

Table 15.8 summarizes the current status of turmeric oleoresin and curcumin in the United States, EU, and Japan, and as can be clearly seen, little common ground exists between the EU, where turmeric oleoresin is not a permitted color, and the United States and Japan.

As previously mentioned, most natural color extracts contain more than one colorant. Commercial curcumin, for example, is primarily a mixture of three curcuminoids, each with its own extinction coefficient. Since the ratio of these three compounds present in a commercial extract varies with

TABLE 15.8   Status of Turmeric/Curcumin in 1999

	Turmeric oleoresin	Curcumin	Minimum specified curcumin content (%)
United States	Yes	Yes	No minimum
EU	No	Yes	90
Japan	Yes	Yes	9.3[a]
JECFA	Yes	—	Not less than declared
JECFA	—	Yes	90

[a]EU calculation.

TABLE 15.9 Curcumin Estimation[a]

Curcumin (%)	Method
93.6	EU method
101.0	Common industrial method
97.9	Proposed new method
96.4	Assuming material is all curcumin
106.0	Using "pure" curcumin as standard

[a]Results obtained on one commercial preparation: E 1%1 cm, 1504 at 428 nm in ethanol; E 1%1 cm, 1616 at 420 nm in acetone.

the variety of turmeric extracted and the processing and storage conditions, there is, therefore, no one correct extinction coefficient that can be used to calculate the pigment content (Table 15.9).

As can be seen from the table, the differences in the calculated curcumin content are very significant when considering that the EC specification for curcumin requires the curcumin content to be in the range of 90–100%. This debate regarding the extinction coefficient is not so important in the United States and Japan, since here the specification does not include a minimum curcumin content.

However, on many occasions too little attention is paid to the detail of the method of analysis and how applicable it is to the colorant in question. For example, the analysis of heavy metals (as Pb), using a hydrogen sulfide method, is extremely unreliable when measuring a strong colorant such as bixin because the test solution, without lead standard, is often as colored as the test solution plus lead standard, thus invalidating the result.

When considering the EC directive relating to the specifications for colorants, it is important to appreciate that most of the parameters specified (such as solvent residues, heavy metals) are not related to the pigment content of the extract. Only a very few purity factors are related to color strength, including the SO_2 content of anthocyanins (E163), the nitrate content of beetroot red (E162), and the total copper content of coppered chlorophyll derivatives (E141).

JAPAN

As recently as April 6, 1999, the Ministry of Health & Welfare in Japan announced the publication of the seventh edition of the Japanese Standards for Food Additives. This document includes newly established specifications for 60 natural additives, 18 of which are natural colorants (Table 15.10), together with new definitions, test methods, as well as a safety assessment of reagents and methods. It is encouraging to note that this seventh edition is

TABLE 15.10   18 Natural Colorants Listed in the Seventh Edition of the
Japanese Standards for Food Additives, 1999

1. Turmeric oleoresin and curcumin	10. Carotene from carrot
2. Caramel a	11. Carotene from palm oil
3. Caramel b	12. Red beetroot
4. Caramel c	13. Grape skin color and extract
5. Caramel d	14. Blackcurrant color
6. Chlorophyll	15. Monascus color
7. Cochineal extract and carminic acid	16. Carthamus red
8. Carotene from Dunaliella	17. Carthamus yellow
9. Paprika color and oleoresin	18. Marigold color

written with reference to JECFA specifications, as demonstrated by the fact
that 17 of the 18 colors included already have JECFA specifications (monascus being the additional one).

One novel aspect of these 18 specifications is the use of a color value as
a measure of minimum color strength. This color value is the calculated absorbance of a 10% solution of the specified colorant at lambda max and was
specifically selected as the means of color estimation because natural colors
contain a number of different colorants, and therefore measurement of a
single compound is not a practical possibility. Thus, for curcumin/turmeric
oleoresin, the minimum color value is stated as 1,500 at lambda max in the
range 420–430 nm when measured in ethanol, whereas the EC minimum
is 14,463 using the same method. For paprika color the Japanese minimum
color value is 300, whereas the EC specification equates to a minimum of
1,470.

It is also interesting to note that in the Japanese regulations the specification for cochineal/carminic acid does not include the lake form, and
thus carmine is still not a permitted colorant in Japan. It is expected that an
eighth edition of the Japanese standards will be published shortly and that
this will include further detailed specifications for additional natural colorants, including annatto and gardenia. It must be remembered that a significant number of other natural colorants, including extracts of gardenia,
perilla, red cabbage, purple yam, and tomato, not included in the seventh
edition are also permitted for food use in Japan.

CONCLUSIONS

If food is to be truly part of global trade, then it is important that food additives should be manufactured to the same standards in all countries.
Thus, the harmonization of colorant specifications and usage is an impor-

tant goal. There are many issues left to address, and it is therefore essential that color manufacturers coordinate their efforts to validate and standardize methods and also work closely with the food industry and regulatory bodies to ensure that consumers continue to receive foodstuffs that are both safe and attractive.

Index